Beef Cattle PRODUCTION

By

Danny D. Simms, Ph.D.

SMS Publishing
P.O. Box 19968
Amarillo, TX 79114
www.smspublishinginc.com

2013

Printed in the United States of America
ISBN: 978-0-615-85327-7

Preface

This book is an attempt to summarize 43 years of experience as a teacher, extension specialist, researcher, consultant, industry nutritionist, partner in a feed company, and manager of a large feed company. The primary audience is intended to be junior/senior level students in a beef production course. It should also be useful for anyone interested in beef cattle production. The goal was to provide good, practical information and recommendations supported by relevant research. The reference section at the end of each chapter should provide more in-depth information if desired.

An additional goal was to provide good reference information on animal requirements and feed composition, which are covered in the Appendices. This information was taken from extension publications and other publications as noted. The generosity of the individuals who allowed the use of this information is appreciated.

The tables in the appendices and examples throughout the book are presented using the English system rather than the metric system. This will bother some researchers, but the fact is that producers still measure feed in pounds, and cattle are weighed for sale in pounds, not kilograms.

While considerable effort has gone into proofing this book, I'm sure that there will still be some typographical errors. Correspondingly, there has been substantial effort to ensure technical accuracy. Please feel free to contact me regarding errors of either type.

Danny Simms
Canyon, Texas

Acknowledgments

I owe a great debt to many people for helping me reach my current level of understanding of the beef industry. In the following paragraphs, I will acknowledge many of them; however, I will undoubtedly omit someone who made a significant contribution. You have my deepest apology.

As a general statement, I must recognize the contributions of the many researchers in the land-grant university system and at USDA experiment stations for their contributions to our understanding of beef cattle production. I can only hope that I have done their work justice in summarizing it in this book.

I owe particular credit to my former colleagues at Kansas State University: Drs. Larry Corah, Gerry Kuhl, Frank Brazle, and Robert Cochran. Their guidance and tutelage contributed greatly to my understanding of this topic.

I am also indebted to the many cattle producers that I interacted with over the years, especially as a beef cattle extension specialist in Kansas. In many cases, I learned far more from them than they learned from me.

I am especially indebted to Dr. Rod Preston (Texas Tech University, Emeritus) for the use of his feed composition tables. Also, I appreciate Dr. Shane Gadberry at the University of Arkansas permitting the use of the animal requirement tables, which were originally developed by Dr. George Davis. In addition, I want to thank the National Academy of Sciences for allowing the use of information taken from several of its publications.

Elanco Animal Nutrition graciously allowed the use of the beef cow body condition score diagrams, and Pearson Education allowed the use of the beef cow digestive system taken from Feeds and Feeding, 6th edition, by Perry, Cullison, and Lowery. The NCBA graciously allowed the use of the quality and yield grade pictures in Chapter 5, and the Kansas Livestock Association provided some of the cover photos.

I owe a special debt of gratitude to the reviewers of the manuscript for technical accuracy:

Dr. Scott Laudert, SBL Consulting

Dr. Sandy Johnson, Kansas State University

Dr. Bob Weaber, Kansas State University

Dr. Walter Fick, Kansas State University

Dr. Jack Whittier, Colorado State University

Dr. Kevin Dhuyvetter, Kansas State University

Dr. David Patterson, University of Missouri

Dr. John Hutcheson, Merck Animal Health

I want to thank Mark Stadtlander for editing the manuscript, and Patrick Hackenberg for formatting the final version and designing the cover. In addition to those previously mentioned, the following researchers warrant special mention for their contributions to our understanding of beef production:

Dr. John Paterson, National Cattlemen's Beef Association

Dr. Dick Pruitt, South Dakota State University (Retired)

Dr. Ivan Rush, University of Nebraska (Retired)

Dr. Keith Lusby, University of Arkansas (Retired)

Dr. Ron Lemenager, Purdue University

Dr. Dan Faulkner, University of Arizona

Dr. Ted McCollum, Texas A&M University

Dr. Bill Kunkle, University of Florida (Deceased)

Dr. Jim Males, Oregon State University

Dr. Robert Bellows, Fort Keogh Research Station, Miles City, Montana (Retired)

Danny D. Simms

Dr. Bob Short, Fort Keogh Research Station, Miles City, Montana

Dr. Don Clanton, University of Nebraska (Retired)

Dr. Jack Whittier, Colorado State University

Dr. Jim Wiltbank, Brigham Young University (Deceased)

Dr. David Patterson, University of Missouri

Dr. Elaine Grings, South Dakota State University

Dr. Cal Ferrell, U.S. Meat Animal Research Center, Clay Center, Nebraska

Dr. Tom Jenkins, U.S. Meat Animal Research Center, Clay Center, Nebraska

Dr. Rick Rasby, University of Nebraska

Dr. Dan Fox, Cornell University (Retired)

Dr. Don Pretzer, Kansas State University (Deceased)

Dr. Jim Mintert, Purdue University

Dr. John Lauchbaugh, Fort Hays Experiment Station, Hays, Kansas (Retired)

Dr. John Brethour, Fort Hays Experiment Station, Hays, Kansas (Deceased)

Dr. Darrell Strobehn, Iowa State University (Retired)

Dr. Dan Loy, Iowa State University

And finally, I must thank my wife, Rachel, for her support throughout this project.

Danny. D. Simms

Table of Contents

Part VI: Evaluating Diet Adequacy

Part VII: Supplementation of Cattle Diets

Part VIII: Additional Cow Herd Management Topics

Part IX: Stocker Management

Part X: Custom Cattle Feeding

Part XI: Special Topics

Chapter 1

Beef Production – An Overview

Beef production is an important industry in much of the United States contributing significantly to the economy in many states and the country. The United States Department of Agriculture (USDA) estimated that the total value of sales from beef cattle was more than $51 billion in 2010. In many areas of the country, beef production is the major source of jobs and income. In addition, it should be noted that the industry takes grasses and other roughages that are unsuitable for human consumption and turns them into high-quality food. As might be expected, cattle are concentrated in areas with abundant forage production.

Segments of the Industry

Traditionally, the industry has been thought of as having three major segments:

1. Commercial Cow/Calf
2. Stocker
3. Feedlot

In addition to these major segments, the seedstock industry is an important part of beef production. As the industry has seen the integration of the packing industry with the other segments, especially with the feedlot industry, it has become a more significant segment in the industry.

Figure 1-1. Beef Cows in the United States in 20011. *(USDA-NASS)*

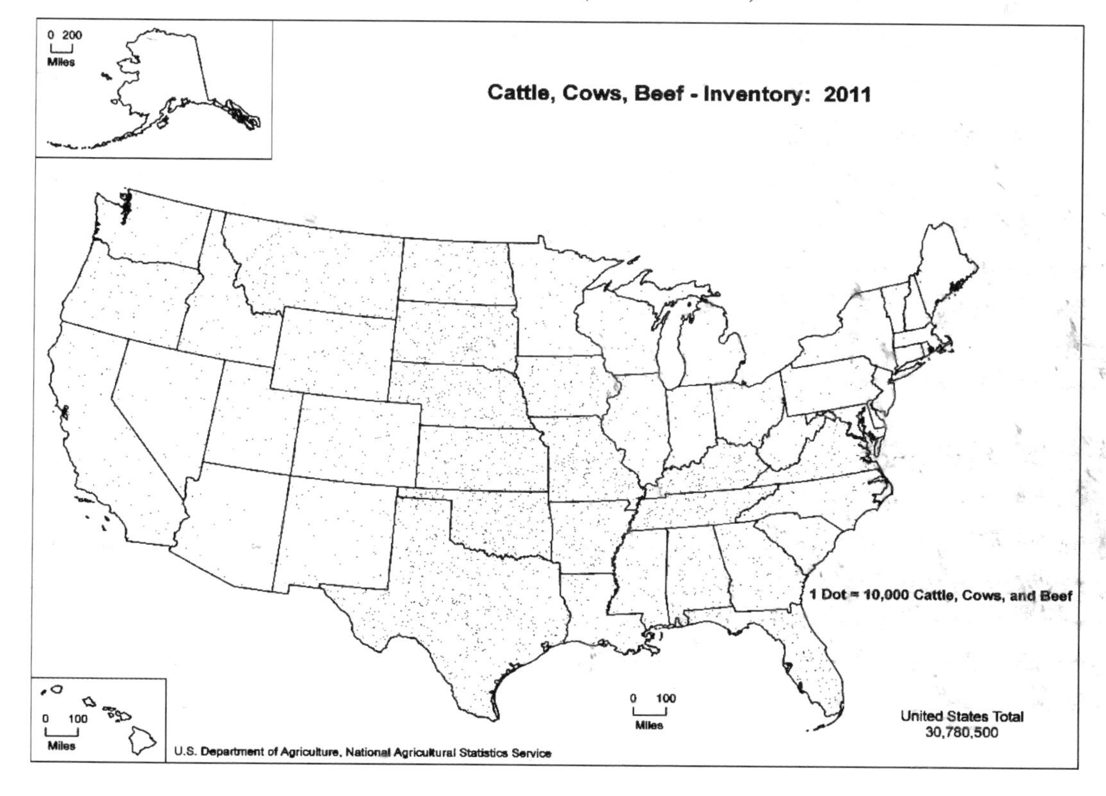

Commercial Cow/Calf Segment

In 2010, there were more than 740,000 cow/calf operations in the United States. While cow/calf production exists in every state, the number of operations and beef cows tends to be higher in areas with significant grassland and in areas with abundant crop residues. Figure 1-1 shows the distribution of beef cows across the United States.

The number of operations with beef cows has been declining in recent years as shown in Figure 1-2. Correspondingly, the number of beef cows in the United States has been declining. This is somewhat puzzling given that calf prices have been at record levels. Historically, high calf prices have caused producers to expand the cow herd; however, to date, that hasn't occurred. There are many reasons for this reduction in beef cow numbers, which will be discussed in more detail in a subsequent chapter.

Most beef cows in the United States are in herds of fewer than 50 head (Figure 1-3). This is especially true in the South and southeastern part of the country. Correspondingly, herds tend to be larger in the western United States. While most herds are smaller than 50 head, larger herds make up more than 50 percent of the total inventory of beef cows with 38 percent in herds of from 100 to 499 head and 16.6 percent in herds of 500 head or more, respectively (Figure 1-4).

Figure 1-2. Number of Beef Cow Operations in the United States by Year.

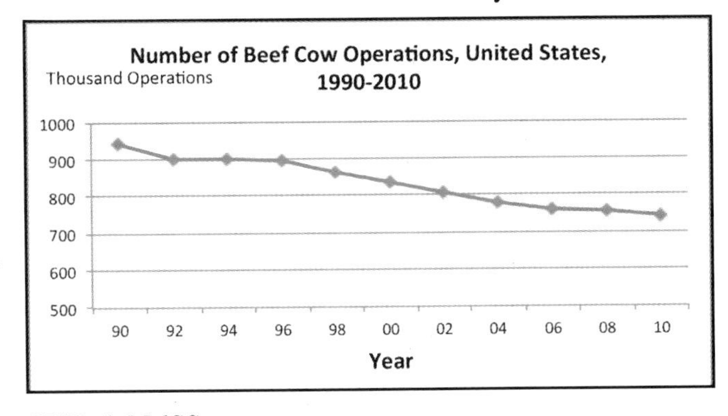

USDA-NASS

Figure 1-3. Beef Cow Operations in the United States by Size in 2010.

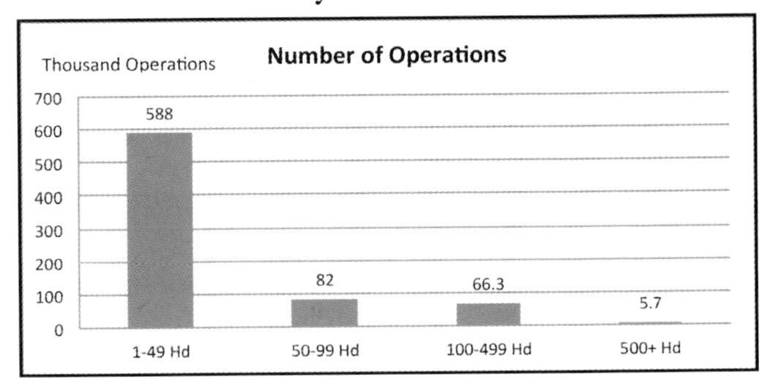

USDA-NASS

Figure 1-4. Percent of Inventory by Cow Herd Size in the United States in 2010.

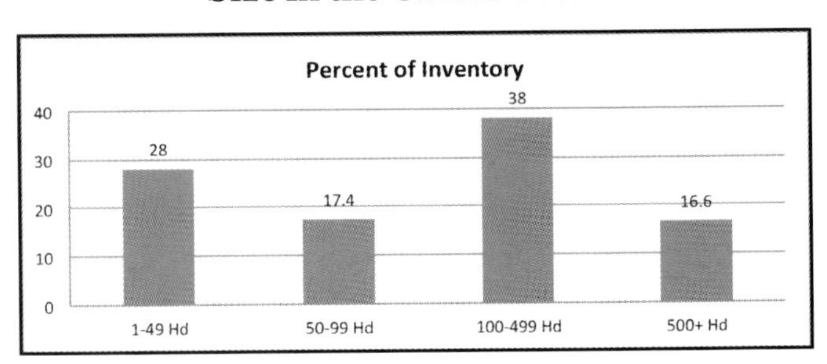

USDA-NASS

Each major cow/calf production area tends to have some competitive advantage that makes cow/calf production viable. For instance, in the western United States, large areas of rangeland provide a forage supply. In addition, many ranches have access to federal or state land for grazing. In the plains region, considerable grassland is available, but the competitive advantage in this region is the availability of crop residues like corn and milo stover, which provide low-cost grazing for a significant part of the year. In the south and southeast, abundant pastureland is available, but the real advantage in this area is the relatively long growing season, which greatly increases the potential forage production per acre.

Stocker Segment

Stocker production consists of growing calves after weaning with the goal of adding weight and in some cases frame before finishing. The growing program may be in a drylot, but more typically, it consists of grazing calves on a pasture like winter wheat. Traditionally, stocker programs existed to put some frame on calves before placing them in a feedlot so

Table 1-1. Cattle on Feed January 1, 2011.

State	1,000 Head	State	1,000 Head
Arizona	258	Minnesota	320
California	470	Nebraska	2,550
Colorado	1,100	Ohio	170
Idaho	250	Oklahoma	380
Illinois	160	South Dakota	410
Indiana	129	Texas	2,850
Iowa	1,380	Washington	215
Kansas	2,400	Wisconsin	250
Michigan	170		

USDA-NASS

they would produce a large enough carcass at slaughter. Improvement in growth rate in all breeds has reduced the importance of growing cattle before finishing with many calves having the potential to produce a desirable carcass weight even if placed directly in the feedlot following weaning. Currently, the high price of weaned calves and high cost-of-gain in feedlots have increased the importance of a stocker phase as a means of providing some low cost gain thereby reducing the overall cost of production.

Major stocker production areas include the wheat growing areas of Kansas, Oklahoma, and Texas where close to two million calves are on wheat pasture during a typical winter. Other major stocker production areas include the Flint Hills in eastern Kansas and the Sand Hills in Nebraska where large numbers of calves and yearlings are placed on grass during the summer. In the Midwest and Great Plains regions, many farmers feed calves in a drylot over winter with the goal of increasing the returns from the grain and forage they produced. With high grain and forage prices, fewer farmers are using this type of stocker production system.

Table 1-2. Major Beef Slaughter Firms in the United States in 2009.

Company	Daily Capacity
Cargill Meat Solutions	29,000
Tyson Foods	28,700
JBS USA	28,600
National Beef Packing Co. LLC	14,000
American Foods Group, LLC	7,000
Greater Omaha Packing Co.	2,900
Nebraska Beef Ltd.	2,600
XL Four Star Beef	2,300
AB Foods LLC	1,600
Sam Kane Beef Processors	1,600

Feedlot Segment

A high percentage of cattle destined for the restaurant and retail trade are placed on a high-energy diet for at least 100 days before slaughter. This finishing period increases the tenderness and palatability of the beef. Until recently, the rations used in this finishing period were more than 80 percent grain; however, the increase in production of ethanol from corn and other grains has resulted in a shift to feeding less grain and more of the by-products resulting from the production of ethanol, starch, and fructose from corn and milo.

Cattle were originally finished in the Corn Belt, but in the 1950s and 60s, cattle feeding shifted to the high plains of Kansas, Oklahoma, and Texas as well as Nebraska. This shift occurred because the milder winters, lower humidity, and drier climates resulted in much better cattle performance. As the cattle moved to the high plains, packing plants were built in that area to process the cattle. Table 1-1 shows the January 1, 2011 cattle on feed for the states that have a significant cattle feeding industry.

In recent years, there has been considerable interest in finishing cattle on grass rather than high-energy diets because of health concerns. While it is certainly possible to finish cattle on grass, it tends to result in higher production costs and actually results in more greenhouse gas. It remains to be seen whether finishing cattle on grass will become more popular.

Packing Segment

As mentioned previously, there has been considerable integration between the cattle feeding industry and the packing industry with numerous types of marketing agreements. Furthermore, some of the largest packers also are major cattle feeders. As with most industries involved in agriculture, there has been considerable consolidation in the packing industry over the last 50 years, which has resulted in five firms (Table 1-2) slaughtering a high percentage (approximately 73 percent) of finished cattle.

Seedstock Segment

The seedstock industry exists to produce the genetics for the other segments of the beef industry. Its primary product is a bull (usually a yearling) for use in the commercial cow/calf industry. Obviously, the genetics of these bulls ultimately ends up in the stocker, feedlot, and packing segments. Over the past 30 years, the seedstock industry has done a tremendous job of improving the growth rate and performance of cattle using Expected Progeny Differences and other selection tools. Historically, most people have considered the seedstock industry as being synonymous with the purebred industry. In fact, the seedstock industry is much broader with cattle breeders developing composites and producing hybrid bulls. In many cases, purebred breeders with two or more breeds have taken the lead in producing F_1 bulls to meet the needs of their customers.

Beef Production and Consumption Trends

Beef Production

Beef production has been relatively stable in recent years as shown in Figure 1-5. This steady supply has been accomplished in spite of the decline in beef cow numbers by significantly increasing carcass weights. This change in carcass weights will be discussed in more detail in Chapter 5.

Per Capita Consumption

Per capita beef consumption in the United States has been declining for many years (Figure 1-6). While there are several reasons for this decline, the major reason is the lower price of competing meats like pork and poultry.

Obviously, the decline in beef consumption by U.S. consumers is not a favorable sign for the industry. Fortunately, exports have been strong in the last few years. For example, in 2010 approximately 8.7 percent of the beef produced in the United States was exported.

Figure 1-5. Total Pounds of Beef Produced in the United States.

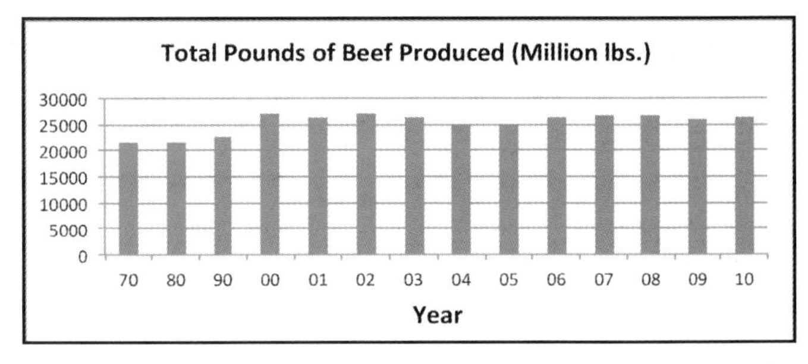

USDA-NASS

Figure 1-6. Per Capita Annual Consumption of Beef in the United States.

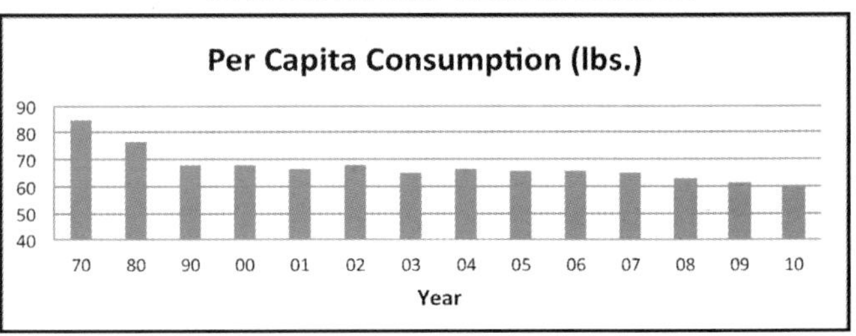

USDA-NASS

References

Beef 2007–08. 2009. USDA.

Galyean, M., C. Ponce, and J. Schutz. 2011. The Future of Beef Production in North America. Animal Frontiers. Vol. 1, No. 2 Pg. 29–36.

USDA National Agricultural Statistics Service Website.

Chapter 2

Commercial Cow/Calf Production

Beef calves are produced across the United States in a wide range of environments from extensive range environments where more than 50 acres may be required per cow unit to intensive environments with less than one acre per cow unit. Obviously, the specific costs associated with these production environments also vary greatly. However, these production systems have one thing in common — the goal to convert pasture/range and low-quality feedstuffs such as crop residues into marketable beef. To illustrate the costs and returns associated with these types of production systems, consider a typical budget for a fairly intensive system as shown in Table 2-1.

Table 2-1. Beef Cow/Calf Enterprise Analysis for the North Dakota Farm Business Management Education Program for 2010 on a Per Cow Basis.

	Average of All Farms	Lowest 20% in Profit	Highest 20% in Profit
Number of Farms	137	27	28
Beef Calves Sold	217.42	133.39	232.41
Transferred out	399.27	430.64	407.52
Cull Sales	115.66	121.92	122.94
Other income	5.88	8.02	3.52
Purchased	-119.00	-193.66	-101.03
Transferred in	-113.17	-86.31	-143.68
Inventory change	72.27	79.70	100.29
Gross Margin	**578.33**	**493.70**	**621.98**
Direct Expenses			
Protein, Vitamins, Minerals	14.85	18.05	14.81
Creep/Starter	7.23	1.40	9.29
Complete ration	3.89	0.42	1.34
Corn silage	22.92	39.93	23.19
Hay, alfalfa	16.73	11.17	31.11
Hay, grass	115.88	119.58	93.09
Pasture	97.72	99.16	83.87
Other feedstuffs	10.84	18.84	9.09
Veterinary	16.19	20.94	16.21
Supplies	16.19	20.94	10.25
Fuel & Oil	23.76	28.48	20.42

	Average of All Farms	Lowest 20% in Profit	Highest 20% in Profit
Repairs	30.98	53.77	20.35
Custom hire	6.25	9.21	5.90
Hired labor	1.43	3.54	1.60
Machinery leases	0.82	5.08	0.10
Operating interest	7.48	10.27	6.11
Total direct expenses	394.34	458.24	346.75
Return over direct expenses	**183.99**	**35.46**	**275.23**
Overhead Expenses			
Hired labor	7.48	15.23	3.82
Farm insurance	8.23	10.22	6.30
Utilities	8.48	11.01	6.53
Interest	12.40	22.68	8.13
Mach & Bldg Depreciation	21.33	21.71	20.29
Miscellaneous	12.88	17.96	9.57
Total Overhead Expenses	70.70	98.81	54.64
Total Direct and Overhead Expenses	465.04	557.05	401.39
Net Return	**113.29**	**-63.35**	**220.59**
Labor and management charge	67.42	82.08	59.47
Net return over labor and management	**45.87**	**-145.43**	**161.12**
Cost of Production Per cwt Produced			
Total direct expense per unit	77.46	95.80	66.66
Total direct & overhead expense per unit	91.35	116.46	77.17
With other revenue adjustments	100.07	133.65	80.84
With labor and management	113.31	150.81	92.27
Estimated labor per cow unit	5.10	6.38	4.59
Other Information			
Number of cows	153.8	118.1	212.3
Pregnancy percentage	97.3	96.6	97.3
Pregnancy loss percentage	2.2	2.7	1.8
Culling percentage	15.0	17.3	14.6
Calving percentage	95.1	94.0	95.6
Weaning percentage	90.3	86.2	91.2
Calf death loss percentage	5.6	8.4	4.5
Average weaning weight	555	521	588
Lbs weaned/exposed female	501	449	536
Feed cost per cow	290.06	308.55	265.79
Average weight of calves sold	580	562	617
Average price per cwt	117.54	109.86	116.54

Adapted from data available on the North Dakota State University Department of Agricultural Economics website.

It is important to note that on average, total feed cost was approximately 74 percent of the total direct cost and 63 percent of total direct and overhead costs. Consequently, the key to a profitable cow/calf operation is for management to focus on keeping feed costs as low as possible while keeping reproductive rates high. Also noteworthy is that more than one-half of the total feed costs were harvested forages (silage and hay).

Table 2-2 shows a budget for a typical range operation on the high plains of Texas. In this situation, the cattle are typically on range all year, whereas in the North Dakota system, the cattle are typically in drylot in the winter. In this extensive production system, much more land is required per cow unit. For example, in this budget, 25 acres were allocated for each cow. Protein and mineral supplements make up the other major feed costs. Feed costs ($225) per cow are lower in this budget compared to the North Dakota budget. However, the expected levels of production are also lower.

The budgets in Tables 2-1 and 2-2 are provided to illustrate the different costs associated with these production systems. In recent years, input costs have risen, and furthermore, they have been extremely variable. Thus, readers are advised to obtain updated budgets rather than using the costs shown in Tables 2-1 and 2-2. Fortunately, the agricultural economics department at the state land-grant university normally prepares budgets for cow/calf enterprises, and it typically updates these budgets on an annual basis.

Factors Influencing Profitability in Cow/Calf Operations

In addition to illustrating typical costs and returns in a North Dakota cow/calf operation, the information in Table 2-1 can be used to identify some key production and cost items that differentiate high- and low-profit producers. For example, if we compare the highest 20 percent of the producers with respect to profit to the lowest 20 percent, we see that the most profitable had lower feed costs per cow unit, especially in terms of harvested feed cost. They also had lower overhead cost per cow,

Table 2-2. Cow/Calf Budget for the High Plains of Texas.

Production Description	Quantity	Unit	$/Unit	Return
Cull cows	0.12 hd @ 1,000 lbs	cwt	63.00	75.60
Heifer calves	0.23 hd @ 474 lbs	cwt	110.00	120.18
Steer calves	0.43 hd @ 525 lbs	cwt	111.00	250.58
Total Gross Income				446.36
		Variable Costs		
Corral repair	1.0	head	1.55	1.55
Fence repair	1.0	head	4.00	4.00
Hay	280	lbs	0.05	14.00
Marketing	0.85	head	5.00	4.25
Miscellaneous	1.00	head	3.00	3.00
Salt and Mineral	50	lbs	0.35	17.50
Supplement	350	lbs	0.16	56.00
Veterinary and medicine	1.0	head	10.00	10.00
Water facility repair	1.0	head	2.50	2.50
Fuel				5.64
Lube				0.56
Repair				0.68
Total Variable Costs				119.69
Residual returns to capital, ownership, labor, land, management, and profit				326.67

Production Description	Quantity	Unit	$/Unit	Return
Capital Investment Description			Rate	
Interest – IT Borrowed	711.16	dollars	0.060	42.67
Interest – OC Borrowed	183.49	dollars	0.048	8.72
Total capital Investment Costs				51.39
Residual returns to ownership, labor, land, management, and profit				275.28
Ownership Cost Description (Depreciation, Taxes, and Insurance)				
Machinery and Equipment				12.89
Livestock				3.04
Total Ownership Costs				15.93
Residual returns to labor, land, management, and profit				259.36
Labor Cost Description				
Machinery and Equipment	1.18	hours	10.59	12.50
Other	6.40	hours	10.60	67.84
Total Labor Costs				80.34
Residual returns to land, management, and profit				179.01
Land Cost Description				
Pasture lease	25.0	acres	5.50	137.50
Total Land costs				137.50
Residual returns to management and profit				41.51

Adapted from a budget prepared by the Texas A&M Department of Agricultural Economics.

partially because they tended to have more cows, which spread their overhead costs over more units. In other words, they had some advantage because of the economies of size.

In terms of production, the highest 20 percent had better reproductive performance and heavier weaning weights resulting in more pounds weaned per female exposed. In addition, they sold their calves for more per pound even though the calves weighed more.

In summary, these data support the findings of other research where high-profit producers are able to keep costs, especially feed costs, low while maintaining high levels of production. In other studies, weaning weight has not been a significant factor with low costs and high reproductive performance being the most significant factors. There are many management decisions that lead to keeping feed costs low while still having relatively high reproductive performance. Subsequent chapters in this book will discuss how to accomplish this combination of low cost and high reproductive performance.

Standardized Performance Analysis (SPA)

On the surface, it would seem relatively easy to perform an economic analysis of a cow/calf operation. However, in reality it is complex, especially in diversified operations. One of the first issues is how will feedstuffs raised on the operation be valued in the analysis. Will they be valued at the actual cost of production or at market value? These values may be quite different in some years. Another issue is how owned land is considered in the analysis. Should it be charged at the current rental rate even though this isn't a true out-of-pocket expense? In fact, the correct approach needs to be based on the questions to be answered by the analysis, and cattle producers typically have several key reasons for performing an economic analysis. Some of the common questions include:

1. Is my cow herd contributing to my net income? This is a common question producers ask when the cow herd is part of a diversified farming/cattle operation.

2. How do I compare with other similar operations?
3. How do my expenses per cow unit compare to other producers?
4. How do I compare to other producers with respect to performance measures like calves weaned per cow exposed and weaning weight?
5. What is my return on investment and how does it compare to other possible investments?

In summary, clearly defined reasons for conducting the analysis are of great value in determining how to value inputs. If the producer is committed to keeping the cow herd, the goal may be to identify weak areas in the operation relative to similar operations. The analysis may then identify areas where management can focus on improvement. On the other hand, if the question is whether to sell the cow herd or not, the inputs should be valued at their opportunity cost. For example, if the pasture currently used by the cow herd can be rented, that rental rate should be its cost in the analysis.

In the 1990s, the Cooperative Extension Service and the National Cattlemen's Beef Association worked together to develop the Standardized Performance Analysis (SPA) program for cow/calf operations. As the name implies, this program developed a standardized approach to performing production and economic analyses of cow/calf operations. Actually, two approaches were developed. The major difference in these approaches is in the way input expenses are established. In the financial analysis, for example, hay is valued at the cost of production, while in the economic analysis it is valued at market value (its opportunity cost). Correspondingly, in the financial analysis, there isn't an expense for owned land while in the economic analysis, it is charged at the current rental rate. Obviously, these two analyses can result in quite different estimates of profitability.

The SPA analysis does provide some useful information, but it doesn't answer all of the questions listed above because only two bases for valuing inputs are considered, and answering some of the questions requires valuing inputs in a different way. The point is that it is important to understand how inputs should be valued based on the questions to be answered.

Table 2-3 on page 12 shows a summary of SPA analysis data from Texas for the years 2004–2008. Conducting a SPA analysis and comparing the results to the averages of these data could be useful in identifying areas of weakness in the operation. For example, comparing raised/purchased feed cost per cow to the average of other herds could indicate a problem area. However, as a caution, comparisons should be made only to herds in a similar environment and production system.

Economic Analysis in Diversified Operations

Performing an economic analysis in a diversified operation is especially difficult because of production from other enterprises being used in the cow/calf enterprise. For example, a cow/calf producer who also raises milo might feed some of the milo to the cattle, and the cattle often graze the milo residue. The question becomes how should the milo be valued — at market value or cost of production? Should the milo stalks be valued at the current rental rate or zero value? As a general rule, these inputs should probably be valued at their opportunity cost.

In addition to the issue of valuing raised feedstuffs used by the cow herd, the allocation of costs associated with shared equipment can be a problem. For example, a tractor that is used to till the farmland and plant crops is often used to move hay bales to the cows during the winter. The allocation of the tractor expenses such as depreciation and interest can be an issue in this situation.

Another important issue is the allocation of overhead costs. For example, a significant expense on many operations is interest on the farm mortgage. This expense must be allocated to enterprises on a fair basis considering the value of the farmland compared to the pastureland. Some other overhead items include insurance and depreciation on vehicles.

It may seem like allocation of these expenses to the cow herd can be straightforward. However, determining the actual contribution of the cow herd to net income in a diversified operation can be difficult. To illustrate this point, consider an example from an analysis conducted by extension specialists at Kansas State University on a diversified farm during the farm financial crises of the 1980s. When an economic analysis of the cow herd enterprise was completed, it showed a net loss of approximately $50 per cow. In this analysis, all inputs from the farming enterprise like grain and residue grazing were

Table 2-3. Texas Cow-Calf SPA Key Measures Summary for the Years 2004–2008.

Production Measures	
Herd Related Measures	**Average**
Pregnancy percentage[a]	89.5
Calving percentage	85.2
Calving death loss based on exposed females	2.9
Calf crop or weaning percentage	82.5
Actual weaning weight, steers and bulls	542.3
Actual weaning weight, heifers	512.6
Average weaning weight	527.4
Pounds weaned per exposed female	437.2
Other Physical Performance Measures	
Raised feed acres per exposed female	0.3
Grazing feed acres per exposed female	17.1
Pounds weaned per acre utilized by the cow/calf enterprise	53.8
Pay Weight Prices Per Cwt	
Weaned calf pay weight price – steers/bulls	116.47
Weaned calf pay weight price – heifers	109.94
Financial Measures	
Investment and Returns (ROA)	
Total investment per breeding cow – cost basis	$3,042
Percent return on assets – cost basis	1.39%
Total investment per breeding cow – market value	$5,083
Percent returns on assets – market value	1.45%
Financial Performance	**$**
Raised/Purchased feed cost per cow	120.05
Grazing cost per cow	107.31
Total cost before noncalf revenue adjustment per cow	557.96
Total cost before noncalf revenue adjusted per cwt	116.64
Net income after withdrawals per cow	-11.57
Economic Performance	**$**
Total cost noncalf revenue adjusted per cow	615.54
Total cost noncalf revenue adjusted per cwt – unit cost	139.49
Net income after withdrawals per cow	-117.44

Adapted from information available on the Texas A&M Ag Economics website. [a]*Based on pregnancy tested herds.*

valued at their opportunity cost and all overhead costs for the operation were also allocated. Using the FINPACK program from the University of Minnesota as a tool, the option of selling the cow herd was considered. When the cow herd was removed from the operation, it resulted in a significant reduction in net farm income. In other words, selling the cow herd made a bad situation even worse. How could cows that were losing $50 per head be making a positive impact on net farm income? On further analysis, it was clear that the contribution of the cow herd to covering overhead costs was the reason. Remember, overhead costs by definition don't change with level of production. For example, interest on the mortgage, which was a very large expense, stayed the same, even if the cows were sold. Other overhead costs like depreciation on the tractor would also stay the same. Without the cow herd, the farming enterprise was less profitable because it had to cover these overhead costs. In addition, some of the land that the cows grazed couldn't be rented.

This example illustrates the complexity of doing enterprise analysis in a diversified operation. Considering a single enterprise without considering its interaction with other enterprises can lead to making bad decisions. Cow herds and the farming operations in a diversified operation often complement each other leading to greater profit.

References

Cow-Calf Enterprise Budget. 2011. Auburn University Extension Service.

Cow-Calf Budgets. 2011. Texas A&M University Agricultural Economics Spreadsheet B-1241.

Dhuyvetter, K., M. Langemeier, and S. Johnson. 2011. Beef Cow-Calf Enterprise. Kansas State University Farm Management Guide MF266.

Dhuyvetter, K. 2011. Differences Between High-, Medium-, and Low-Profit Producers: An Analysis of the Kansas Farm Management Association Beef Cow-Calf Enterprise. Kansas State University Ag Manager Info Website.

Dunn, B. 2002. Measuring Cow-Calf Profitability and Financial Efficiency. Proceedings of Beef Improvement Federation.

Hughes, H. 2007. Cattle Economics – Assessing 2005 IRM Herds.

Lalman, D. and D. Doye, Eds. 2008. Beef Cattle Manual, 6th Ed. Oklahoma State University Extension Service.

Livestock Enterprise Budgets for Iowa – 2010. Iowa State University Department of Agricultural Economics.

Livestock Enterprise Analysis North Dakota Farm Business Management Education Program State Report 2010. Beef Cow-Calf. North Dakota Department of Agricultural Economics.

Ramsey, R., D. Doye, and C. Ward. Economic Factors Affecting Cow Herd Performance. Oklahoma State University Extension Bulletin AGEC-595.

Ramsey R., D. Doye, C. Ward, J. McGrann, L. Falconer, and S. Bevers. 2005. Factors Affecting Beef Cow-Herd Costs, Production, and Profit. Journal of Agricultural and Applied Economics 37:91–99.

Chapter 3

The Stocker Industry

Historically, stocker production has been based on using high-quality native pasture, small-grain pasture, or harvested feedstuffs to put additional frame and weight on weaned calves. This additional frame was required by the smaller-framed cattle prevalent in the 1960s and 1970s because they produced carcasses that were too light at harvest if they were placed directly on a finishing program following weaning. The increase in frame and growth potential through improved genetics over the past 30 years has reduced the need for a stocker program before finishing solely for adding frame.

The significant increase in feed costs caused by development of the ethanol industry has had an impact on the importance of the stocker phase. This increase in feed costs has meant that the gains in a typical stocker program may be the cheapest gains during the animal's life. Thus, many producers try to maximize the weight put on in a stocker program as a means of reducing the total cost of production through finishing.

Stocker programs are of two types: 1) Grazing of pastures or crop residues of various types, or 2) Drylot programs where cattle are kept in a pen and fed harvested feedstuffs. Drylot stocker programs are often referred to as backgrounding programs. Stocker programs emphasizing grazing are prevalent in many areas in the country while drylot programs are prevalent in the northern High Plains and upper Midwest where farmers background cattle as a means of adding value to their crops.

Factors Influencing Profitability in Stocker Production

The key factors in the profitability of stocker cattle are:

1. The buy/sell margin — the difference between the purchase price per pound and sale price per pound.
2. Performance (gain).
3. Length of ownership.

Additionally, health problems and death loss can significantly influence profitability. For example, many stocker programs in the high plains are based on purchasing lightweight calves from the South and Southeastern United States. These calves are often

Table 3-1 Effect of Wintering Rate of Gain on Performance and Profitability.

Winter Gain	Slow	Slow	Fast	Fast
Grazing Season	**Short**	**Long**	**Short**	**Long**
Winter, ADG, lb	0.78	.80	2.01	2.04
Grazing ADG, lb	2.45	2.01	1.44	1.29
Finishing ADG, lb	3.69	3.28	3.55	3.06
Feed/Gain	6.45	7.13	6.70	7.77
Final Weight, lb	1,211	1,276	1,277	1,310
Breakeven, $/Cwt	69.85	69.94	69.27	71.41

University of Nebraska, 1996.

Table 3.2 Projected Wheat Pasture Budget for 2011 Prepared by the Agricultural Economics Faculty at Oklahoma State University.

Assumptions: October purchase at 500 lbs, February sale at 747 lbs • Average Daily Gain = 2.0 lbs, 1.5% death loss • Stocker Phase = 135 days

Production	Wt., lbs	Price/Cwt	Quantity	$/Head
Feeders	747	$136.00	0.985 Head	$1,000.55
Total Receipts				**$1,000.55**
Operating Inputs	**Wt., lbs**	**Price**	**Quantity**	**$/Head**
Stockers	500	$142.00	1	$710.00
Pasture		76.46	1	76.46
Hay		12.83	1	12.83
Salt		0.15	1	0.15
Minerals		0.19	1	0.19
Vet Services/Medicine		3.88	1	3.88
Vet Supplies		0.71	1	0.71
Marketing		7.84	1	7.84
Mach/Lube, Fuel, Repairs		14.59	1	14.59
Mach/Equip Labor		10.25	1.11	11.38
Other Labor		10.25	1.50	15.38
Operating Capital		6.5%	251.28	16.33
Total Operating Expenses				**$869.74**
Returns Above Operating Expenses				**$130.81**
Fixed Costs		**Rate**		**$/Head**
Machinery/Equipment				
Interest at		6.00%		1.40
Taxes at		1.00%		0.41
Insurance		0.60%		0.14
Depreciation				3.53
Total Fixed Costs				5.48
Total Costs (Operating + Fixed)				$875.22
Returns Above All Specified Costs				$125.33
Break-Even Analysis				
Break-Even Purchase Price ($/Cwt)				
Above Operating Costs				$168.16
Above Total Costs				$167.07

highly stressed when they arrive, resulting in high levels of sickness, high medication costs, and high death loss. Producers purchasing these calves usually get them bought at a significant discount to local cattle. Thus, they're trying to keep the buy/sell margin as small as possible.

The Buy-Sell Margin – Stocker production is considered a high-risk enterprise primarily because the difference between the purchase price of the calf (or the value of the calf raised at home) and the sale price, which is referred to as the buy/sell margin. It is the single most important factor in the profitability of

Table 3-3. Projected Budget for a Drylot Stocker Program in 2011-2012 Prepared by the Agricultural Economics Faculty at Auburn University.

Assumptions: 60 Head with a purchase weight of 450 lbs 150 day feeding period • 2.60 lbs ADG 1.5% death loss 10.00 lbs feed conversion • 823 lbs ending weight with 2% shrink

Item	Head	Unit	Quantity	Price or Cost/Unit	Total Value/Cost	$ /Head Sold	% of Total
Gross Receipts							
Feeder cattle	59	Cwt	8.23	$127.00	$61,667.39	1,045.21	100.00
Variable Costs							
Stocker calves	60	Cwt	4.50	130.00	35,100.00	594.92	53.36
Starter feed		Ton	2.10	210.00	441.00	7.47	0.67
Ration		Ton	109.80	195.00	21,411.00	362.90	32.55
Vet & Med		Head	60	25.00	1,500.00	25.42	2.38
Hay		Ton	18.00	85.00	1,530.00	25.93	2.33
Labor		Hours	300.00	8.25	2,475.00	41.95	3.76
Marketing Expenses		Head	59.00	15.00	885.00	15.00	1.35
Beef Promotion Fee		Head	59	1.50	88.50	1.50	0.13
Equipment – Repair		$			121.17	2.05	0.18
Interest on Op. Cap.		$	21,868.99	0.0550	1,202.79	20.39	1.83
Total Variable Costs					**$64,754.46**	**1,097.53**	**98.4%**
Income Above Variable Costs					$-3,087.07	-52.32	
Fixed Costs							
General Overhead		Head	60	2.50	150.00	2.54	0.23
Int. on Bldg. and Equip.		$	3,652.50	0.065	237.41	4.02	0.36
Depre. On Bldg Equip		$			579.08	9.81	0.88
Other Bldg. and Equip.		$			52.78	0.89	0.08
Total Fixed Costs					1,019.28	17.28	1.55
Total Expenses					65,773.74	1,114.81	100.00
Net Return Above Total Costs					$-4,106.35	$-69.60	
Total Production Cost Per Head ($/Head Sold)					$519.89		
Value of Gain Per Cwt ($/Cwt)					123.24		
Cost of Gain Per Cwt: To Cover Variable Costs ($/Cwt)					137.56		
To Cover Total Costs ($/Cwt)					142.29		
Net Returns Per Head Sold: Above Variable Costs ($/Hd)					-52.32		
Above Total Costs ($/Head)					-69.60		
Breakeven Feeder Price: To Cover Variable Costs ($/Cwt)					133.36		
To Cover Total Costs ($/Cwt)					135.46		
Maximum Stocker Purchase Price: To Cover Variable Costs					124.82		
To Cover Total Costs ($Cwt)					121.04		

stocker production. Historically, the margin in stocker programs has been negative, and at times extremely negative. In other words, typically stockers have been priced higher than the sell price of the feeder animal produced. However, high grain and roughage prices tend to make the margin smaller. The importance of the buy/sell margin on profitability means that getting stocker cattle bought at a good price is a key to profitability.

Rate of Gain – As a general statement, the higher the rate of gain in a stocker program, the greater the profitability because input costs like pasture and mineral are spread across more pounds. Additionally, a higher percentage of feed inputs go to gain rather than just to maintenance. Stocker producers have some control over rate of gain through supplementation programs.

While in general higher rates of gain are more profitable, there are some exceptions to this rule. For example, if the rate of gain is high enough to promote excessive fat deposition in the stocker phase, the cattle may be discounted when sold. Secondly, if ownership of the cattle is to be retained after the stocker phase, the overall performance of the cattle must be considered because rate of gain in one phase of production influences performance in subsequent phases of production. Table 3-1 shows a University of Nebraska study that illustrates this concept. The calves wintered at the low rate of gain exhibited compensatory gain when grazed the following year, and they performed better even in the finishing phase.

If one only considers the wintering program, the calves that gained more than 2 pounds per day would probably have been more profitable than the calves that gained less than 1 pound per day. However, compensatory gain and performance by the calves wintered at the low rate of gain resulted in similar breakevens after finishing. High feed prices and buy/sell margins that are much smaller than historical levels make it crucial for producers to consider overall performance if the cattle are owned for more than one phase of production.

Length of Ownership – Generally, the longer the period of ownership, the higher the chance for profitability because more pounds are produced if the cattle are gaining significantly. With short periods of ownership, changes in the market (buy/sell margin) have a greater impact on profitability.

Table 3-2 is presented to illustrate the type of costs associated with a typical wheat pasture stocker program in the High Plains. Obviously, the buy/sell margin and the costs are constantly changing. The Cooperative Extension Service in many states has spreadsheets available that make calculating breakevens with up-to-date prices and costs easy.

The projected budget in Table 3-2 clearly shows the importance of the investment in the stocker animal in the overall budget. In this example, the stocker animal represents ($710.00 ÷ $875.22) = 81 percent of the total costs of the stocker program. Obviously, this points out the importance of the buy/sell margin and death loss.

Table 3-3 shows a projected budget for a drylot stocker program prepared by Auburn University Cooperative Extension faculty. In this scenario, the percentage of total expenses represented by the stocker animal is much lower than in the wheat pasture example because the expenditures for feed are relatively higher at 32.6 percent of the total. This illustrates the impact that high feed costs and relatively poor feed conversion (10.0 lbs of feed per pound of gain) have on profitability. Consequently, even though the buy/sell margin is small ($-3.00), these stockers were projected to produce a significant loss.

Health Program – Because morbidity and mortality can be major problems in stocker programs, a sound health program including vaccinations and treatment protocols is essential for profitability. Receiving and health programs for stocker and feeder cattle are covered in a subsequent chapter.

References

Alabama Enterprise Budget Summaries. Website: *www.ag.auburn.edu/agec/*

Dhuyvetter, K. and M. Langemeier. 2011. Summer Grazing of Steers in Eastern Kansas. Kansas State University Farm Management Guide MF-1008.

Dhuyvetter, K. and M. Langemeier. 2011. Drylot Backgrounding of Beef. Kansas State University Farm Management Guide MF-600.

Feedlot Management Primer Chapter 2. Shipping and Receiving Cattle. Beef Information Website. Ohio State University.

Lawrence, J. and C. Ostendorf. How Profitable is Backgrounding Cattle? Iowa State University Extension Service.

Morris, C., I. Rush, B. Weichenthal, and B. Van Pelt. 1996. Beef Production Systems for Weaning to Slaughter in Western Nebraska. NE Beef Report.

Self, H. and N. Gay. 1972. Shrink During Shipment of Feeder Cattle. JAS 35:489-494.

Chapter 4

The Cattle Feeding Industry

The cattle feeding industry is a dynamic industry. It has gone through numerous changes during its history and is going through significant changes today. To illustrate, remember that cattle feeding was once centered in the Corn Belt, where many corn farmers fed cattle as a means of marketing some of their corn. In the 1960s, large feedlots were built in the High Plains regions of Texas, Oklahoma, and Kansas. At that time, both of these areas developed significant irrigation based on drawing water from the Ogallala aquifer. This irrigation allowed production of corn to supply these feedlots. The High Plains had the advantage of better weather for feeding cattle compared to the Corn Belt. Over time, cattle feeding declined in the Corn Belt with Texas, Kansas, and Nebraska becoming the major cattle feeding states. The infrastructure required to support cattle feeding, such as packing plants, followed the cattle to these states while many packing plants closed in the Corn Belt.

The development of cattle feeding in the High Plains was accompanied by the development of custom cattle feeding where the feedlot fed cattle owned by someone else. In fact, in many cases, these custom feedlots actually owned few, if any, of the cattle they fed. They simply supplied a service. The owners of the cattle were cattle producers, farmers, or investors looking for a way to make money feeding cattle without investing in a feeding facility.

In the last 15 years, there have been other significant changes in this dynamic industry including:

Consolidation – In the 1970s and 1980s, there were hundreds of independent feedlots across the High Plains with 10,000 to 20,000 head capacity. Gradually large corporations bought these yards and consolidated them. For example, JBS has a number of feedlots with a combined one-time capacity of more than 900,000 head. Correspondingly, Cactus has feedlots with a one-time capacity of more than 500,000 head. These feeding companies were built by purchasing independent feedlots in the 1980s and 1990s and by expanding the feeding capacity of these feedlots.

Increase in the Size of Feedlots – During the past 30 years, the size of feedlots has increased markedly to take advantage of increased efficiency. These larger feedlots spread their fixed costs, such as a feed mill, over more cattle, reducing the cost per head. Currently, there are a number of feedlots with a capacity of more than 100,000 head.

Marketing Arrangements – In the 1970s and 1980s, feedlots marketed their cattle independently with almost all of the fed cattle sold live on a cash basis. Over the past 20 years, cattle feeding companies have developed marketing arrangements with packers. In most cases, the cattle are sold on some type of a grid or on a carcass basis. In addition to this type of arrangement, some independent yards have formed marketing groups to pool enough cattle to have market influence. This trend developed in the 1990s in response to the fact that an independent 10,000 to 30,000 head feedlot had little marketing influence, especially when fed cattle were abundant.

Loss of Custom Cattle Feeders – One of the most significant trends in the last 20 years has been the decline in the number of people who want to feed cattle in a custom feedlot. This has occurred because profits from feeding cattle have not been good for the last 15 years. This loss of customers has forced custom feedlots to own more and more of the cattle in the lot and assume the risk of loss on the cattle. Owning a custom feedlot full of cattle owned by someone else was a profitable business. Conversely, acquiring the capital to own a high percentage of the cattle in the lot and assuming the risk of ownership of these cattle has put a tremendous financial strain on many feedlots.

More Cattle Being Fed by Packers – A higher percentage of cattle are being fed by companies that also own packing plants than in the past. The best examples are JBS and Cargill. These companies have increased the number of cattle they feed to ensure a supply for their packing plants. Obviously, these company-owned cattle have the potential for allowing packers to pressure the fed cattle market. This type of integration has a number of implications for the industry. It will be interesting to see the effect of this trend in the future.

Rations with By-Products – In the past 15 years, the ingredients in finishing rations have changed significantly. For example, 15 years ago, a typical finishing ration was 85 percent grain, 10 percent roughage, and 5 percent supplement. Today it is common for a finishing ration to contain 20 to 40 percent of a by-product of some type. In areas where ethanol plants and corn processing plants are prevalent, the percentage of by-products may be even higher.

Capital Requirements – The investment required to feed cattle has increased markedly in the past few years. For example, in late 2011, it required approximately $1,600 in investment in the feeder and feed to produce a 1,300-pound finished steer compared to around $750 in the 1980s. This requirement for capital has put considerable financial stress on many cattle feeders.

Increased Volatility – Feed costs, feeder cattle prices, and fed cattle prices have been more volatile than any time in history. This increase in volatility has resulted from a dwindling supply of feeder cattle and the increased demand for corn because of the ethanol industry. This volatility has made many decisions regarding purchasing inputs and selling cattle much more important and subsequently much more stressful. Cattle feeders have been forced to manage price risk by using the futures markets and forward contracts more than any time in history.

Structural Changes in Marketing – Volatility has forced cattle feeders to use alternative marketing strategies for fed cattle like grids, formulas, and forward contracts reducing the percentage of fed cattle sold live as shown in Table 4-1.

Table 4-1. Percentage of Fed Cattle Marketed Using Alternative Methods and Percent of Out Front Beef Sales.

	Percent of Cattle Marketed Using Alternative Methods (Grids, Formulas, and Forward Contracts)	Percent of Beef Sold for Future Delivery (Formula Sales and Forward Contracts)
2005	53	50
06	55	55
07	58	54
08	62	58
09	64	59
10	66	63
11YTD	71	63

USDA

Economics of Cattle Feeding

Factors Affecting Cattle Feeding Profits

Price Margin – The difference between the purchase price of the feeder and the price received for the finished animal is the price margin. The price margin is the major factor affecting profit in a cattle feeding enterprise. Typically, the price margin is negative with the price received for the finished animal per pound lower than the purchase price paid for the feeder. In recent years, this has been especially true. The excess feeding capacity in the feedlot industry has led cattle feeders to compete aggressively for feeder cattle keeping the feeder price high relative to the market value of the finished animal. The net result has been low profits in the cattle feeding sector over the last 15 years. In fact, cattle feeding has produced a net loss during this period.

Cost of Gain – The cost to put on a pound of gain is the second most important factor affecting cattle feeding profits. Feeding costs are typically divided into two categories: 1) Feed costs, and 2) Nonfeed costs. The major nonfeed costs in a custom feedlot include yardage, medicine, and other incidental charges. Generally, published cost-of-gain quotes include feed and nonfeed costs, but interest on the investment in the animal and feed is not included. The major factors affecting cost of gain include:

Table 4-2. Cattle Feeding Budget Prepared by the Agricultural Economic Department at Iowa State University.

Income	Price	Unit	Quantity	Unit	Total
Steer sales		per lb	1,250	lbs	
Variable Costs					
Yearling feeder cost	$1.45	per lb	750	lbs	$1,087.50
Interest	9%		5.5	months	44.86
Feed Costs					
Corn	$7.25	per bu	50	bu	$362.50
Fair quality hay	$125.00	per ton	0.25	tons	31.25
Modified distiller grain	$120.00	per ton	0.95	tons	114.00
Supplement & minerals	$0.16	per lb	100	lbs	16.00
Corn silage	$55.00	per ton	0	tons	0.00
Total Feed Costs					$523.75
Other Variable Costs					
Veterinary and health					$8.00
Machinery and equipment					7.00
Marketing and miscellaneous					16.00
Interest on variable costs	9%		2.75	months	11.44
Labor	$14.00	per hour	2	hours	28.00
Death loss	1%	head			14.29
Total Variable Costs					$1,754.85
Income Over Variable Costs					($1,754.85)
Fixed Costs					
Machinery, equipment, housing					$14.00
Total Costs					$1,754.85
Income Over All Costs					($1,754.85)
Breakeven selling price for variable costs (per lb)					$1.39
Break-even selling price for all costs (per lb)					$1.40

Adapted from a spreadsheet available on the agricultural economics website from Iowa State University.

- **Feed Prices** – This item has the greatest influence on the variability in cost of gain over time, and the development of the ethanol industry has significantly increased feed prices in recent years.
- **Feed Conversion** – Over time, feed conversion has the next greatest influence on cost of gain. In a feedlot at any time, feed conversion is the major factor in profit differences between pens of cattle. Consequently, feedlot nutritionists focus diligently on improving feed conversion.
- **Daily Gain** – There is a relationship between daily gain and feed conversion. Cattle with good feed conversion tend to be cattle that have a higher-than-average rate of gain. However, daily gain also is an independent factor because other costs like medicine and yardage are spread over more pounds if rate of gain is high.
- **Health Problems and Death Loss** – Abnormally high levels of morbidity result in high costs of gain because of the cost of

medicine and veterinary services. For example, feedlots typically charge a chute charge every time they treat an animal, and many of the newer medicines are very expensive. Higher-than-normal death losses increase the cost of gain. Many feedlots run their close-out analysis when the pen is sold with deads out and deads in. From a profitability standpoint, every dead animal increases the true cost of gain and decreases profitability.

Cattle Feeding Budget

Table 4-2 shows a typical budget for a finishing operation in the Midwest. As shown, the cost of the feeder is the major expense emphasizing the importance of the purchase price. For example, reducing the purchase price to $1.30 per pound reduces the break-even selling price to cover all costs to $1.31. The other major expense is the cost of the feed. For example, reducing the cost of corn to $5 per bushel also reduces the break-even selling price to $1.31 per pound. This budget also illustrates that expenses are typically separated into variable expenses (those that vary with the number of animals being fed) and fixed expenses (those that do not vary with the number of animals). Interest is charged for the entire feeding period on the cost of the feeder. However, interest is usually only charged for one-half of the feeding period for feed costs. As noted previously medicine expenses and death loss can significantly influence break-even selling prices. For example, increasing the death loss to 3 percent increases the break-even selling price by $0.03 per pound.

This budget also indicates the large investment currently required to feed cattle. With $1,755 invested in each animal at the end of the feeding period, feeding many head requires considerable capital.

The budget shown in Table 4-2 was generated using a spreadsheet on the agricultural economics website from Iowa State University. This type of spreadsheet is available from many land-grant universities. These spreadsheets are useful in preparing a break-even analysis for cattle feeding. Because the cost of inputs like cattle and feed are constantly changing, they are even more useful.

Breakeven Analysis

Spreadsheets and other software make calculating break-even selling prices easy; however, breakeven selling prices also can be calculated by hand as illustrated below.

Calculating the Breakeven Sale Price

Assumptions: Purchase 100 head of 850 lb yearling steers for $120/cwt = $1,020/head

Freight to feedlot = $5.00 per head

Expected finished weight of 1,325 lbs = 475 lbs of gain

Projected ADG = 3.4

Projected days on feed = 140

Projected cost of gain = $106/cwt or $1.06 per pound

Typical death loss = 1.0%

Total feed cost = 475 × $1.06/lb = $503.50

Total costs per head without interest: $1,020 + 5 + 503.50 = $1,528.50

Total costs for 100 head = $152,850

Projected pounds to be sold:

Cattle to be sold live with a 4% pencil shrink

1,325 less shrink (1,325 × 0.04 = 53) = 1,272 lbs (pay weight)

Total pounds = 99 head (100 less 1% death loss) × 1,272 = 125,928 lbs

Breakeven Sale Price = $152,850 ÷ 125,928 lbs = $1.21/lb

Note: This breakeven selling price does not include interest on the animal and feed

Interest on the steers at 5%:

$1,020 + $5 (freight) = 1,025 × 0.05 = $51.25 × (140 ÷ 365) = $19.66 per head

Interest on the feed, yardage, etc., @ 5%:

Since feeding costs are paid monthly, the average amount invested is one half of the total over the feeding period 4.75 cwt of gain @ $106 = $503.50 ÷ 2 = $251.75 (average investment) × 0.05 = $12.59 × (140 ÷ 365) = $4.83 per head

Total interest on cattle and feed = ($19.66 + $4.83) = $24.49 per head × 100 = $2,449

Breakeven sale price with interest:

$152,850 + $2,449 = $155,299 ÷ 125,928 = $123.32 per cwt

Calculating the Breakeven Purchase Price for the Feeder

Assumptions: 100 head of 850 lb steers

Freight to feedlot = $5 per head

Expected finished weight of 1,325 lbs = 475 lbs of gain

Projected ADG = 3.4

Projected days on feed = 140

Projected cost of gain = $106/cwt or $1.06 per pound

Typical death loss = 1.0%

Futures market shows a projected sale price of $120/cwt

Feeding Costs = $1.06 × 475 = $503.50 + 5 freight charge = $508.50

Projected lbs at sale = 1,325 less 4% pencil shrink (1,325 × 0.04) = 1,272 lbs

Projected receipts = 99 head × 1,272 × $1.20 (projected market) = $151,113.60

Projected receipts per head = $151,113.60 ÷ 100 = $1,511.14

Projected receipts minus feeding costs = $1,511.14 – $508.50 = $1,002.64

Breakeven purchase price = $1,002.64 ÷ 850 lbs = $117.96/cwt

Breakeven purchase price to cover interest as calculated above:

($1,002.64 – $24.49) = $978.15 ÷ 850 = $115.08/cwt

References

Albright, M., M. Langemeier, J. Mintert, and T. Schroeder. Factors Affecting Cattle Feeding Profitability and Cost of Gain. Beef Cattle Handbook. BCH-8050.

Dhuyvetter, K. and M. Langemeier. 2011. Finishing Beef. Kansas State University Farm Management Guide MF592.

Feeding Beef Cattle. Land O' Lakes Feed.

Feedlot Management Primer Chapter 2. Shipping and Receiving Cattle. Beef Information Website. Ohio State University.

Galyean, M., C. Once, and J. Schutz. 2011. The Future of Beef Production in North America. Animal Frontiers. Vol. 1, No. 2. Pg 29-36.

Lawrence, J., S. Shouse, W. Edwards, D. Loy, J. Lally, R. Martin. Beef Feedlot Systems Manual. Iowa Beef Center Website. Iowa State University.

Lawrence, J., D. Loy, D. Busby, and J. Sellers. Alternative Cattle Finishing Systems to Reduce Feed Cost. Beef Feedlot Systems Manual. Iowa Beef Center Website. Iowa state University.

Schulz, L. Finishing Yearling Steers Spreadsheet. Iowa State University Ag Decision Maker Website.

Chapter 5

The Product

The goal of beef production is to produce a high-quality protein source that also provides a good eating experience. The target is to give consumers beef that is lean, juicy, tender, and flavorful. In most cases, beef has all of these attributes, but unfortunately beef is too variable to provide a consistent eating experience. Specific problems with beef will be discussed in Chapter 59. Obviously, the other major goal is to make a profit on the beef produced.

Beef Products

The primary beef products are the cuts of meat as shown in Figure 5-1. However, only about 40 percent of live weight ends up as retail product. Much of the remaining weight provides numerous by-products, some of which are used in human food. Other inedible by-products are used in many ways (Table 5-1). For example, consider the many uses of leather, such as clothing and automobile seats.

Market Segments

The markets for beef are diverse, but they can be categorized into three types as follows:

Retail – This segment is composed of supermarkets and other retail outlets. This segment can use a wide range of carcass weights and quality grades from select to prime. Beef products sold include retail cuts such as steaks and hamburger, and processed products such as hot dogs.

Food Service – This segment includes a wide range of restaurants from fast food establishments to steak houses. Correspondingly, this segment uses a diversity of beef products from hamburger to steaks. This segment uses a wide range of carcass weights and quality grades.

High-end Restaurants – This segment is differentiated by the emphasis on using high-quality beef (average choice to prime) to provide a pleasing eating

Table 5-1. Examples of the Many Uses of Beef By-products.

By-Product	Uses
Edible By-products	
Gelatin from bones	Gelatin candy (gummy bears), marshmallows, gelatin desserts, yogurt
Fatty acid-base from fats	Chewing gum, oleo margarine and shortening
Plasma protein from blood	Cake mixes, deep fry batters
Inedible By-products	
Horns and hooves	Combs, pet chews, piano keys
Fats and fatty acids	Crayons, candles, floor wax, detergent, bar soap, shaving cream
Gelatin from bones	Paper and cardboard glues
Hides and hair	Leather goods, paint brushes, sports equipment

Adapted from There's More to Cattle than Just Beef

Figure 5-1 Shows a Diagram of a Beef Animal and the Source of the Various Cuts of Meat. These Cuts of Beef are the Primary Product, but it Should be Noted that only about 40 percent of the Live Weight is Lean Meat. The Remainder Termed "By-product," has a Wide Range of Uses as Shown in Table 5-1.

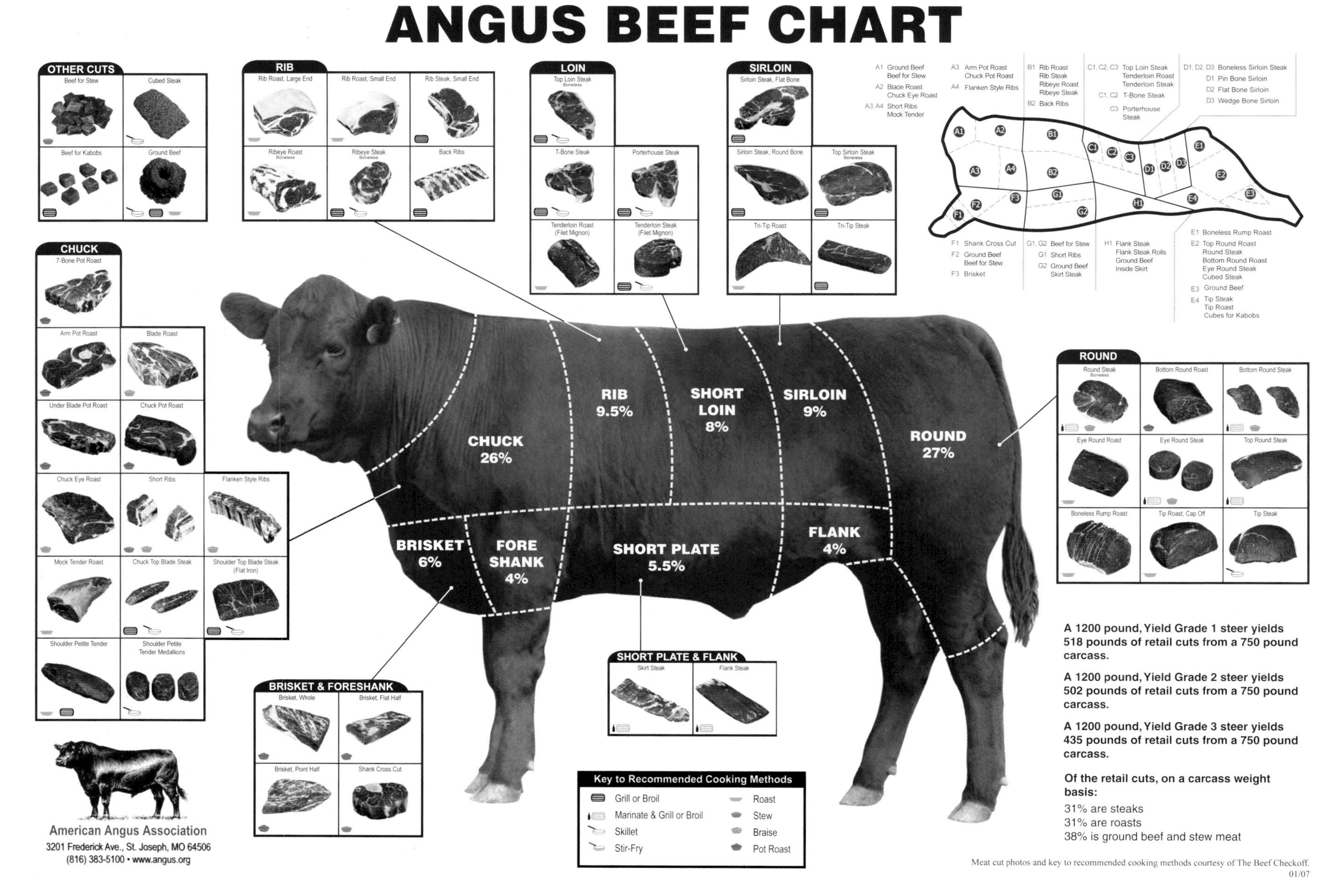

Table 5-2. Beef Skeletal Maturity, Maturity Class, and Approximate Age.

Maturity	% Thoracic Ossification	Approximate Age (Months)
A	0	9–30
B	10	30–42
C	35	42–72
D	70	72–96
E	90	>96

Table 5.3 Relationship between Marbling Score, Intramuscular Fat, and Quality Grade.

Marbling Score	Percent Intramuscular Fat	Quality Grade
Slightly Abundant	8.96	Low Prime
Moderate	7.14	High Choice
Modest	7.02	Average Choice
Small	5.21	Low Choice
Slight	3.76	Select
Traces	2.54	Standard

Iowa State University. 1996 Beef Research Report

experience. This segment has grown in recent years. It appears there is still growth potential, because many people are willing to pay more for a consistent, quality product.

To aid in defining differences in beef carcasses, the USDA developed grading standards concentrating on both quality (emphasizing marbling) and product yield. Grading is voluntary, but a high percentage of beef carcasses are graded using the USDA system.

Quality Grading

The two primary considerations in the quality grading system are maturity and marbling, which is the amount of intramuscular fat in the carcass. This fat supplies flavor and a sense of juiciness. Fat is also more tender than muscle, and as marbling increases, tenderness increases to some degree. And finally, fat provides some protection against overcooking.

Maturity – Older cattle tend to be less tender than younger cattle because their lean has a coarser texture, and they tend to have more connective tissue. An assessment of maturity is made during grading by evaluating skeletal maturity. As the animal ages, the cartilage in the spine ossifies, starting in the posterior of the animal and moving forward. The amount of ossification in the vertebral buttons near the thorax is used to assess maturity. Table 5-2 shows the relationship between ossification and the maturity group and approximate age. Using the degree of ossification gives some estimate of the age of the animal, but it is not perfect because cattle differ in when ossification occurs. Meat graders visually assess the degree of ossification and assign the carcass to the appropriate maturity group such as A, B, etc.

Marbling – As noted above, marbling is the amount of intramuscular fat as shown in Table 5-3. Historically, the amount of marbling was visually assessed on the ribeye between the 12th and 13th ribs using a scale from practically devoid to moderately abundant. This marbling scale is used to determine the quality grade of prime, choice, et cetera, as shown in Figure 5-2. This grade is stamped on the carcass.
In recent years, instrument-grading systems have been developed and accepted for use in cattle grading. These systems should be more consistent than the visual appraisal approach. There is a relationship between maturity and

Figure 5-2. Examples of Quality Grade.

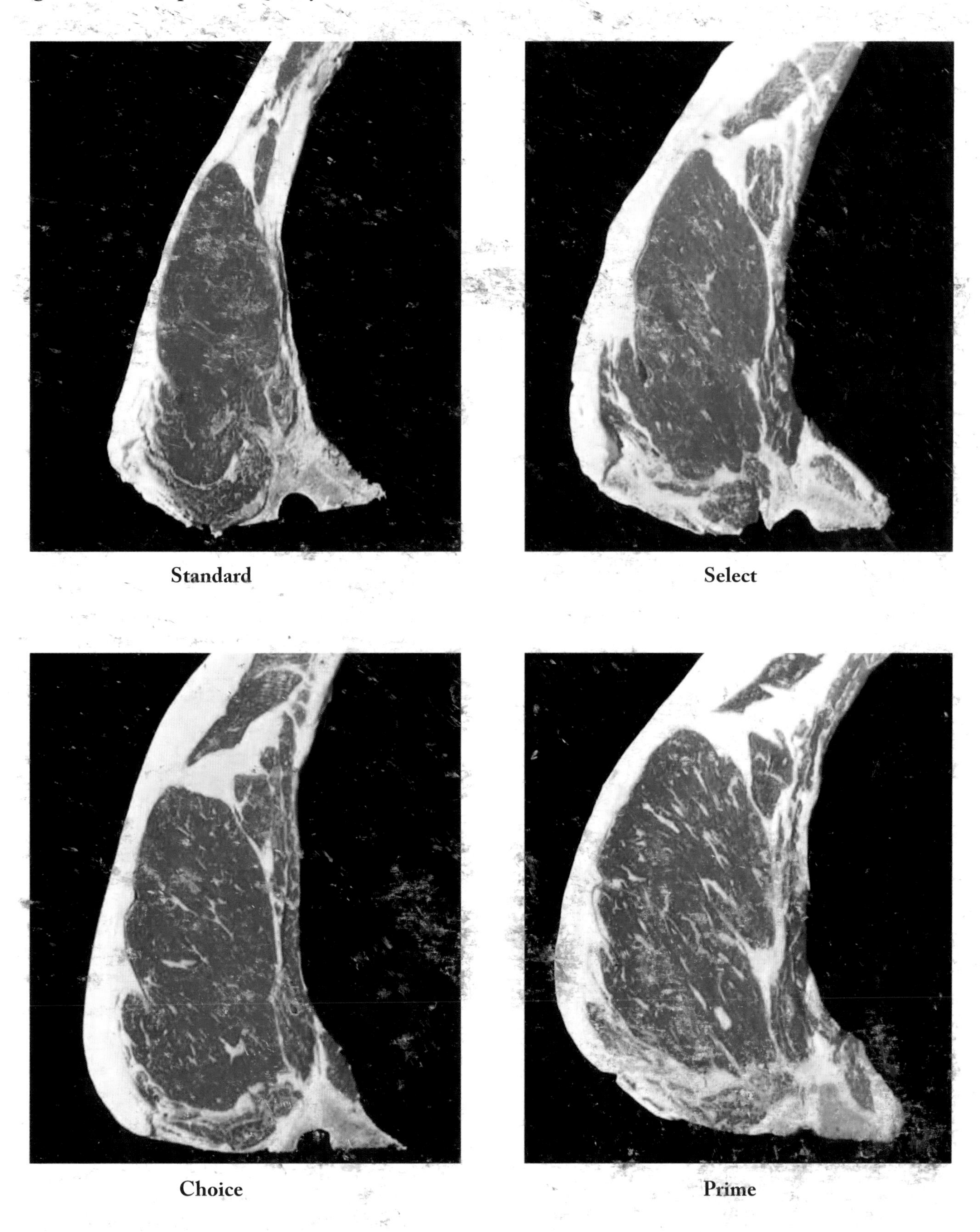

Standard **Select**

Choice **Prime**

Figure 5-3. Relationship between Marbling, Maturity, and Carcass Quality Grade*.

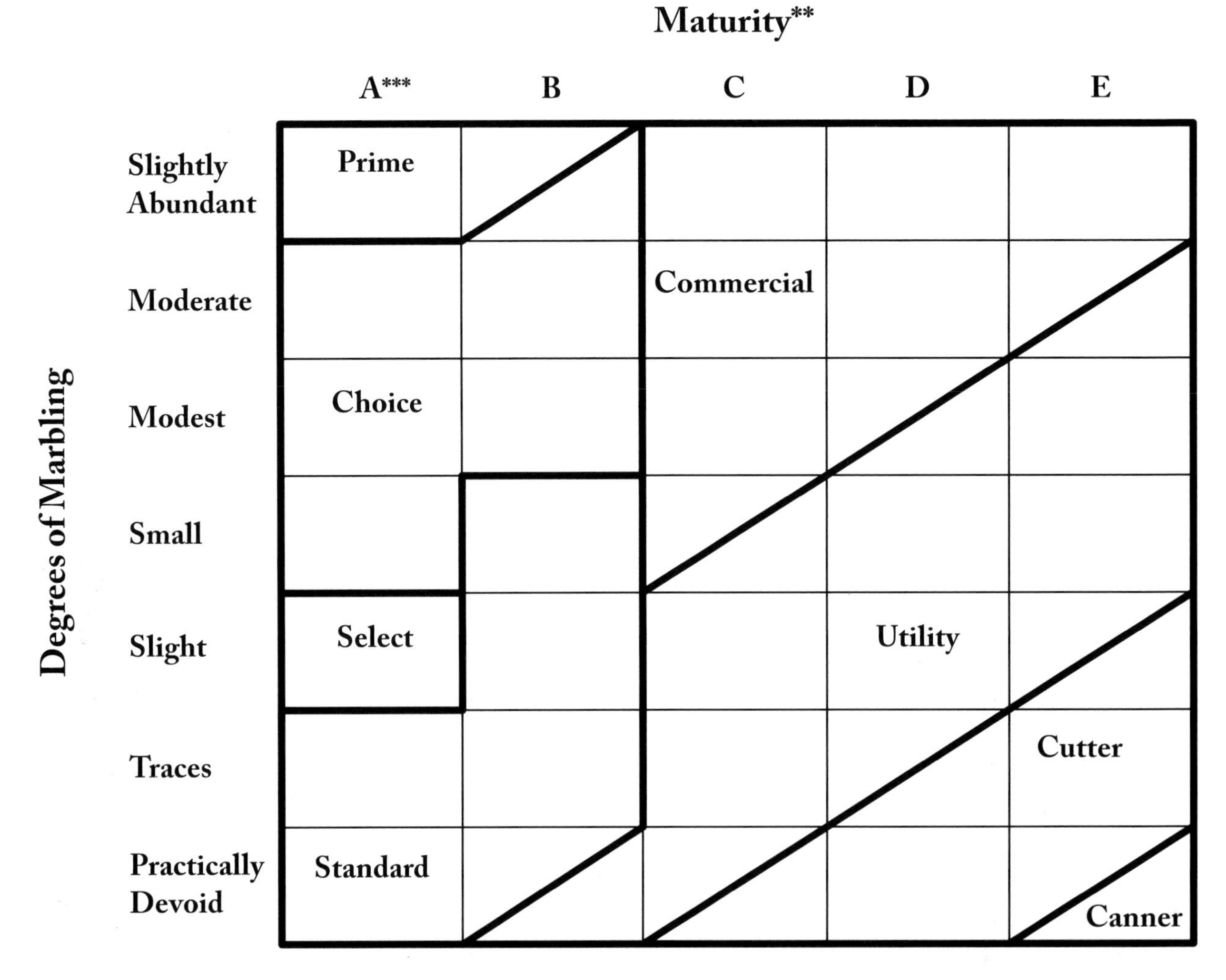

**Assumes that firmness of lean is comparably developed with the degree of marbling and that the carcass is not a "dark" cutter. **Maturity increases from left to right (A through E). ***The A maturity portion of the figure is the only portion applicable to bullock carcasses.*

Figure 5-4. Frequency Distribution of USDA Quality Grade for the National Beef Quality Audit – 2011.

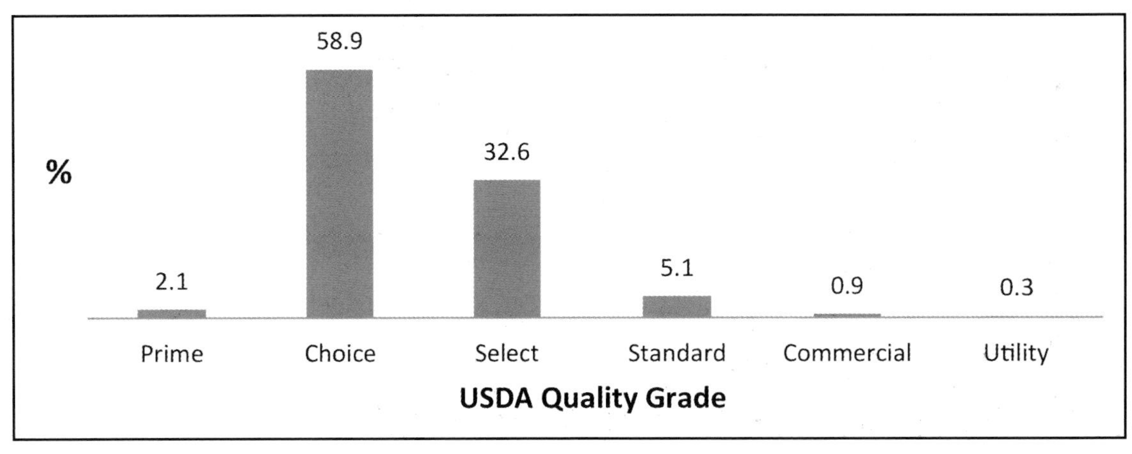

Source: NBQA 2011

Figure 5-5. Comparison of Percent Prime and Choice from USDA in 1974 and the National Beef Quality Audits of 1991, 1995, 2000, 2005, and 2011.

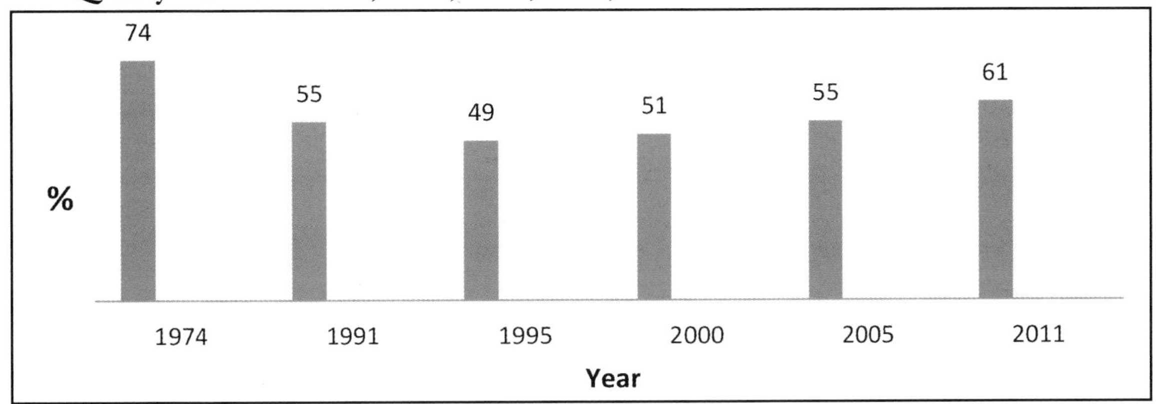

NBQA 2011

Figure 5-6. Examples of Yield Grades.

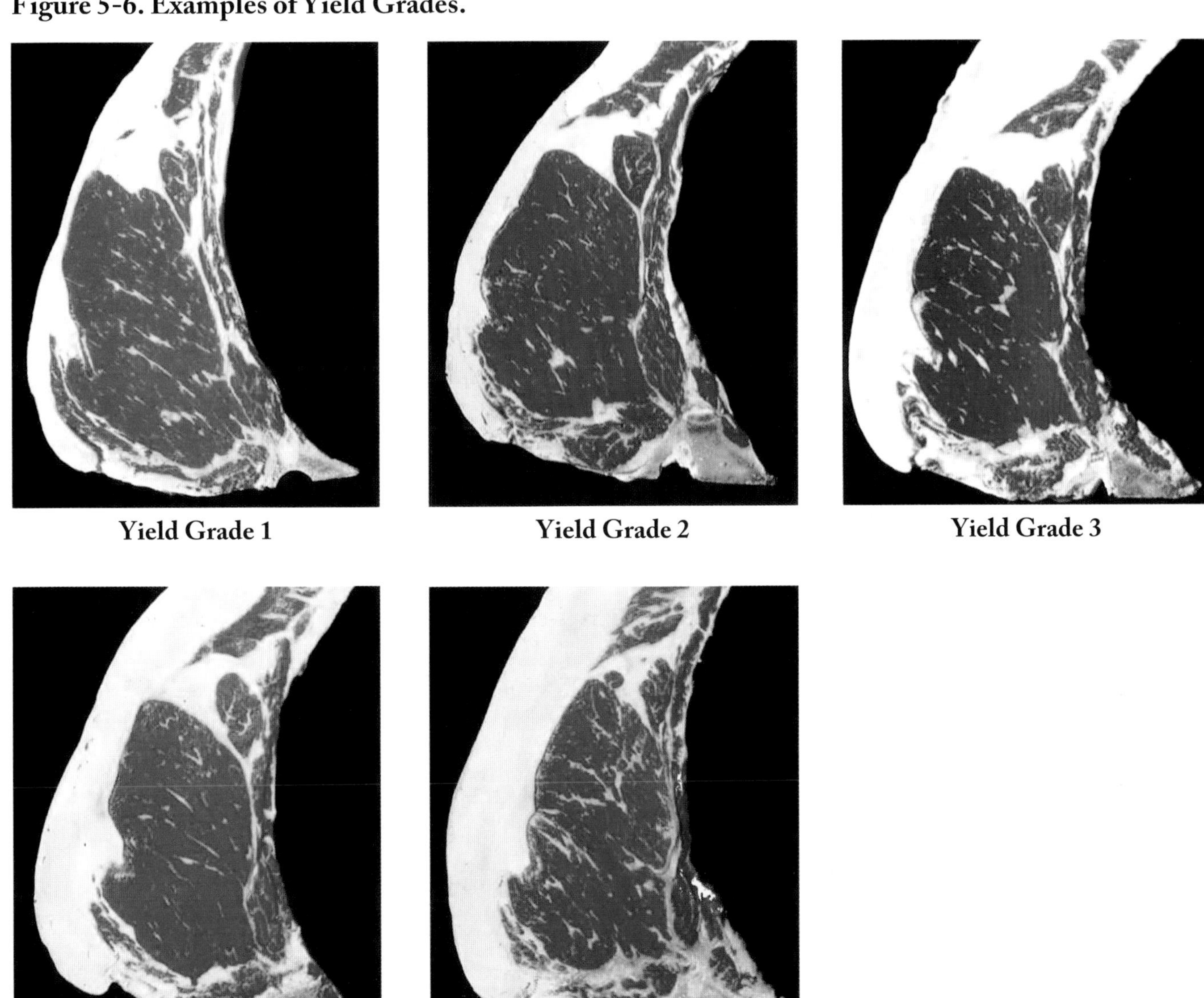

marbling shown in Figure 5-3 whereby it takes more marbling for a B maturity carcass to grade choice than an A maturity carcass. In addition to maturity and marbling, other factors affect the quality grade. For example, if the color of the lean is too dark, the grade will be reduced.

Current Quality Grade Distribution and Trends

Figure 5-4 shows the quality grade distribution of cattle assessed in the National Beef Quality Audit (NBQA) in 2011. As shown, more than 60 percent graded choice or higher and slightly more than 6 percent failed to qualify for at least the select grade. These carcasses are often referred to as "no-rolls" because they do not receive a USDA quality grade stamp. Figure 5-5 shows the percentage of carcasses grading choice or higher in the USDA analysis in 1974 and the NBQA audits conducted in 1991, 1995, 2000, 2005, and 2011. The marked decline in the percentage of cattle grading choice or higher from 1974 to 1995 probably resulted from the introduction of the continental breeds. As shown, the industry made a 10 percent improvement in the percentage of carcasses grading choice or higher from 2000 to 2011.

Yield Grade

The other grade applied to beef carcasses is the yield grade, which is an attempt to differentiate carcasses on the amount of closely trimmed retail cuts they will produce. Table 5-4 shows the yield grades and the approximate percentage of carcass weight that will be in closely trimmed retail cuts. Meat graders assign the yield grade based on:

1. **Amount of external fat** – measured three-fourths around the ribeye muscle away from the chine bone.
2. **Hot carcass weight** – weight taken right after slaughter.
3. **Amount of kidney, pelvic, and heart fat** (internal carcass fat).
4. **Area of the ribeye muscle** between the 12th and 13th rib.

Figure 5-6 shows examples of the ribeyes of carcasses representing the various yield grades. The official USDA yield grade formula is: Yield grade = 2.5 + {(2.50 × adjusted fat thickness, inches) + (0.2 percent of kidney, pelvic, and heart fat) + (0.0038 × hot carcass weight, pounds) – (0.32 × ribeye area, square inches)}. In summary, as fat thickness or kidney, pelvic, and heart fat, or carcass weight increase, yield grade increases. Conversely, as ribeye area increases, yield grade decreases. Fat thickness has the greatest influence on yield grade.

Current Yield Grade Distribution

Figure 5-7 shows the frequency distribution of yield grade as assessed by the NBQA conducted in 2011. Probably the most disturbing finding was that almost 10 percent of the carcasses were yield grade 4 or higher, which indicated a significant amount of waste fat.

Table 5-5 shows the averages for various carcass traits for each of the National Beef Quality Audits. As shown,

Figure 5-7. Frequency Distribution of USDA Yield Grade in the NBQA Conducted in 2011.

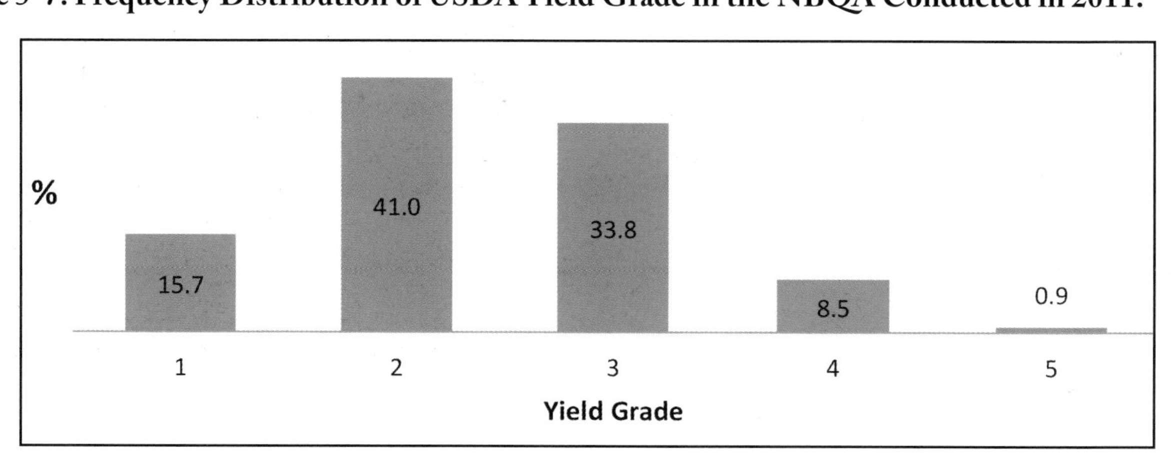

NBQA 2011

Table 5-4. Relationship between Yield Grade and the Percentage of Closely Trimmed Retail Cuts from a Beef Carcass.

Yield Grade	Percent of Carcass Weight in Closely Trimmed Retail Cuts
1	More than 52.3
2	50.1 to 52.3
3	47.8 to 50.0
4	45.5 to 47.7
5	Less than 45.5

Table 5-5. Means for USDA Carcass Traits from the NBQA in 1991 through the NBQA in 2011.

Trait	1991	1995	2000	2005	2011
USDA Yield Grade	3.2	2.8	3.0	2.9	2.6
Adj. Fat Thickness, in.	0.6	0.5	0.5	0.5	0.5
Hot Carcass Weight, lbs	760.6	747.8	786.8	793.4	824.6
Ribeye Area, sq. in.	12.9	12.8	13.1	13.4	13.8
Marbling Score[a]	424	406	423	432	440

[a]100 = Practically Devoid^{00}, 300 = Slight^{00}, 500 = Modest^{00}, 700 = Slightly Abundant^{00}. NBQA 2011.

yield grade has decreased and there has been a slight decrease in fat thickness. The most significant change has been in carcass weight with carcasses averaging almost 64 pounds more in 2011 than in 1991. The other significant change has been in ribeye area with an increase of almost one square inch from 1991 to 2011. It appears marbling score increased slightly in the 2011 audit.

Production of Retail Product Relative to Live Weight

The average finished animal will dress approximately 63 percent, which results in a 1,300-pound animal producing a 1,300 × 0.63 = 819 pound carcass. If the carcass is a yield grade 2, it will yield an average of about 51 percent in closely trimmed retail product (Table 5-4). Thus, the 819-pound carcass yields about 819 × 0.51= 418 pounds of retail cuts.

Beef from Cull Cows

Cull animals account for 15 to 20 percent of the annual income for beef and dairy producers. In addition, it is a major source of beef. For example, in 2011, there were 26.7 million steers and heifers slaughtered in federally inspected plants. In the same year, 6.8 million cull beef and dairy cows were slaughtered. Most people think that all of the beef from these cull animals ends up in hamburger. While most of it does, almost all facilities that slaughter cull animals produce whole muscle cuts. These cuts are primarily from the rib, loin, round, and chuck. They are marketed primarily to the food service trade, such as lower-end restaurants, buffets, and cafeterias. This trend to produce cuts from cull cows rather than just hamburger is growing in the industry.

Classification of Cull Cows

While a few young, cull cows may fit into the A or B maturity classes, most will fall into the C, D, or E maturity classes. Cull cows are procured based on body condition scores (illustrated in Chapter 23), which reflect fatness. Categories reported are:

- **Breakers** – defined as cows, normally with a yield grade range of 2 to 4 (estimated red meat yield of 75 to 80 percent), which are processed into various cuts.
- **Boners** – defined as cows (estimated red meat yield of 80 to 85 percent), normally boned for processed beef after removal of the merchandisable cuts.

Table 5-6. Approximate Associations between Cull Cow Marketing Classification, Carcass Quality Grade, and Cow Body Condition Score for Young Cows.

Marketing Class	Red Meat Yield	Dressing %	Approximate Carcass Quality Grade	Body Condition Score
Breaker	75–80	High	Commercial	8–9
		Average	Commercial	8
		Low	Com/Utility	7–8
Boner	80–85	High	Utility	6–7
		Average	Utility	6
		Low	Utility	5.5–6
Lean	85–90	High	Utility/Cutter	4.5–5.5
		Average	Cutter	4–4.5
		Low	Cutter	3–4
Light	75–90	High	Cutter	2–3
		Average	Cutter/Canner	2
		Low	Canner	1–2

Oklahoma State University Cooperative Extension Bulletin AGEC-613.

- **Lean** – defined as cows with an estimated red meat yield of 85 to 90 percent, which will yield at most a few merchandisable cuts with the majority of the carcass used for boneless processing beef.
- **Light** – is the term used for cows that vary in estimated red meat yield from 75 to 90 percent but always produce fewer pounds of boneless beef because the animal is small in overall size and weight, light muscled, and/or extremely thin.

Table 5-6 shows the relationship between body condition scores, carcass quality grade, and market class for young cows.

Palatability of Beef from Cull Cows

Beef from cull cows is tougher, leaner, and less juicy than beef from conventionally produced grain-finished animals. However, as noted, packers are merchandizing cuts to low-end restaurants and other food service operations. In many cases, this beef is partially to blame for complaints about inconsistency in beef quality. One practice that has improved the quality of beef from cull cows is feeding cull cows a high-energy diet before slaughter. This practice will be discussed in a subsequent chapter.

References

Izquirdo, M., D.E. Wilson, G.P. Rouse, and V. Amin. 1996. Relationship Between Carcass Endpoints and USDA Marbling Quality Grades: A Progress Report. Iowa Beef Research Report. Pg 49.

Lalman, D. and D. Doye, Eds. 2008. Beef Cattle Manual, 6th Ed. Oklahoma State University Extension Service.

Meat Evaluation Handbook. 1977. National Livestock and Meat Board.

Mikel, B., D. Bullock, and G. Anderson. The End Product. University of Kentucky Beef Book.

Morgan, J. 1997. Factors Affecting Dressing Percentage in Cattle. Proceedings American Feed Industry Association Symposium. 1997.

Peel, D.S. and D. Doye. Cull Cow Grazing and Marketing Opportunities. Oklahoma Cooperative Extension Service Bulletin AGEC-613.

United States Standards for Grades of Carcass Beef. 1997. USDA Agricultural Marketing Service.

Wagner, W. and P. Osborne. 1997. Quality and Yield Grades in Beef Cattle. University of West Virgina Extension Website.

Woerner, D.R. 2010. Beef from Market Cows. National Cattlemen's Beef Association White Paper. *www.beefusa.org* Website

Chapter 6

Basic Genetic Principles

Successfully breeding beef cattle requires an understanding of a few basic genetic principles. The physical expression of a trait is termed the animal's *phenotype*. It is a combination of the genetic makeup of the animal (its *genotype*) and environmental effects. The genotype is coded by deoxyribonucleic acid (DNA) and stored in units called chromosomes. The DNA codes for proteins that influence the animal's phenotype.

Cattle have 30 pairs of chromosomes with one member of each pair coming from the sire and one from the dam. Genes, the basic unit of inheritance, occur along these 30 chromosomes with each gene coding for a specific protein. The location of the gene on the chromosome is called the locus for that gene. Each animal has two copies of each gene, which may be the same or different. Alternative forms of a gene located at a specific locus are referred to as alleles. These alleles may interact with one being dominant to the other, which then dictates the phenotype, or both alleles may produce proteins that influence the phenotype.

Qualitative vs. Quantitative Traits

Beef cattle traits can be defined as either qualitative or quantitative based on the number of genes controlling the trait. An example of a qualitative trait in beef cattle is coat color, with only a few genes determining the color of the coat. Conversely, weaning weight is an example of a quantitative trait with numerous genes impacting the phenotype. Another way of looking at the difference is that qualitative traits tend to exhibit distinct groups, while quantitative traits are expressed across a range. Quantitative traits usually exhibit a normal distribution around the average of the group. Another typical distinction is that quantitative traits are influenced by the environment while qualitative traits are not. For example, weaning weight is greatly influenced by the environment while coat color is rarely influenced by the environment.

As noted, coat color is a trait controlled primarily by a single gene with several alleles possible at this locus on the chromosome. Two of the most common alleles are black and red with black being dominant to red. This explains why breeding an Angus bull to Hereford cows results in all black calves. These calves, however, will have both a copy of the black allele and a copy of the red allele at the color gene locus. A purebred Angus bull is termed homozygous because it carries two alleles that are the same (both black) while the calves of an Angus/Herford cross are termed heterozygous because the alleles are different (one black, one red). While the primary color of cattle is determined by this one gene, there are genes at other loci that can also influence color. For example, another gene tends to dilute the color to some extent. When present, this gene dilutes black to gray and red to yellow. Even though more than one gene influences coat color, it is still a qualitative trait because color is controlled by only a few genes and cattle can be put into distinct color groups.

Typically, quantitative traits in beef cattle are influenced by many genes, each with some positive or negative effect. The phenotype is essentially a sum of the effects of all of the genes impacting the trait plus the environmental affects. The genetic effects on a quantitative trait are of two types, additive and non-additive. With additive gene action, the gene has the same influence on the trait regardless of the other genes present. For example, a certain allele produces a 5-pound positive impact on weaning weight regardless of other genes present in the animal. With non-additive effects, the influence of the gene depends on interactions with other genes. Additive gene action is passed from generation to generation with a consistent influence on the phenotype. The sum of all of the genes that have an additive effect on a trait is termed the animal's breeding value for the trait. Figure 6-1 illustrates the concept of breeding value.

Figure 6-1. Example of Additive Effects of Genes and the Resulting Breeding Value.

Allele	Effect on Weaning Weight
A	+5 lbs
a	-2 lbs
B	+10 lbs
b	-6 lbs
C	+3 lbs
c	-2 lbs

Bull A Genotype: AABbcc Breeding Value = (+5)+(+5)+(+10)+(-6)+(-2)+(-2) = 10 lbs

Bull B Genotype: AaBBCc Breeding Value = (+5)+(-2)+(+10)+(+10)+(+3)+(-2) = 24 lbs

In the formation of gametes, the sperm in the male and egg in the female, one allele of each gene will be present. These gametes receive a random sampling of the genes in the respective parent. To illustrate, consider Bull B in Figure 6-1. The best sperm that this bull can pass on for weaning weight would be the combination ABC with a breeding value of (+5)+(+10)+(+3) = +18 lbs The poorest sperm would be the combination aBc with a breeding value of (-2)+(+10)+(-2) = +6 lbs

Heritability

As noted previously, the phenotype that an animal expresses is a combination of the genotype and environmental effects. For each trait, the percentage of the variability in phenotype controlled by additive genetics is termed its heritability. To illustrate this concept, consider a group of calves with an average weaning weight of 550 pounds, with a low of 450 pounds and a high of 625 pounds. The question is, how much of the difference in weaning weight in these calves is due to genetics and how much is due to the environment? The heritability of weaning weight is approximately 25 percent, which means that 25 percent of the difference

Table 6-1. Heritability Estimates for Traits in Angus and Charolais Cattle.

Trait	Angus	Charolais
Birth Weight	0.42	0.49
Calving Ease Direct	0.18	0.24
Weaning Weight	0.20	0.25
Post-Weaning Gain	0.20	0.31
Residual Average Daily Gain	0.31	
Yearling Height	0.45	
Scrotal Circumference	0.43	0.31
Docility	0.37	
Calving Ease Maternal	0.12	0.16
Maternal Milk	0.14	
Mature Weight	0.55	
Mature Height	0.82	
Hot Carcass Weight	0.30	0.21
Ribeye Area	0.28	0.34
Fat	0.25	0.26
Marbling	0.36	0.36

Source: Websites for the American Angus Association and the International Charolais Association, respectively.

is due to additive genetics and 75 percent is due to non-additive genetics and environmental influences. Table 6-1 shows the heritability of traits in beef cattle.

Obviously, selection for traits with high heritability is more successful because more of the differences between animals are due to additive genetics and less to non-additive genetics and environmental effects. Unfortunately, with traits low in heritability, most of the differences between animals are due to environmental effects, which won't be passed on. As shown, many economically important traits such as weaning weight and post-weaning gain are lowly to moderately heritable. This makes within-herd-selection less accurate and increases the value of tools like Expected Progeny Differences (EPD), which will be discussed in a later chapter.

The heritability of a trait is important because it indicates the expected response to crossbreeding. Traits with low heritability typically show greater heterosis due to crossbreeding than traits high in heritability. This relationship means that to some extent, selection pressure should be focused on traits high in heritability and crossbreeding used to improve performance in lowly heritable traits.

Genetic Correlations

Often, when one selects for a trait in beef cattle, other traits show a correlated response either positively or negatively. For example, there is a positive genetic correlation between birth weight and yearling weight primarily because some of the same genes influence growth rate in the womb and after birth. Moreover, there is a positive correlation between preweaning and postweaning growth rate and growth rate and mature size. There is a negative genetic correlation between growth rate and calving ease. There is also a negative correlation between growth rate and milking ability. One of the most significant genetic correlations is between birth weight and direct calving ease with a strong negative correlation. In other words, as birth weight increases, calving ease decreases. Fortunately, the development of Expected Progeny Differences has greatly improved the ability to avoid the impact of genetic correlations. For example, within several breeds, there are bulls that have high pre- and postweaning growth rate but moderate birth weight. This has allowed breeders to increase weaning and yearling weights without significantly increasing calving problems.

Selection

Selection is the practice of giving some individuals a greater chance to breed and pass on their genetics than others. The goal in selection is to increase the breeding value of the herd for desired traits. Many criteria have been used to select the parents of the next generation. Historically, visual appraisal was the primary basis for selection. In the 1960s and 1970s, on-farm performance programs and central bull tests contributed information to be used in selection. In the 1980s, EPD were developed. They have had a major impact on the accuracy of selection for many cattle traits. Recently, genomics (DNA evaluation) has been used to increase the accuracy of EPD. It is clear that this technology will have a significant impact in the future.

Genetic Progress

The genetic progress for a trait resulting from selection is a function of:

Heritability of the Trait – genetic change is directly related to the percentage of the variability in a trait that is due to additive genetics.

Selection Intensity – selection intensity is the difference between the animals that reproduce and the average of the group. For example, if the average weaning weight of a group of bull calves was 550 pounds, and the bulls selected for breeding averaged 650 pounds at weaning, the selection intensity is 100 pounds. The larger the selection intensity is, the greater the progress. The overall selection intensity is the average of the selection intensity for the bulls and the replacement females.

Generation Interval – this is essentially the average age of the parents when the replacements are born. In most beef cattle herds, the generation interval is about 5 years. Some purebred breeders turn generations over quickly to increase the rate of genetic progress. For example, if the best yearling heifers are superovulated and the embryos placed in other females, the generation interval will be greatly reduced.

The genetic progress for a trait per year is calculated by the formula: (Heritability × Selection Intensity) ÷ Generation Interval.

Example:

Assumptions:
Heritability of Weaning Weight = 0.20
Selection Intensity in Bulls – Selected group averaged 100 lbs above average
Selection Intensity in Heifers – Selected group averaged 20 lbs above average
Generation Interval = 5 years
Genetic Progress = (100 + 20) ÷ 2 × 0.20 ÷ 5 = 2.4 pounds per year.

References

Beef Sire Selection Manual. National Beef Cattle Evaluation Consortium.

Guidelines for Uniform Beef Improvement Programs, 9th Ed. 2010. Beef Improvement Federation.

Lalman, D. and D. Doye. Editors. Beef Cattle Manual. 6th Edition Oklahoma Cooperative Extension Service Publication E-913.

Chapter 7

Important Beef Cattle Traits

Numerous traits affect profitability in beef cattle production. The relative importance of these traits varies depending on the environment, marketing program, and current weaknesses in the herd. Typically, major traits in beef cattle have been categorized into three groups: Growth, Reproduction, and Carcass.

Growth Traits

Birth Weight

Birth weight can have both a positive and negative affect on production. For example, low birth weight calves cannot handle cold stress as well as heavier calves. On the other hand, heavy birth weights tend to increase calving difficulty and calf death loss, and cause delayed rebreeding of the dams. Calving difficulty can result in injury and even death of the dam. Birth weights are affected by several factors:

- **Genetics of the Sire** – Birth weight is a moderately heritable trait (heritability of 40 to 45 percent), and the sire has a major influence on birth weights.
- **Genetics of the Dam** – The dam contributes one-half of the genetics for birth weight to the calf, and thus she has a major influence on birth weight.
- **Maternal Environment** – In addition to contributing one-half of the genetics to the calf, the dam also provides the maternal environment during gestation. Some cows restrict calf growth in the womb, while others tend to produce heavy calves. There is a genetic component to this maternal effect, which is separate from the genetics for growth or birth weight. Several breed associations provide EPD for maternal calving ease, which estimates differences in the maternal genetic effect on calving difficulty.
- **Nutrition** – Inadequate nutrition, especially a deficiency of energy and/or protein, will reduce birth weights to some extent. It appears that providing nutrients to the developing fetus has a high priority in beef females because birth weights are not reduced significantly unless there is a severe deficiency of energy and/or protein. A slight deficiency may reduce birth weight by only a couple of pounds. A severe deficiency will reduce birth weights by several pounds. In summary, one can't starve out calving difficulty.
- **High Protein Levels** – A few research trials have shown high protein levels fed before calving result in heavy birth weights; however, this result has been inconsistent.
- **Weather** – Cold weather increases birth weights because it results in an increased blood flow to the internal organs and thus to the fetus. This explains why the same genetics result in lower birth weights in southern climates compared to northern climates. This also explains the tendency for the incidence of calving difficulty to be higher following a cold winter compared to a mild winter.
- **Age-of-Dam** – Actual birth weight of a calf is also affected by the age of the dam. Beef Improvement Federation guidelines for adjusting birth weights are shown in Table 7-1. Most breeds use age-of-dam adjustments based on their own data in their genetic evaluations. The

Table 7-1. Beef Improvement Federation Guidelines for Age-of-Dam Adjustments for Birth Weight and Weaning Weight.

		Weaning Weight	
Age-of-Dam at Birth of Calf (years)	Birth Weight Adjustment (lbs)	Male	Female
2	+8	+60	+54
3	+5	+40	+36
4	+2	+20	+18
5 to 10	0	0	0
11 and older	+3	+20	+18

BIF Guidelines

data in Table 7-1 indicate that a 2-year-old will have a calf 8 pounds lighter than she will as a mature cow (5 to 10 years old) on average.

Weaning Weight

Since most producers sell their calves at weaning, this trait typically has a major affect on profitability. Weaning weights are influenced by:

- **Genetics from Sire and Dam** – Weaning weight is low to moderate in heritability ($h^2 = 0.20 - 0.25$).
- **Milk Production by the Dam** – Weaning weight also is influenced greatly by the amount of milk provided by the dam. In fact, in a group of calves in the same environment, differences in milk production by the cows account for a significant amount of the variation in weaning weights.
- **Quantity and Quality of Forage** – Both the amount of forage available and its nutrient content influence weaning weights.
- **Age-of-Dam** – The age of the dam has a significant affect on weaning weight. Table 7-1 shows the age-of-dam adjustments suggested by the Beef Improvement Federation. For example, a 2-year-old dam will produce a calf 60 pounds lighter at weaning than she will as a mature cow. As with birth weight, each breed association uses its own adjustment factors based on its data. Age-of-dam effects on weaning weight result from lower milk production by a cow in her first couple of lactations compared to subsequent lactations.
- **Age at Weaning** – Actual weaning weight is also a function of age with older calves typically being heavier. For genetic comparisons such as are typical in seedstock operations, weaning weights are adjusted to 205 days-of-age. While this adjustment is relevant in assessing genetic differences, from a practical standpoint, older, heavier calves typically return higher revenues and are more profitable. Thus, in commercial cow herds, more emphasis should be placed on actual weaning weights rather than adjusted 205-day weights when evaluating cows.
- **Sex of the Calf** – Bull or steer calves are usually 30 to 50 pounds heavier than heifer calves at weaning.

Yearling Weight

Yearling weight is the animal's weight at approximately one year of age. For genetic comparisons, the actual yearling weight is adjusted to 365 days-of-age. Yearling weight is an especially important trait for producers who retain ownership through a stocker and/or finishing program. It is an important trait even for producers who sell at weaning because good performance in a finishing program results in repeat buyers willing to give a premium for the calves at weaning. Actual yearling weight is influenced by:

- **Weaning Weight** – Heavier calves at weaning tend to be heavier as yearlings because many of the same genes influence both preweaning and postweaning growth rate.

Table 7-2. Relationship between Yearling Hip Height, Frame Score, Mature Weight of Cows, and Steer Weights.

	Yearling Hip Height (in)		Expected Weight (lbs)		
Frame Score	Bulls	Heifers	Mature Cows	Steer Live	Steer Carcass
3	45	43	1,025	950	600
4	47	45	1,100	1,050	660
5	49	47	1,175	1,150	725
6	51	49	1,245	1,250	785
7	53	51	1,320	1,350	850
8	55	53	1,395	1,450	915
9	57	55	1,465	1,550	975

Beef Improvement Federation Guidelines

- **Nutrition Postweaning** – The quality of the nutrition program significantly influences the actual yearling weight.
- **Sex** – Bulls and steers are usually heavier than heifers at a year of age.

Yearling Height

In the 1970s and 1980s there was considerable emphasis on increasing the frame size of cattle, especially in the British breeds. Frame score (based on hip height adjusted for age) was widely used as a selection tool. During this period, cattle varied significantly in frame score, and larger framed cattle did tend to gain faster. Currently, hip height is not an important selection criterion in most breeds. Table 7-2 shows the relationship between frame score, hip height, expected mature cow weight and carcass weight of steers.

Mature Height

Another trait that once received considerable emphasis was mature height, and again as with yearling height, it is not a major emphasis in most breeding programs today.

Mature Weight

Mature weight is important because of the increased nutrient requirements required to support the additional weight. Some experts believe mature weights are too heavy in many cattle today, meaning that negative pressure should be applied in many breeding programs.

Milk

Milk production is important because of the influence it has on weaning weight; however, milk production requires significant amounts of protein and energy. In many situations, negative pressure should be applied for milk production.

Reproductive Traits

Fertility

Given the importance of this trait, it is unfortunate that it is very low in heritability (less than 10 percent in most studies). Response to selection for fertility will be negligible, but response to crossbreeding will be significant because of heterosis. Obviously, there is some natural selection for fertility because only those animals that reproduce pass their genes on to the next generation. There is another relationship between genetics and fertility that is important in that if the genetics, especially for cow size and milk production, does not fit the environment, reproductive performance is typically reduced.

Calving Difficulty

Calving difficulty causes significant losses in the cattle industry because calves that experience a difficult birth have a lower survival rate. Additionally, females that experience a difficult birth take longer to return to estrus often resulting in a younger, smaller calf in the following year. Table 7-3 shows the level of calving difficulty in first-calf heifers reported in the Beef 2007-08 survey conducted by the USDA. It appears that calving problems have been reduced based on a

Table 7-3. Percentage of First-Calf-Heifers Requiring Various Levels of Assistance During Calving.

Assistance Level	1992/93 CHAPA*	Beef 97 Comparable	Beef 97	Beef 2007–08
	Percent Replacement Heifers			
Easy pull	9.4	10.6	11.2	7.7
Hard pull	7.4	4.7	5.1	3.4
Cesarean section	0.4	0.3	0.4	0.5
No assistance	82.8	84.4	83.3	88.4

* *Cow/Calf Health and Productivity Audit*

comparison of the levels reported in 1992/1993 to those reported in the Beef 2007-08 survey. It is likely that this reduction is the result of selection pressure for calving ease applied by seedstock producers over the period between these surveys.

As a note, a survey conducted in Kansas by this author in the early 1980s indicated an overall assistance level of 25 percent in first-calf-heifers. At that time, many producers were shifting from British breeds to the European breeds with much higher growth rates and frame size. This resulted in significant levels of calving difficulty in many herds.

Scrotal Circumference

Research has shown a relationship between the scrotal size of a bull and puberty in his daughters with larger scrotal size resulting in earlier puberty in daughters. However, scrotal circumference should probably be considered a threshold trait, which means that once adequate scrotal size has been reached, there is little reason to apply selection pressure. The minimum scrotal circumference as a yearling should be 12.5 inches (32 cm). Adequate scrotal size is not a problem in most breeds, thus little emphasis needs to be placed on this trait in these breeds.

Carcass Traits

Over the past 20 years, carcass traits have become more economically important with the development of branded-beef programs. Producers that retain ownership through slaughter have been rewarded for breeding cattle that produce a higher-quality carcass. The high-quality beef market is still growing, which means carcass traits may increase in economic importance.

Marbling

Marbling (intramuscular fat) is related to tenderness, juiciness, and palatability of beef. With the growth of the branded beef programs, this trait has received considerable selection pressure. It has a moderate heritability (h^2=0.35) so it responds to selection. Furthermore, most breeds have done a good job of identifying bulls significantly higher than average for marbling, which means there can be significant selection intensity. Because of significant premiums for cattle with a grade of choice or better, the potential economic impact of improving this trait can be important for producers that retain ownership. Conversely, this trait is much less important for producers that sell their calves at weaning.

Carcass Weight

Carcass weights have increased dramatically over the past 20 years because of the economics of production rather than a need to make carcasses larger. In fact, the current carcass size is too large in many cases when one considers the size of retail cuts from these carcasses. For example, a ribeye steak from a carcass with a 15 square inch ribeye cut one inch thick is too large for many consumers. The economics of production, however, may make carcasses even larger in the future. The food service industry is developing innovative approaches to fabricating carcasses to obtain acceptable portion sizes.

Ribeye Area

Ribeye area has a moderate heritability (h^2=.37), and it is an indicator of the relative amount of muscle in the carcass. Some breeds need to increase ribeye area while others need to moderate it to some extent because of the points made in the previous paragraph. For example, the continental breeds such as Simmental and Charolais have adequate muscle, while some lines of Angus would benefit from an increase in ribeye area. Ribeye area is used to arrive at yield grade with larger ribeye area resulting in lower numeric yield grade, which is an advantage in most grid pricing formulas.

Fat Thickness

Fat depth is also moderately heritable (h^2=0.44), and it is used in the yield-grade formula. Consequently, it has economic significance in many pricing formulas. The goal should be to produce carcasses with high levels of marbling without excessive external fat.

Yield Grade

As with the other carcass traits, yield grade is moderately heritable (h^2=0.36). Carcass weight, ribeye area, and fat are all major factors in the yield-grade formula. Yield grade is a good predictor of the percentage of lean, retail product that the carcass will produce. Yield grade 1 and 2 carcasses are rewarded in most pricing grids, while yield grade 4 and 5 carcasses are discounted, as will be illustrated in a subsequent chapter.

Other Important Traits

In addition to the traits listed above, there are several other important traits that do not fall into one of the major categories.

Soundness

This trait covers a broad range of physical characteristics that can lead to premature culling. Probably the most common problem is poor udders, especially enlarged teats. Other soundness issues include a tendency to prolapse following calving and a tendency to get cancer eye. In addition, sound feet and legs are important. Some breed associations are calculating a stayability EPD to aid in selecting for general soundness.

Temperament

The heritability of temperament in cattle is quite high (h^2 = 0.3 to 0.5). Cattle that are difficult to handle increase the time required for general processing and in some cases are simply dangerous. Research shows that cattle with bad temperament grow slower in growing and finishing programs, and reproductive performance is lower in heifers with bad temperament. High-strung cattle should be culled whenever possible. Some breeds are calculating a docility EPD as a tool for applying selection pressure for temperament.

Feed Efficiency

This is an extremely important trait both in the cow herd and in the feedlot. Research studies have shown that there are significant differences between animals in their maintenance requirements and ability to convert feed to gain. Unfortunately, it is expensive and time consuming to identify these differences, and thus it has not been done on a large scale. Many seedstock producers, however, are placing increased emphasis on this trait. A few breeds are collecting data, which may make it possible to delineate efficiency differences in the future. Furthermore, genomics may be a powerful tool in selecting for this trait in the future.

Relative Trait Emphasis

Obviously, many traits can influence the profitability of a beef cattle operation. Consequently, considerable thought should be given to the relative importance of traits in a specific operation. It is important to remember that as the number of traits considered in the selection program increases, the improvement in each trait is reduced. This makes identification of the economically important traits in an operation essential. Trait emphasis should be based on:

1. **Production System and Feed Resources** – In production systems where the quantity and/or quality of the basal feedstuffs is restricted, it may be more important to place emphasis on traits that affect feed requirements such as cow weight and level of milk production. Conversely, in production systems with abundant, high-quality feed throughout most of the year, less emphasis should be placed on these traits.
2. **Marketing Program** – If the calves will be sold at weaning, considerable emphasis should be placed on weaning weight and milk production with little consideration of yearling weight and carcass characteristics. Conversely, if the calves will be retained, even for a stocker program,

there should be more emphasis on postweaning growth as indicated by yearling weight.

3. **Current Performance in the Herd** – To illustrate this concept, consider two herds, Herd A with a high level of calving difficulty in first-calf-heifers and Herd B with minimal calving difficulty. Herd A should probably put significant emphasis on selecting for calving ease while Herd B should place minimal emphasis on this trait.

Economically Relevant Traits and Indicator Traits

Obviously, from the coverage of traits in this chapter, many traits can be considered in a breeding program. However, since the goal is profitability, it is crucial to identify the traits that directly affect profitability. Traits that directly influence profitability are referred to as Economically Relevant Traits (ERT). For example, weaning weight is an ERT for producers that sell their calves at weaning, and marbling score is an ERT for producers who retain ownership and market on a high quality-grid. Traits that may or may not influence profitability are termed Indicator Traits. For example, birth weight is considered an important trait because of its relationship to calving difficulty. However, a slight increase in birth weight typically will have little influence on calving difficult or profitability. An exception to this rule would be in bulls mated to yearling heifers where even a few pounds difference in birth weight can significantly influence calving problems and profitability.

There are numerous examples of indicator traits like selecting for increased scrotal circumference in bulls to increase the percentage of the bull's daughters that cycle early and get pregnant. In some cases, it is effectively impossible to select directly for a trait because there is not an ERT for the trait. For example, cow maintenance feed requirement has a significant influence on profitability, but it is impossible to measure in a commercial operation. Consequently, producers are forced to use cow weight and level of milk production (indicator traits) to estimate differences in feed requirements.

In some cases, a trait may be both an ERT and an indicator trait. For example, cow weight is an ERT because heavier cows increase income when culled and sold. And as noted previously, cow weight is an indicator trait for cow maintenance feed requirements.

When an ERT is available for a trait, it should be emphasized over any indicator traits. For example, when selecting bulls, both a direct calving ease EPD and birth weight EPD are usually available. The calving ease EPD is a better predictor of potential calving difficulty. Furthermore, in most breed genetic evaluations, the birth weight EPD is factored in to the calving ease EPD.

References

Beef 2007-08. 2009. USDA.

Beef Sire Selection Manual. National Beef Cattle Evaluation Consortium.

Guidelines for Uniform Beef Improvement Programs, 9th Ed. 2010. Beef Improvement Federation.

Marshall, D.M. 1994. Breed Differences and Genetic Parameters for Body Composition Traits in Beef Cattle. JAS 72:2745-2755.

Chapter 8

Breeds of Beef Cattle

One of the most important decisions a cattle producer makes is the breed or breeds that will comprise the herd. This decision is complicated by the fact that there are more than 60 recognized breeds of beef cattle in the United States. In the next chapter, the concept of matching the genetics to the resources is discussed. A good match is crucial for a profitable cow/calf operation. Clearly, making good decisions regarding breeds requires some understanding of the relative strengths and weaknesses of each breed. It should be noted, however, that even though each breed exhibits certain general characteristics, there is tremendous variation in performance levels within each breed. This means that selecting the source of genetics within a breed may be as important as the choice of breed(s).

Table 8-1. Breed Characteristics Based on U.S. Meat Animal Research Center Data (1970s Evaluation).

Breed	Growth Rate and Mature Size	Percent Retail Product	Age at Puberty	Milk Production
Angus	XXX	XX	XX	XXX
Brahman	XXX	XXX	XXXXX	XXX
Brangus	XXX	XX	XXXX	XX
Braunvieh	XXXX	XXXX	XX	XXXX
Charolais	XXXXX	XXXX	XXXX	X
Chianina	XXXXX	XXXX	XXXX	X
Devon	XX	XX	XXX	XX
Galloway	XX	XXX	XXX	XX
Gelbvieh	XXXX	XXXX	XX	XXXX
Hereford	XXX	XX	XXX	XX
Holstein	XXXX	XXXX	XX	XXXXX
Jersey	X	X	X	XXXXX
Longhorn	X	XXX	XXX	XX
Maine Anjou	XXXXX	XXXX	XXX	XXX
Nellore	XXX	XXX	XXXX	XXX
Piedmontese	XXX	XXXXX	XX	XX
Pinzgauer	XXX	XXX	XX	XXX
Red Poll	XX	XX	XX	XXX
Sahiwal	XX	XXX	XXXXX	XXX
Salers	XXXXX	XXXX	XXX	XXX

Breed	Growth Rate and Mature Size	Percent Retail Product	Age at Puberty	Milk Production
Santa Gertrudis	XXX	XX	XXXX	XX
Shorthorn	XXX	XX	XXX	XXX
Simmental	XXXXX	XXXX	XXX	XXXX
South Devon	XXX	XXX	XX	XXX

Adapted from Beef Cattle Handbook BCH-1410. Increasing number of Xs indicate higher performance.

Breed Characterization

In the 1960s and 1970s, many new breeds of cattle entered the United States, especially from Europe. This introduction of new genetics stimulated a germplasm evaluation program at the Meat Animal Research Center at Clay Center, Nebraska. This program has evaluated more than 30 breeds of cattle from all over the world. The data from this research have given the industry a general characterization of these breeds for growth, maternal performance, age at puberty, and carcass characteristics. Table 8-1 shows a summary of characteristics for the breeds evaluated through 1993.

Does the Breed Characterization Information in Table 8-1 Reflect Current Breed Characteristics?

As discussed previously, the breed evaluation project at the Meat Animal Research Center was initiated in the late 1960s. The genetics selected for evaluation reflected a good representation of the breeds at that time. However, selection programs in all breeds have changed their characteristics since the 1970s. In 1999, the Meat Animal Research Center initiated a project to compare the major breeds again to make a more current comparison. Seven breeds were selected for evaluation: three British breeds (Hereford, Angus, and Red Angus) and four Continental breeds (Simmental, Gelbvieh, Limousin, and Charolais). Approximately 20 bulls were selected from each breed to represent the genetics in the breed in 1999. Tables 8-2 through 8-9 show some of the results from that project. It is important to note that the differences shown represent just the sire effect. Thus the differences between breeds are actually twice as large. For the maternal traits shown in Tables 8-7, 8-8, and 8-9, the differences shown represent one-quarter of the direct and one-half of the maternal breed effects. The LSD (Least Significant Difference) in each table indicates the difference that is significant at the 5 percent level. Thus, when comparing two breeds, there is a statistically significant difference at the 5 percent level if the difference in their performance is greater than the LSD. For example, in Table 8-2, there is a significant difference in the weaning weight of

Table 8-2. Breed of Sire Means for Preweaning Traits of F_1 Calves Produced in Cycle VII of the GPE Program.

Sire Breed of Calf	No. of Calves	Calving Unassisted, %	Calving Difficulty Score	Birth Wt., lbs	200-Day Wt., lbs
Hereford	190	95.6	1.24	90.4	524
Angus	189	99.6	1.01	84.0	533
Red Angus	206	99.1	1.06	84.4	526
Simmental	201	97.7	1.10	92.2	553
Gelbvieh	209	97.8	1.10	88.6	534
Limousin	200	97.6	1.13	89.5	519
Charolais	199	92.8	1.40	93.7	540
LSD<.05		3.6	.21	3.3	14.1

Germplasm Evaluation Program Progress Report No. 22

Table 8-3. Least Square Means for Breed of Sire for Postweaning Growth and Carcass Traits of F_1 Steers (Slaughtered at 445 Days-of-Age).

Sire Breed of Calf	No.	Post Wn. ADG, lbs	Final Wt., lbs	Carcass Wt., lbs	Dress. %	Marb. Score	% Choice	YG	Fat In.	REA Sq. in.
Hereford	97	3.23	1,322	803	60.5	526	65.4	3.19	.50	12.3
Angus	98	3.32	1,365	836	61.1	584	87.6	3.44	.58	12.9
Red Angus	93	3.26	1,333	811	60.7	590	89.9	3.44	.53	12.1
Simmental	92	3.26	1,362	829	60.8	527	65.7	2.72	.37	13.6
Gelbvieh	90	3.12	1,312	800	61.0	506	57.7	2.60	.35	13.4
Limousin	84	3.12	1,285	795	61.8	504	56.9	2.43	.37	13.9
Charolais	95	3.21	1,348	826	61.3	517	61.9	2.54	.34	13.7
LSD<.05		0.13	40	24	0.6	28	10.7	0.31	0.08	0.52

Germplasm Evaluation Program Progress Report No. 22

Table 8-4. Least Square Means for Breed of Sire for Estimated Retail Product, Fat Trim, and Bone Yields and Weights of F_1 Steers in Cycle VII of the Germplasm Evaluation Program at the U.S. Meat Animal Research Center.

		Retail Product		Fat Trim		Bone	
Sire Breed	**No.**	**%**	**lbs**	**%**	**lbs**	**%**	**lbs**
Hereford	86	60.7	480	26.0	213	14.3	113
Angus	83	59.2	488	27.7	235	13.7	113
Red Angus	81	59.1	474	27.6	226	13.8	111
Simmental	80	63.0	522	23.6	197	14.3	119
Gelbvieh	81	63.8	509	22.7	182	14.6	116
Limousin	73	63.7	504	23.5	188	14.1	111
Charolais	85	63.5	523	22.9	190	14.5	119
LSD<.05		1.3	16	1.6	16	.4	4

Germplasm Evaluation Program Progress Report No. 22

calves out of Hereford sires compared to calves out of Charolais sires because there is more than 14.1 pounds difference in the 200-day weights of their calves. Correspondingly, there is not a significant difference in the weaning weights of calves sired by Hereford and Angus bulls.

As shown in Table 8-2, Charolais sires caused more calving difficulty as would be expected given their heavier birth weights.

Table 8-3 shows the postweaning performance and carcass characteristics of the F_1 steers. These data indicate that the British breeds have made significant improvement in postweaning gain since the original evaluation of these breeds. In fact, there was little difference in postweaning gain, slaughter weights, and carcass weights between these breeds. There were, however, some major differences in carcass characteristics. In making comparisons amongst breeds, it is important to note that these cattle were slaughtered on an age constant basis. This resulted in the carcasses from the British breeds being considerably fatter. Correspondingly, they had higher yield grades, marbling scores, and percent choice. On the other hand, the carcasses from the Continental breeds were heavier

Table 8-5. Least Square Means for Breed of Sire for Warner–Bratzler Shear Force and Sensory Characteristics of Rib Steaks Aged for 14 Days.

Sire Breed	No.	WB Shear Force, lbs*	Sensory Panel**		
			Tenderness Score	Flavor Score	Juiciness Score
Hereford	86	9.1	5.6	4.9	5.3
Angus	83	8.9	5.8	4.9	5.4
Red Angus	82	9.2	5.7	4.9	5.4
Simmental	80	9.5	5.6	4.9	5.3
Gelbvieh	81	10.0	5.3	4.8	5.2
Limousin	73	9.5	5.7	4.9	5.3
Charolais	85	9.6	5.5	4.9	5.2
LSD<.05		0.7	0.32	0.12	0.12

** Lower shear values reflect greater tenderness. ** Sensory scores: 1 = extremely tough, bland, or dry through 8 = extremely tender, intense, or juicy. Germplasm Evaluation Program Progress Report No. 22*

Table 8-6. Breed of Sire Means for Estimates of Feed Efficiency (Live Weight Gain Per Unit of Metabolizable Energy Consumes, lb/Mcal) for Alternative Intervals and Endpoints.

Sire Breed	Time 187 Days	Weight 750–1,300 lbs	Marbling Small 35	Fat Thickness 0.43 in.	Fat Trim 24.8%	Retail Product 456 lbs
Hereford	.1306	.1275	.1289	.1334	.1331	.1256
Angus	.1248	.1236	.1344	.1312	.1319	.1214
Red Angus	.1227	.1202	.1329	.1273	.1291	.1159
Simmental	.1265	.1271	.1250	.1226	.1233	.1330
Gelbvieh	.1221	.1201	.1182	.1185	.1182	.1246
Limousin	.1281	.1230	.1223	.1243	.1248	.1300
Charolais	.1202	.1202	.1170	.1151	.1158	.1264
LSD<.05	.0018	.0049	.0076	.0075	.0074	.0095

Germplasm Evaluation Program Progress Report No. 22

Table 8-7. Sire Breed Least Square Means for Growth and Puberty Traits of Heifers.

Sire Breed of Female	18 Month Wt., lbs	Frame Score*	Puberty Expressed %	Puberty Wt., lbs	Age at Puberty Days	Preg. Rate %
Hereford	950	5.5	79.2	733	357	94
Angus	936	5.3	97.2	750	343	88
Red Angus	953	5.2	97.4	744	342	91
Simmental	961	5.9	91.6	757	342	90
Gelbvieh	922	5.6	91.5	711	329	83
Limousin	933	5.8	80.3	785	377	87
Charolais	950	5.8	88.5	758	358	91
LSD<.05	32	.3	9.9	35	16	13

** Based on hip height adjusted for age with larger frame scores indicating taller animals. Germplasm Evaluation Program, Progress Report No. 22 U.S. Meat Animal Research Center.*

muscled and had a higher retail product yield as is shown in Table 8-4. It appears that while the differences in growth rate between breeds has narrowed, differences in muscling and marbling ability still exist.

Table 8-5 shows the palatability characteristics of steaks from steers representing the sire breeds evaluated. Carcasses from Angus sired steers tended to be more desirable, probably because of their higher fat content. Again, if these steers had been fed to the same compositional endpoint (the same external fat thickness), there would probably be less difference in carcass palatability.

Table 8-6 shows breed of sire estimates for feed conversion. The British breeds were more efficient when the target was marbling, fat thickness, or fat trim percentage. Conversely, the Continental breeds were more efficient when the goal was producing 456 pounds of retail product.

The differences in carcass characteristics between breeds are extremely relevant when one considers the number of cattle that are being marketed on grids that reward quality and/or retail product yield. These grids will be discussed in a subsequent chapter.

Sire breed influence on heifer puberty and growth are shown in Table 8-7. Differences in weight at 18 months of age were relatively minor, but the Continental breeds were slightly taller as shown by their larger frame scores. Weight at puberty was much heavier than in the data from the 1970s. Gelbvieh expressed puberty at the youngest age while Limousin were the oldest when they expressed puberty. There were not any significant differences in pregnancy rates.

Table 8-8 shows the calving and maternal performance of the F_1 females as two-year-olds. Notably, the Simmental sired heifers calved the easiest, while the Gelbvieh sired heifers exhibited the most difficulty, which makes sense because they had the heaviest birth weights. Simmental and Gelbvieh sired heifers produced the heaviest calves at weaning indicating higher milk production as well as higher growth rate.

Heights, condition scores, and weights for these F_1 cows as 5-year-olds are shown in Table 8-9. The Continental breeds were still slightly taller and had a tendency to be in a slightly lower body condition compared to the British breeds. Differences in body weights were insignificant with the exception of the Gelbvieh sired cows, which were lighter than most of the other breeds. This similarity in mature weights is quite different than was observed in the 1970s evaluation where the cows sired by the British breeds were smaller than those sired by the Continental breeds. The common perception that the British breeds have lighter mature weights simply is not true any more.

The data in Table 8-10 show a comparison of breed of sire means for a number of breeds, including several breeds that were not included in Cycle VII. Again, the values shown represent just the sire effect, and the actual differences between breeds would be twice as large. Table 8-10 provides estimates of calf performance from average bulls of each breed.

A review of Tables 8-2 through 8-10 indicates that there are still some differences between breeds, but the differences are much less than in the 1970s

Table 8-8. Least Square Means by Breed of Dams Sire for Reproduction and Maternal Traits of F_1 Females Producing Their First Calf at 2 Years of Age.

Sire Breed of Female	No.	Calf Crop		Calving Diff. Score	Unassist. Births, %	Birth Wt., lbs	200-Day Wt. Per Calf, lbs
		Born, %	Weaned, %				
Hereford	80	92	70	1.9	74	81.6	413
Angus	84	83	76	2.0	72	79.8	424
Red Angus	104	86	76	2.2	68	78.3	415
Simmental	98	86	69	1.5	86	79.6	442
Gelbvieh	109	79	68	2.2	64	83.6	447
Limousin	109	85	73	2.0	68	80.2	429
Charolais	97	87	73	2.1	69	81.6	430
LSD<.05		7	8	.21	3.7	3.5	19.2

Germplasm Evaluation Program Progress Report No. 22

Table 8-9. Breed of Sire Means for Height, Condition Score, and Weight of F_1 Cows (Adjusted for Condition Score) at 5 Years of Age.

Sire Breed of Female	No.	Height at % Years of Age, in.	Condition Score at 5 Years of Age*	Weight at 5 Years of Age, lbs**
Hereford	56	53.3	7.00	1,420
Angus	62	52.9	7.22	1,411
Red Angus	68	52.9	7.40	1,409
Simmental	70	54.1	6.98	1,404
Gelbvieh	68	53.1	6.77	1,323
Limousin	80	54.1	6.95	1,391
Charolais	69	54.1	7.09	1,371
LSD<.05		0.8	0.28	57

** Condition scores were assigned on a scale from 1 = very thin to 9 = very fat. ** Because the condition scores were abnormally high as a result of early weaning the calves because of drought, the weights shown are the result of adjusting the condition scores to 5.5. Germplasm Evaluation Program Progress Report No. 22*

Table 8-10. Breed of Sire Means for 2010 Born Animals Under Conditions Similar to USMARC.

Breed	Birth Wt., lbs	Weaning Wt., lbs	Yearling Wt., lbs	Maternal Milk	Marb Score*	REA Sq. in.	Fat in.
Angus	89.8	582	1,037	570	5.92	13.0	0.59
Hereford	94.3	576	1,005	549	5.19	12.8	0.53
Red Angus	90.3	566	999	563	5.59	12.6	0.54
Shorthorn	96.3	566	1,016	568	5.34	12.9	0.42
South Devon	94.8	579	1,021	569	5.84	13.0	0.48
Beefmaster	95.0	578	997	558			
Brahman	100.8	592	980	577			
Brangus	92.4	571	1,007	566			
Santa Gertrudis	96.0	578	993		4.82	12.5	0.46
Braunvieh	92.1	557	977	582	5.23	13.6	0.39
Charolais	97.2	599	1,041	561	5.05	13.8	0.36
Chiangus	93.2	557	989		5.32	13.1	0.45
Gelbvieh	93.3	581	1,013	579			
Limousin	93.3	580	1,000	559	4.75	14.2	
Maine-Anjou	93.8	561	995	563	4.92	13.7	0.37
Salers	91.6	573	1,017	571	5.58	13.4	0.37
Simmental	93.9	591	1,031	571	5.11	13.8	0.38
Tarentaise	91.6	584	1,002	572			

** Marbling score units: 4.00 = Sl^{00}; 5.00 = Sm^{00} (Choice). USMARC Website*

(Table 8-1) because of the selection programs implemented by purebred breeders in all breeds. At this time, there is almost as much variation within each breed as there is across breeds. This means that rather than think in terms of generalized breed differences, one must identify seedstock breeding programs within a breed that fit the resources and goals of the commercial operation.

References

Cundiff, L. Biological Types of Cattle. Beef Cattle Handbook BCH-1410.

Cundiff, L., T.L. Wheeler, K.E. Gregory, S.D. Shackelford, M. Koomaraie, R. M. Thallman, G.D. Snowder, and L.D Van Vleck. 2004. Germplasm Evaluation Program Progress Report No. 22.

Cundiff L. and L.D. Van Vleck, Proceeding of Beef Improvement Federation, 2006.

Cundiff, L.V., R.M. Thallman, L.D. Van Vleck, G.L. Bennett, and C.A. Morris. 2007. Cattle Breed Evaluation at the U.S. Meat Animal Research Centre and Implications for Commercial Beef Farmers. Proceedings of the New Zealand Society of Animal Production Vol 67.

Wheeler, T.L., L.V. Cundiff, L.D Van Vleck, G.D. Snowder, R. M. Thallman, S.D. Shackelford, and M. Koomaraie. 2006. Germplasm Evaluation Program, Progress Report No. 23 U.S. Meat Animal Research Center.

U.S. Meat Animal Research Center Website. (*www.marc.usda.gov*).

Chapter 9

Matching Genetics to the Resources

An important characteristic of profitable cow/calf operations is a match between the genetics of the cattle and the resources. The importance of matching genetics to the resources has increased in recent years because of the tremendous increase in animal performance and because of high feed costs resulting in part because of the development of the ethanol industry. In other words, with current genetics, it is fairly easy to exceed the level of production that will result in the greatest profit.

Economic vs. Biological Efficiency

Comparisons of different types of cattle are often made strictly in terms of biological efficiency such as the pounds of calf weaned per cow exposed. Biological efficiency ignores the costs of feed inputs and value per unit for the final product. Therefore, using biological efficiency to make comparisons may result in erroneous conclusions about profitability. To illustrate, consider two herds, one having a high level of milk production and the other a low level. If both herds are fed to allow the cattle to reach their genetic potential, the herd with the high level of milk production should have a higher weaning weight per cow exposed — a greater biological efficiency. However, the economic efficiency ranking may be just the opposite, because of much higher feed costs in the herd with high milk production.

Another example of comparisons based on biological efficiency leading to a misleading conclusion is crossbreeding. Research has shown that crossbreeding increases pounds of calf weaned per cow exposed by as much as 20 to 25 percent. When the comparison is made on an economic basis, however, the increase in profitability as a percentage is lower because crossbred cows and crossbred calves consume more feed to produce the extra pounds of calf.

Too often, the assumption is made that if a technology or practice increases productivity, it will increase profitability, and this often is not the case. If the goal is to maximize profits, both the cost and return side of the equation must be considered. Furthermore, this cost/return analysis should be on a total farm/ranch basis rather than a per head basis. To illustrate why, compare running high-producing cows on a ranch to running cows with a more moderate level of production. On a per head basis, the higher-producing cows may be more profitable, but the ranch can produce feed for fewer of them, which may result in lower profitability for the operation. The goal should be to maximize gross profit for the entire operation, not profit per cow.

The Resources

When thinking about the resources in a cow/calf operation, the first one that usually comes to mind is the feed resource, because it is the most important factor determining the most profitable level of production. However, other resources must be considered including labor, capital, and marketing options.

Feed Resource Considerations

There is a wide range in feed resources used for cattle production across the United States. As a general statement, areas with abundant feed supplies can support cattle with higher productivity than areas with limited quantity or quality of feed. This relationship is shown in Table 9-1.

Even in an environment where feed availability is high, the ability to store fat during periods of abundant feed can be a valuable trait because there are few environments where high quality feed is abundant all

Table 9-1. Matching the Genetic Potential for Beef Cattle Traits to the Production Environment.

Production Environment	Traits				
Feed Availability	Milk Production	Mature Size	Ability to Store Energy*	Calving Ease	Lean Yield
High	M to H	M to H	L to M	M to H	H
Medium	M to H	M	M to H	M to H	M to H
Low	L to M	L to M	H	M to H	M

*L = Low, M = Medium, and H = High *Ability to store fat during times of abundant feed for use when feed quantity or quality is limited. Mature Size – Low = < 1,000 lbs, Medium = 1,000 to 1,200 lbs, and High = > 1,200 lbs*
Modified from Beef Improvement Federation Guidelines, 9th addition.

year. And stored fat can be used as an energy source to reduce the need for supplementation during periods when the available forage is low in quality and/or quantity.

Labor Resource Considerations

Availability of quality labor has become a problem for many ranching operations in recent years. This has forced cattle producers to prioritize the use of labor and minimize the need for labor as much as possible. As a general rule, cattle with high productivity require more labor inputs because of more calving difficulty and a greater need for supplementation.

Capital Resource Considerations

On many cattle operations, the cattle enterprise must compete with other enterprises for capital. For example, in diversified farming/ranching operations, producers often have limited operating capital and a decision must be made to spend the capital on fertilizer for crops or on supplement for the cows. Limited capital also can be a problem on a pure ranch operation when the capital available will not allow for the purchase of the optimal amount of supplement or other inputs. Thus, if capital is limited, a cow with lower productivity may be most profitable.

Marketing Option Considerations

Another resource that influences the optimal cow for an operation is the marketing options available. For example, if the calves must be sold at weaning, more emphasis should be placed on traits that influence reproductive performance, total weaning weight, and the value of the calves per pound. Conversely, if the calves can be retained through finishing, more emphasis should be placed on postweaning growth and carcass quality with less emphasis on preweaning growth and possibly milking ability. In many operations, capital availability may dictate the available marketing options. For example, bankers often force cow/calf producers with significant debt to sell their calves at weaning to reduce their debt.

Assessing the Match between Genetics and the Resources

Because there is such a strong relationship between the nutrition program and reproductive performance in the herd, reproductive performance is a good indicator of the match. If reproductive performance is good without excessive supplementation, then the cattle fit the resources. If excessive supplementation is required to obtain good reproductive performance, the cattle probably exceed the optimal level of performance. However, the cost of the supplementation must be considered. In making this assessment of the match, it is important to evaluate both the calf crop and calving sequence. In addition, this evaluation should be done by age group of the females. To illustrate why, consider that the newest genetics in the herd is represented by the yearling heifers and two-year-olds. One of the first indications of a poor match is poorer reproductive performance in these younger females. For example, a reduced percentage of two-year-olds rebreeding in the first 21 days of the breeding season may be an indication of a problem. This illustrates the importance of monitoring reproductive performance annually.

References

Beef Sire Selection Manual. National Beef Cattle Evaluation Consortium.

Davis, K.C., M.W. Tess, D.D. Kress, D.E. Doornbos, and D.C. Anderson. 1994. Life Cycle Evaluation of Five Biological Types of Beef Cattle in a Cow-Calf Range Production System: Biological and Economic Performance. JAS 72:2591-8.

Gosey, J. 1984. Matching Genetic Potential to Feed Resources. Proceedings Cornbelt Cow-Calf Conference, Ottumwa, IA.

Guidelines for Uniform Beef Improvement Programs, 9th Ed. 2010. Beef Improvement Federation.

Joandet, G.E. and T.C. Cartwright. 1975. Modeling Beef Production Systems. Journal of Animal Science 41:1238.

Lalman, D. and D. Doye, Eds. 2008. Beef Cattle Manual, 6th Ed. Oklahoma State University Extension Service.

Chapter 10

Crossbreeding Beef Cattle

As a general statement, commercial cattle producers should be crossbreeding because crossbreds have two major advantages over straightbreds:

- **Crossbreds exhibit hybrid vigor or heterosis** – crossbreds are more vigorous and exhibit higher levels of production than the average of the straightbred parents. This is particularly true with respect to calf vigor and survival, and reproduction, especially in harsh environments.
- **Crossbreeding can take advantage of breed complementarity** – for example, if a breed excels in growth rate but is below average in carcass quality, it can be crossed with a breed that excels in carcass quality. The resulting crossbred may not be superior to either parental breed in any trait but superior in overall performance. Selection of breeds for a crossbreeding program is key to a successful program — the breeds must complement each other for the program to produce significant improvements in performance.

What is Heterosis?

Heterosis is the phenomenon where a crossbred exceeds the average performance of the straightbred parents. It is important to note that heterosis is measured relative to the average of the straightbred parents rather than to the best straightbred parent. To illustrate this concept, cross breed A with an average weaning weight of 530 pounds to breed B with an average weaning weight of 580 pounds. The average of the two breeds is 555 pounds. If the weaning weight of the crossbreds averages 575 pounds, the percentage of heterosis is (575 – 555) = 20 ÷ 555 = 3.6 percent. Note that the average weaning weight of the crossbreds is lower than the superior parental breed (Breed B). Then why make this cross rather than just use Breed B? Remember that heterosis will be exhibited for numerous traits, not just weaning weight. Thus, we would expect that hybrid vigor would result in better calf survival and a higher percentage of cows weaning a calf. In some research trials, crossbreeding has increased the pounds of calf weaned per cow exposed by 25 percent.

What Determines the Amount of Heterosis?

The amount of heterosis exhibited by crossbreds is influenced by:

- **Heritability of the Trait** – There is an inverse relationship between the heritability of a trait and the amount of heterosis exhibited. In other words, traits high in heritability exhibit little heterosis, while traits low in heritability exhibit high levels of heterosis. This makes sense when considering that heritability is a measure of the additive genetic component for a trait, while heterosis is a function of the non-additive genetic component.
- **Breed Differences** – The greater the differences between two breeds, the higher the level of heterosis. For example, crossing *Bos taurus* breeds with *Bos indicus* breeds results in more heterosis than crossing a *Bos taurus* breed with another *Bos taurus* breed. The amount of heterosis exhibited is a function of the number of different alleles in the crossbred, and there are more alternate forms of alleles that are different between *Bos taurus* and *Bos indicus.*
- **Percentage of Each Breed in the Crossbred** — The amount of heterosis exhibited is also a function of the percentage of each breed in the

Table 10-1. Examples of the Percentage of Maximum Heterosis in Various Breed Crosses.

Breed Combination	Percent of Maximum Heterosis in Progeny
Angus × Hereford	100
Angus × Simmental	100
1/2 Angus, 1/4 Simmental, 1/4 Hereford	100
3/4 Hereford, 1/4 Angus	50
3/4 Simmental, 1/4 Angus	50
3/8 Hereford, 5/8 Angus	75

cross with 100 percent of maximum heterosis exhibited as long as the highest percentage of any breed does not exceed 50 percent in the cross as shown in Table 10-1.

Maximizing the Benefits of Crossbreeding

Maximizing the benefits of crossbreeding requires having both a crossbred cow and calf. Crossbred cows tend to produce more pounds of calf per cow exposed, while crossbred calves are more vigorous with higher survival and growth rates. Furthermore, crossbred cows exhibit increased longevity and earlier puberty. As a general statement, about two-thirds of the benefit of crossbreeding comes from increased performance of the crossbred cow with the other one-third coming from the increased performance of the crossbred calf.

Crossbreeding Programs

Taking maximum advantage of heterosis requires a systematic approach that fits the overall management scheme. Over the years, many systems have been proposed. It has been this author's experience that most of these systems are too complex for most commercial cattle operations because too few breeding pastures are available or other constraints. First consider the traditional systems and then consider systems that are more practical.

Traditional Crossbreeding Programs

- **Two-Breed Rotation** – This system is one of the simplest of the traditional systems where cows sired by breed A are then mated to bulls of breed B and visa versa. At least two breeding pastures are required and some means of identifying the breed of sire for all replacements. This system works best if the bulls selected in each breed are similar for mature size, birth weight, and milk production. A two-breed rotation system is illustrated in Diagram 10-1.
- **Three-Breed Rotation** – This system is essentially the same as the two-breed rotation but with a third breed added. In this system, a female is always bred to the breed to which she is least related. At least three breeding pastures are required as well as a good system to identify breed groups. Diagram 10-2 illustrates this system.
- **Terminal Sire System** – In this system, cows of a specific two-breed cross are mated to bulls

Diagram 10-1. A Two-breed Rotational Crossbreeding System.

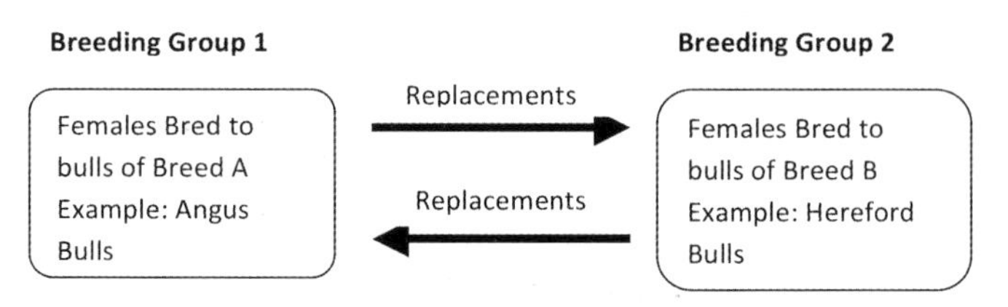

Diagram 10-2. Three-breed Rotational Crossbreeding System.

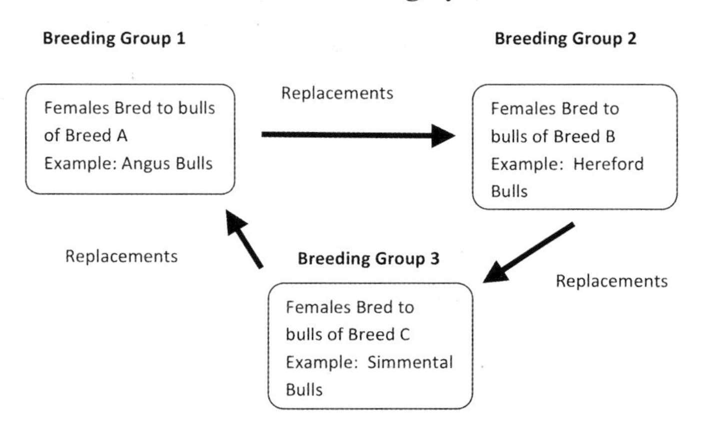

of a third breed with all offspring marketed. Obviously, this allows the use of a two-breed cross that is well suited to the environment, which is then mated to a breed with high growth rate. Only one breeding pasture is required, which makes this system relatively simple. Unfortunately, this system does not produce its own replacements — they must be purchased or produced in another crossbreeding program. If straightbred cows are maintained to produce the desired crossbred cows to be bred by the terminal sire, the overall level of heterosis is greatly reduced. Purchasing the desired crossbred cows is a good option, but they may not be available with the quality desired or at a reasonable price. A terminal sire system with the breeding groups required to produce the desired crossbred female is shown in Diagram 10-3.

- **Rota-Terminal System** – This crossbreeding system is essentially a combination of a rotational system and a terminal sire system with the rotational groups producing the replacements for the terminal system as shown in Diagram 10-4. Although this system requires a high level of management, it does maintain high levels of heterosis, and it does produce its own replacements. The breeds in the rotational portion can be selected to match the resources and for complementarity. The terminal sire should be selected for growth rate and carcass desirability.

Diagram 10-3. Terminal Sire Crossbreeding System with Breeding Groups Required to Produce Replacements.

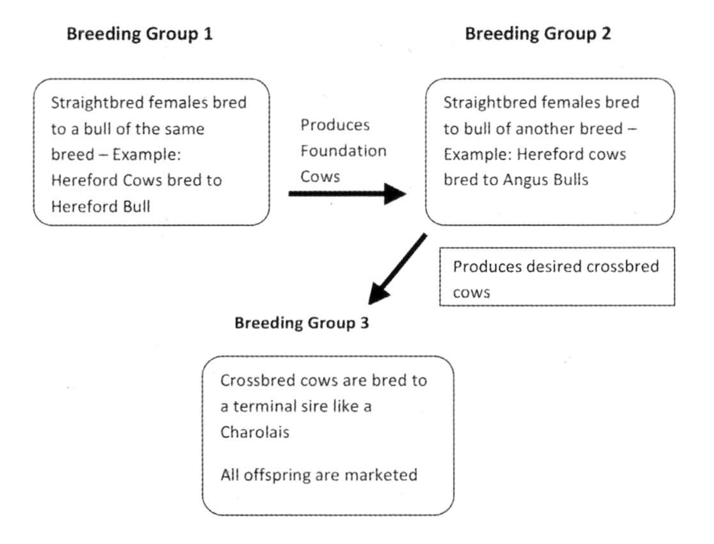

Diagram 10-4. Rota-terminal Crossbreeding System.

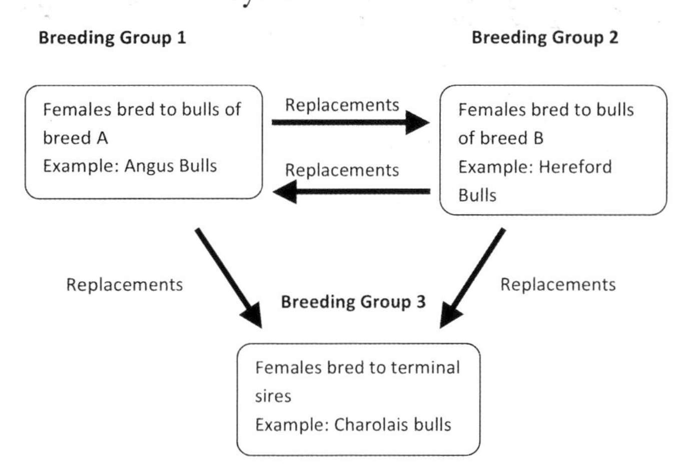

Simple Crossbreeding Systems

While the crossbreeding systems outlined in the previous section are quite effective, few cattle producers actually use them because they do not have enough breeding pastures or find them too difficult to manage. The crossbreeding systems outlined in the following section are simple to implement and take advantage of the benefits of heterosis and complementarity to a high degree. There has been an increase in the use of these systems in the past few years.

- **Purchase Crossbred Females and Breed to a Terminal Sire** — This is an effective way to take advantage of crossbreeding if quality crossbred females can be purchased at a reasonable price. Unfortunately, this often is not possible. However, some commercial cattlemen have developed arrangements with other cattle producers to have them produce the desired crossbred female. For example, some purebred Hereford breeders will breed a percentage of their cows to Angus bulls and sell the Angus × Hereford females to commercial cattle producers.

- **Rotate Sire Breeds** – This system consists of simply using a breed of bull for a few years and then switching to another breed. To keep the level of heterosis relatively high, the breed should be rotated every 4 years. In small herds or herds with only one breeding pasture, it will

Table 10-2. Percent Advantage in Calf Weaning Weight Per Cow Exposed for Common Crossbreeding Systems.

Crossbreeding System		Advantage* %	Retained Heterosis, %**	Minimum # of Breeding Pastures
Two-Breed Rotation	A×B	16	67	2
Three-Breed Rotation	A×B×C	20	86	3
Terminal Cross with Purchased F_1 Females	T × (A×B)	24	100	1
Terminal Cross with Straight Females	T × A	8.5	0	0
Terminal Sire with Rotational (2 Breeds)	A×B T × (A×B)	21	90	3
Terminal Sire – Buy F_1 Females	T X (A×B)	24	100	1
Rotate Sire Breeds Every 4 years	A×B A×B×C	12–16 16–20	50–67 67–83	1
Composite	2 Breeds	12	50	1
	3 Breeds	15	67	1
	4 Breeds	17	75	1
Rotating Unrelated F_1 Bulls	A×B × A×B	12	50	1
	A×B × A×C	16	67	1
	A×B × C×D	19	83	1

*Adapted from Ritchie, et al. * Percent increase in pounds of calf weaned per cow exposed.*
*** Relative to the F_1 with 100% heterosis.*

be necessary to replace the bulls after 2 years to avoid having a bull breed his daughters.

- **Composites** – A number of composites have been developed by seedstock producers. Research has shown that these composites retain a high level of heterosis for many generations. Developing a good composite requires a large number of cattle and a good understanding of the strengths and weaknesses of each breed included in the composite.
- **F_1 Bulls** – Many seedstock breeders are producing crossbred bulls for use in commercial herds. Using these bulls can be an effective way of crossbreeding in small herds or when a simple system is preferred in a large herd. One might think that this system would result in tremendous variability in the calves, but the variability is similar to that exhibited when purebred bulls are used.

Relative Advantage of Crossbreeding Systems

As noted, the benefit from crossbreeding over straightbreeding is a function of the amount of heterosis in the cow and the calf. Table 10-2 shows the advantage over the straightbreds in calf weaning weight per cow exposed for common crossbreeding systems. As with any increase in production, there is typically an associated cost. For example, crossbred cows require more feed inputs than the average of the parental breeds because the crossbred cows will be larger and give more milk. In addition, crossbred calves will grow faster requiring more feed. Research has shown, however, that the increase in pounds weaned per cow exposed more than offsets the cost of additional feed.

The Importance of the Crossbred Cow

While crossbred calves exhibit heterosis, the major benefit of crossbreeding comes from the crossbred cow. She tends to be more fertile resulting in more calves in her lifetime and greater longevity reducing the cost of replacements. Furthermore, she produces more milk increasing weaning weights.

The Importance of Sire Selection in a Crossbreeding Program

Crossbreeding is a powerful tool because it allows a producer to take advantage of heterosis and breed complementarity. However, sire selection is just as important in a crossbreeding program as it is in a straightbreeding program. Remember: Heterosis is the advantage of the crossbred over the average of the parent breeds. Crossbreeding cannot make up for poor genetics — the additive genetic component is still extremely important.

References

Beef Cattle Manual, 6th Ed. 2008. Oklahoma State University Extension Service.

Beef Sire Selection Manual. National Beef Cattle Evaluation Consortium.

Buchanan, D.S. and S. Northcutt. The Genetic Principles of Crossbreeding. Beef Cattle Handbook BCH-1400.

Bullock, D. Planning the Breeding Program. University of Kentucky Beef Book.

Gosey, J. 2005. Crossbreeding The Forgotten Tool. Proceedings, The Range Beef Cow Symposium XIX.

Guidelines for Uniform Beef Improvement Programs, 9th Ed. 2010. Beef Improvement Federation.

Kress, D. 1989. Practical Breeding Programs for Managing Heterosis. The Range Beef Cow Symposium XI.

Kress, D. 1994. Crossbreeding with a New Target. Beef Improvement Federation Proceedings.

Lamberson, B., J. Massey, and J. Whittier. 1991. Crossbreeding Systems for Small Herds of Beef Cattle. University of Missouri Extension Bulletin G2040.

Notter, D.R., J.O. Sanders, G.E. Dickerson, G.M. Smith, and T.C. Cartwright. 1979. Simulated Efficiency of Beef Production for a Midwestern Cow-Calf-Feedlot Management System. III. Crossbreeding Systems. JAS 49:92.

Ritchie, H., D. Banks, J. Cowley, and D. Hawkins. 1999. Crossbreeding Systems for Beef Cattle. Michigan State University Extension Bulletin E-2701.

Ritchie, H., B. Banks, D. Buskirk, J. Cowley, and D. Hawkins. 1999. Development and Use of Composite Breeds: A Summary. Michigan State University Extension Bulletin E-2702.

Simms. D.D., K.O. Zoellner, and R.R. Schalles. 1990. Crossbreeding Beef Cattle. Kansas State University Extension Bulletin C-714.

Spangler, M.L. 2007. The Value of Heterosis in Cowherds: Lessons from the Past that Apply to Today. Proceedings, The Range Beef Cow Symposium.

Chapter 11

Expected Progeny Differences

As noted previously, the development of expected progeny differences (EPD) greatly improved the accuracy of selection and the performance of cattle for many traits. In the early years of their use, many cattle breeders questioned the accuracy of EPD in predicting breeding value, and in fact, there were problems with the reliability of EPD, especially in some breeds. However, as the technology used to calculate EPD improved and the databases used to calculate them increased in size, the usefulness of EPD as predictors of true genetic merit increased significantly. They are the major tool for making selection decisions in today's industry.

What Are Expected Progeny Differences?

Expected Progeny Differences represent the average transmitting ability of an individual for a trait expressed in the units of measure for the trait. It is an estimate of the breeding value or true genetic merit of the animal. For example, a bull with a +30 lb weaning weight EPD would be expected to produce calves 30 pounds heavier at weaning than the base bull (0 EPD) for the breed. Expected Progeny Differences allow direct comparisons of the expected performance of the progeny from bulls and cows. Thus, the bull with a +30 lb weaning weight EPD would be expected to produce calves 10 pounds heavier at weaning than a bull with a +20 lb weaning weight EPD. It should be noted that the average EPD for any trait is not zero. In other words, a bull with a +30 lb EPD is not going to produce calves 30 pounds heavier than the average of the breed. This bull will produce calves 30 pounds heavier than the base animal. However, this still allows for direct comparisons in actual units of the trait. The average EPD for breeds can be found in their sire summaries or on the breed association websites.

How Are Expected Progeny Differences Calculated?

Expected progeny differences are based on the performance information submitted to breed associations. This information is analyzed by sophisticated computer programs that make comparisons of performance of animals within each herd and across the breed to calculate an estimate of the breeding value of each individual. In making this calculation, the program considers the performance of relatives in other herds compared to their contemporaries. Clearly, the reliability of this estimate of the breeding value in predicting the true breeding value is a function of the amount of data submitted to the breed association and the accuracy of the data.

In calculating the EPD for a bull calf following weaning, for example, the calf's performance in the herd compared to his contemporaries will be considered as well as the EPD of his ancestors. As more information becomes available, the EPD will be recalculated. It is important to note that the bull should only be compared to his contemporary group (calves raised in the same environment).

How Reliable Are Expected Progeny Differences in Predicting Breeding Value?

While EPD have had a significant influence on beef breeding, they are not perfect in predicting the true genetic merit of an individual animal. This inaccuracy has caused some cattle breeders to question the value of EPD. To illustrate why there is some inaccuracy in EPD, consider a weaning weight EPD calculated for a bull calf at birth. The EPD for weaning weight will initially be the average of his parent's weaning weight EPD. As noted in a previous chapter, the breeding value of his ancestors represents the average contribution of the genes they carry for the trait in question. In the formation of gametes, these genes will segregate with some gametes getting a better than average set of genes and others getting a poorer than average set of genes. Consequently, our bull calf may have received a better than average or poorer than average set of genes from one or both parents. This gene segregation causes the EPD on unproven animals to be inaccurate in many cases.

It is important to note, however, that while the EPD for any unproven animal may not be reliable, on average, EPD do represent genetic merit. To illustrate this point, consider 50 bull calves from the same bull and cow. Initially, all 50 calves would have the same EPD. If we then progeny tested these bulls, we would expect the EPD for 25 of these bulls to be higher than the initial value and 25 lower, with an average EPD value close to the original estimate. In other words, on average EPD are reliable in estimating genetic merit in beef cattle.

What are Accuracy Values?

The accuracy value for an EPD is the reliability that can be placed on the EPD. Accuracy values range from 0 to close to 1.0 with the accuracy value increasing as the reliability increases with more information such as progeny performance. To illustrate, consider a bull calf. His initial EPD (also called the pedigree estimated EPD) will be based on an average of the EPD of the parents. Then as the bull's own weaning weight is included in the evaluation, his EPD for weaning weight would be adjusted (the interim EPD). The accuracy would also be increased substantially. Subsequently, if the bull has progeny and their performance information was entered in the breed evaluation, his EPD would again be adjusted and the accuracy increased. Bulls that have been widely used across the breed will have accuracies of 0.95 or higher. These bulls are often referred to as "proven" bulls — their breeding value is known, and their EPD will change little in the future.

Accuracy values can be thought of as a measure of risk. There is a far greater chance of a bull with a low accuracy value having a "true" breeding value well below his current EPD. Conversely, with low accuracy, a bull's "true" breeding value may be much greater than his current EPD predicts. There is just much less risk with bulls having high accuracies — their EPD are not going to change much.

Use of Genetic Correlations in Calculating EPD

Genetic correlations between traits are considered in the calculation of EPD. For example, birth weight and weaning weight have a positive genetic correlation. In the calculation of the birth weight EPD, the

Table 11-1. Accuracy and Associated Possible Change Values for Hereford Cattle in the Spring 2012 Sire Summary.

	Production			Maternal	Carcass	
Accuracy	BW	WW	YW	Milk	Marb*	REA**
.10	2.5	11.7	19.5	8.5	.14	.23
.20	2.2	10.4	17.4	7.5	.12	.21
.30	1.9	9.1	15.2	6.6	.11	.18
.40	1.6	7.8	13.0	5.6	.09	.16
.50	1.4	6.5	10.9	4.7	.08	.13
.60	1.1	5.2	8.7	3.8	.06	.10
.70	0.8	3.9	6.5	2.8	.05	.08
.80	0.5	2.6	4.3	1.9	.03	.05
.90	0.3	1.3	2.2	0.9	.02	.02

** Marbling score with 4.00 = slight (select) and 5.00 = Small (choice) ** Ribeye area in square inches. American Hereford Association Website: Hereford.org.*

weaning weight would be factored into the analysis to calculate the birth weight EPD. Correspondingly, the birth weight would be considered in the calculation of the weaning weight EPD.

Possible Change

The accuracy value has an associated possible change value that represents the expected deviation in the EPD. To illustrate, consider a Hereford bull with a weaning weight EPD of 45 and an accuracy of 0.30. Table 11-1 shows a possible change value of 9.1, which means that the bull's EPD would range from 35.9 (45 – 9.1) to 54.1 (45 + 9.1) two-thirds of the time. If this bull had an accuracy of 0.70, the expected range would be 41.1 (45 – 3.9) to 48.9 (45 + 3.9). The possible change values will be different for each breed. Figure 11-1 shows the relationship between the accuracy of the EPD and the possible change for a Hereford bull with a weaning weight EPD of 30 pounds. Clearly, as the accuracy value increases, the confidence that can be placed in the EPD increases.

Types of EPD

Growth Traits

- **Birth Weight** – Expressed in pounds, it is the expected difference in birth weight compared to the base animal of the breed. The difference between the birth weight EPD of two bulls, for example, would be the expected difference in the weight of their calves at birth on average.
- **Weaning Weight** – Expresses the expected difference in an animal's offspring at weaning in pounds compared to the base animal. As with birth weight EPD, the difference between the weaning weight EPD of two bulls, for example, would be the expected difference in their progeny. Since most producers sell their calves at weaning, this is an important EPD.
- **Milk** – Expresses the difference in the expected weaning weight of the daughter's offspring resulting from differences in milk production. It is expressed in pounds of calf weaning weight, not in pounds of milk. In other words, if bull A has a milk EPD of +20 and Bull B has a milk EPD of +30, the expected difference in weaning weight of the calves from their daughters would be 10 (30 – 20) pounds because of differences in milk production. This EPD also indicates to some extent the differences in feed requirements that might be expected. It will take a significant amount of energy and protein to produce the milk required for the daughters of Bull B to produce the additional 10 pounds of calf at weaning. Thus, if feed quantity or quality is limited, an upper limit should be placed on milk EPD to avoid reducing reproductive performance.
- **Yearling Weight** – Expresses the expected differences in the weight of progeny at 1 year of age. Typically, a larger value is better, but yearling weight EPD is positively correlated with mature weight. Thus, a bull with a high yearling weight EPD would be expected to produce daughters that are larger at maturity, which means they would require more feed. This EPD is a good predictor of differences in performance in the finishing phase of production.

Figure 11-1. Relationship between the Accuracy Value and Possible Change Value.

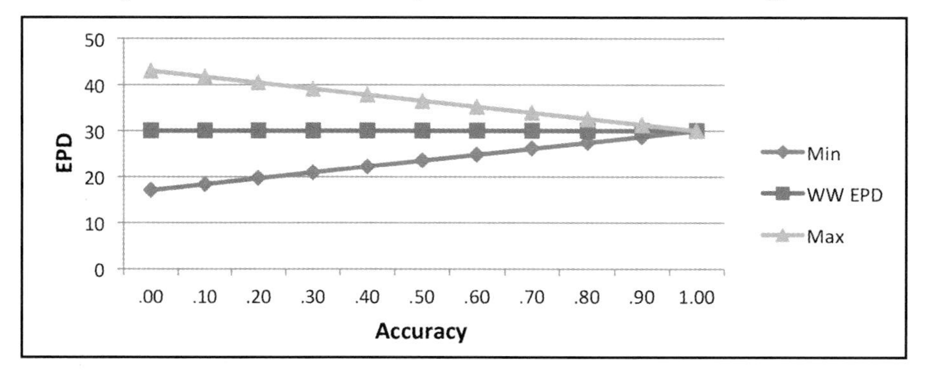

American Hereford Association Website: Hereford.org.

- **Total Maternal** – Expresses the expected difference in weaning weight in calves from the daughters of an animal. For example, if Bull A has a Total Maternal EPD of +60 and Bull B has a total maternal EPD of +50, we would expect the daughters of Bull A to produce calves 10 pounds heavier at weaning. This EPD is calculating by taking one-half of the Weaning EPD and all of the Milk EPD.
- **Yearling Height** – Expresses expected difference in hip height at one year of age.
- **Mature Height** – Expresses the expected differences in mature height of progeny. Bigger values might be useful in herds desiring to increase cow size, while smaller values might be desirable in herds needing to decrease cow size.
- **Mature Weight** – Expresses the expected differences in the weight of progeny at maturity in pounds. Since to some extent, feed requirements increase as mature weight increases, this EPD predicts differences in feed requirements. Thus, it can be used to assist in matching cattle to a specific environment.

Reproductive Traits

- **Scrotal Circumference** – Expresses the expected difference in scrotal circumference of male progeny. As noted previously, this EPD was useful in breeds with a tendency for small testicles, but it probably has limited value today.
- **Gestation Length** – Expresses the expected difference in days that an animal's progeny will stay in utero. It is an indicator trait for potential calving difficulty since the calf is growing rapidly in late pregnancy. In addition to shorter gestation length reducing calving difficulty, heifers and cows with short gestation lengths have a little longer to start cycling after calving before the breeding season starts. This gives them a better chance of getting rebred in the next breeding season.
- **Calving Ease Direct** – Expresses the expected differences in calving difficulty as a percentage. For example, if Bull A has a calving ease direct EPD of 105 and Bull B of 95, we would expect 10 percent more calving problems in first-calf-heifers with Bull B. In most breeds, higher values indicate more unassisted births. This EPD is based on actual levels of calving difficulty in first-calf heifers as reported by breeders. It is useful because a bull may produce calves that have heavier birth weights than desired, but the shape of the calves and other factors allows them to be born easier than one would expect based on their birth weight.
- **Calving Ease Maternal** – Expresses expected differences in calving difficulty in the daughters of an animal. This EPD recognizes that some cows calve easier than others do and that there is a genetic component to this difference. It is usually expressed as a percentage with a higher value meaning more unassisted births.
- **Heifer Pregnancy** – Expresses the expected differences in pregnancy in yearling heifers. It is expressed as a percentage with higher values indicating more pregnancies in the progeny.

Carcass Traits

With the move to more branded-beef programs, there has been a major focus on developing carcass EPD by the major breed associations. These EPD are calculated on an age constant basis as if all cattle were slaughtered at the same age. Some breeds base their EPD only on actual carcass data while others base carcass EPD on ultrasound measurements. Some breeds use a combination of these data.

- **Carcass Weight** – Expresses expected differences in carcass weight when the progeny are slaughtered at the same age.
- **Ribeye Area** – Expresses the expected difference in the ribeye area measured between the 12th and 13th ribs on the carcass. As with other EPD, bigger is not always better because in heavier muscled cattle the ribeyes are often too large.
- **Fat Thickness** – Expresses the differences in external fat on the carcass in inches. A lower value is usually considered better, but some external fat is desirable.
- **Marbling** – Expresses the expected differences in marbling (intramuscular fat) in the progeny. It is expressed in terms of differences in marbling score.

Table 11-2. Adjustment Factors to Add to EPD of 18 Breeds to Estimate across Breed EPD – 2012 Evaluation.

Breed	BW	WW	YW	Milk	Marb.	REA	Fat
Angus	0.0	0.0	0.0	0.0	0.00	0.00	0.000
Beefmaster	6.7	35.3	32.5	7.8			
Brahman	11.1	42.5	4.8	22.4			
Brangus	3.7	13.0	13.5	6.8			
Braunvieh	1.2	-19.2	-38.5	30.0	-0.67	0.23	-0.095
Charolais	8.6	40.1	46.8	5.7	-0.46	0.92	-0.222
Chiangus	3.3	-14.9	-31.3		-0.42	0.40	-0.157
Gelbvieh	4.0	5.7	-13.5	13.6			
Hereford	2.7	-2.8	-20.1	-16.7	-0.34	-0.11	-0.053
Limousin	3.8	-0.9	-34.7	-9.2	-0.70	1.07	
Maine-Anjou	4.1	-13.0	-34.5	-4.7	-0.79	0.88	-0.210
Red Angus	2.4	-0.6	-12.0	-3.1	0.03	-0.10	-0.034
Salers	1.8	-3.1	-14.3	2.4	-0.11	0.75	-0.210
Santa Gertrudis	7.8	34.2	24.8		-0.64	-0.18	-0.146
Shorthorn	5.9	17.9	41.7	19.6	-0.10	0.24	-0.151
Simmental	5.2	24.9	22.4	19.8	-0.55	0.92	-0.215
South Devon	4.2	3.2	-6.3	-2.3	0.05	0.15	-0.111
Tarentaise	1.7	33.1	21.2	23.4			

U.S. Meat Animal Research Center

- **Retail Product** – Expresses the expected difference in saleable meat from the progeny. It is expressed as a percent with higher values indicating more meat that is saleable. It considers the same carcass characteristics as the yield grade EPD.
- **Yield Grade** – Expresses the expected differences in yield grade of the carcasses of progeny. The major factors in yield grade are fat thickness, carcass weight, and ribeye area. This EPD is expressed in terms of yield grade units with lower values indicating lower yield grades and more saleable meat.
- **Tenderness** – Expresses the expected differences in tenderness of the carcass of the progeny in terms of pounds of Warner Bratzler Shear Force required to cut through the meat. Higher values indicate that more pounds were required. Thus, lower values indicate more tender meat.

Other Traits

- **Stayability** – Expresses the expected differences in the longevity of the progeny in the herd. It is expressed as a percent, and it predicts the probability that a bull's daughter will stay in the herd through six years, for example. The higher the value is, the higher the probability of longevity.
- **Docility** – Expresses expected differences in behavior in the progeny. It is expressed as a percentage with higher values indicating a higher percentage of calm animals in the progeny.

Across Breed Expected Progeny Differences

As noted, EPD are calculated on a breed-by-breed basis, and thus they can't be compared directly. In other words, a weaning weight EPD of 50 for an Angus bull is not the same as a weaning EPD of 50 for a Charolais bull. However, many times commercial producers with a crossbreeding program want to select bulls of several breeds that have similar genetic merit as a way of reducing variation within the herd. Fortunately, the U.S. Meat Animal Research Center (USMARC) has calculated adjustment factors to allow comparisons of EPD between cattle of different breeds as shown in Table 11-2.

To illustrate the use of this table, compare a Tarentaise bull with an EPD for weaning weight of 35 to a Gelbvieh bull with a weaning weight EPD of 45. The across breed adjustment factor for weaning weight in Tarentaise is 34.8, which results in an across breed EPD of 69.8 (35 + 34.8). Correspondingly, the across breed adjustment for weaning weight EPD for Gelbvieh is 3.9, which results in an across breed EPD of 48.9 (45 + 3.9). Thus, the Tarentaise bull would be expected to produce calves 20.9 pounds (69.8 – 48.9) heavier at weaning than the Gelbvieh bull. It is important to note that Table 11-2 should not be used to make a direct comparison of breeds — remember each breed has a different base. In addition, Table 11-2 shows adjustments based on the calculations in 2012. These adjustments are updated annually and available on the Beef Improvement Federation website (*www.beefimprovement.org*).

Genomic Enhanced EPD

Over the past 10 years, there have been tremendous advances in our ability to analyze DNA. Current technology providing genomic evaluations use a panel of DNA markers, commonly referred to as SNPs (pronounced "snips"; single-nucleotide polymorphisms). Each SNP does not provide much information, but a panel of critically selected SNPs for multiple traits provides molecular breeding values (MBV). These MBV are being included in the calculation of the EPD for most traits by several breed associations. Thus, the EPD for a trait is based on the individual's performance, pedigree information, progeny information, and the MBV. As the databases on MBV increase in size, the importance of the genomic evaluation will increase. This technology allows the identification of superior genetics early in life.

To illustrate the potential for genomic-enhanced EPD, consider 10 bulls produced by embryo transfer from the same bull and cow. At birth, they would all have the same EPD for all traits. However, genomic evaluation could be used to identify the superior bull(s) in the group. Obviously, genomics has the potential to make selection much more accurate.

Multi-Breed EPD

Recently, the American Simmental Association and the Red Angus Association pooled their datasets, which already contained data from the Canadian Angus and some other Angus data. The result was the world's largest multi-breed dataset with more than 10 million records. In 2012, Red Angus and Simmental adjusted their EPD to the same base and scale meaning that their EPD will be directly comparable. Hopefully, this trend of cooperation between breeds continues because it increases the accuracy of genetic evaluation and makes it easier for commercial cattle producers to identify genetics that fit their operation.

$Value Indexes

Because there are so many EPD to consider in making selection decisions, several breeds are calculating selection indexes that take into consideration both the production traits and the traits that influence the cost of production. The goal in calculating these value indexes is to consider economics in the selection process and make selection easier. This approach allows commercial producers to directly compare the expected profitability of the progeny of sires under consideration. Each breed calculates different indexes with the goal of providing the most useful indexes for commercial breeders.

Angus Association Indexes

The following discussion of $Values represents the Angus Association approach to calculation of dollar value indexes. This information was taken from the Angus website (*www.angus.org*) in January 2013.

Trying to consider the large number of EPD currently available through the American Angus Association and individual performance information in making selection decisions can become overwhelming. To make this process easier the Angus Breed Association calculates several $Value Indexes that consider EPD that influence the expected returns and EPD that influence the cost of production at a specific marketing endpoint. The major $Value indexes are Weaned Calf Value ($W), Feedlot Value ($F), Grid Value ($G), and Beef Value ($B). These indexes are bioeconomic values, expressed in dollars per head. Also, a Cow Energy Value ($EN) is available for fine-tuning the cow herd.

$Values encompass the revenue generated from genetically derived outputs and associated costs (expenses) from required inputs. $Values only have meaning when used in comparing the relative merit or ranking of two individuals. Each sire listed in the Sire Evaluation Report is comparable to every other sire. The $Values are sensitive to the assumptions for the industry-relevant components used in calculating the indexes. As with EPD, variation in $Values between animals indicates expected differences in the relative value of progeny if random mating is assumed. Averages and percentile breakdowns also are provided for $Values as reference points for the Angus database. A $Value of 0 does not correlate to the lowest ranking or to an average animal.

Weaned Calf Value ($W)

Weaned Calf Value ($W) considers four primary economic impact areas:

1. **Birth Weight** — birth weight influences on calf death losses related to dystocia, weaned calf crop percentage, and resulting revenue per cow.
2. **Weaning Weight** — direct growth impact on weaning weight revenue (preweaning growth and pounds of calf sold) and energy requirements and related costs necessary to support preweaning calf growth.
3. **Maternal Milk** – revenue from calf preweaning growth and pounds of calf sold as influenced by varying cow milk levels related to lactation energy requirements.
4. **Mature Cow Size** – expense adjustments are made for maintenance energy as related to differing mature cow size, including mathematical linkages between mature weight and yearling weight.

$W provides the expected dollar-per-head difference in future progeny preweaning performance in a multi-trait fashion, within a typical U.S. beef cow herd. If Bull A has a $W of +25.00 and Bull B has a $W of +15.00, and these sires were randomly mated to a comparable set of females and the calves were exposed to the same environment, and a normal number of replacement females were saved from both sires, on average you could expect Bull A's progeny to have a +$10.00 per head advantage in preweaning value over Bull B's progeny (25.00-15.00 = 10.00 per head).

As with other $Values, the Weaned Calf Value includes assumptions, as listed below:

Base Calf Price	$140 per cwt
Cow/Heifer Mix	80%/20%
Cow weight	1,300 lbs
Feed energy cost	$.095 per Mcal NEm

Cow Energy Value ($EN)

A Cow Energy Value ($EN) is available to assess differences in cow energy requirements, expressed in dollars per cow per year, as an expected dollar savings difference in future daughters of sires. A larger value is more favorable when comparing two animals (more dollars saved on feed energy expenses). Components for computing $EN savings difference include lactation energy requirements and energy costs associated with differences in mature cow size.

Cow Energy ($EN), Savings, $/cow/year	+15.75
Cow Energy ($EN), Savings, $cow/year	+4.68

The expected difference in cow energy savings per cow per year for future daughters of the two animals is +$11.07 (15.75 – 4.68 = +11.07).

$Values ($Feedlot, $Grid, and $Beef)

- Feedlot Value ($F), Grid Value ($G), and Beef Value ($B) are postweaning bio-economic $Values, expressed in dollars per head. The $Values were developed primarily to serve as selection tools for commercial bull buyers.
- Although feedlot and carcass merit are important components of the beef production chain, it should be stressed to producers that the $Values ($F, $G, $B) are not to be used as a single selection criterion, since the indexes only encompass postweaning and carcass performance.
- $Values have meaning when used in comparing the relative merit or ranking of two individuals. For example, Bull A has a $B value of +26.00 and Bull B has a $B value of +16.00. If these bulls were randomly mated to a comparable set of females and the calves were exposed to the same environment, on the average you would

expect Bull A's progeny to have a $10 per head advantage in postweaning performance and carcass merit over Bull B's progeny (+26 minus +16 = +10 per head).

- $Feedlot, $Grid, and $Beef Values incorporate available carcass EPD, converted into economic terms, incorporating industry-relevant components for feedlot performance and carcass merit. These base components used to calculate $Values for any registered animals are:

Feedlot assumptions:

Time on feed	160 days
Ration cost	$315 per ton of dry matter
Fed market	$110 per cwt, live

Grid assumptions:

Quality components:		**Yield components:**	
Prime premium (above choice)	$12.00	YG 1 premium	$4.50
CAB premium (above choice)	$4.00	YG 2 premium	$2.25
Choice-Select spread	$10.00	YG base	$0.00
Standard discount	-$20.00	YG 4 & 5 discount	-$20.00
		Average carcass wt., lbs	816
		Heavyweight discount	-$20.00

Feedlot Value ($F)

Feedlot Value ($F), an index expressed in dollars per head, is the expected average in future progeny performance for postweaning performance compared to progeny of other sires. $F incorporates weaning weight (WW) and yearling weight (YW) EPD along with trait relationships. Typical feedlot gain value, feed consumption and cost differences are accounted for in the final calculations, along with a standard set of industry values for days on feed, ration costs and cash cattle price.

Grid Value ($G)

Grid Value ($G) is the expected average difference in future progeny performance for carcass grid merit compared to progeny of other sires expressed in dollars per head. The $G combines quality grade and yield grade attributes, and is calculated for animals with carcass EPD.

- **Quality Grade ($QG)** represents the quality grade segment of the economic advantage found in $G. $QG is intended for the specialized user wanting to place more emphasis on improving quality grade. The carcass marbling (Marb) EPD contributes to $QG.
- **Yield Grade ($YG)** represents the yield grade segment of the economic advantage found in $G. $YG is intended for the specialized user wanting to place more emphasis on red meat yield. It provides a multi-trait approach to encompass ribeye, fat thickness, and carcass weight into an economic value for red meat yield.

Beef Value ($B)

Beef Value ($B) facilitates what most every beef breeder is already seeking — simultaneous multi-trait genetic selection for feedlot and carcass merit, based on dollars and cents. $B represents the expected average dollar per head difference in the progeny postweaning performance and carcass value compared to progeny of other sires. The $B value is comprised of two pieces: $F and $G. To align $B with market place realities and approximately value carcass weight in Angus cattle, the following factors are incorporated into the final calculation for $B.

$B is not simply the sum of $F and $G. Projected carcass weight and its value are calculated along with production cost differences. $B takes into consideration any discount for heavyweight carcasses. Final adjustments are made to prevent double-counting weight between feedlot and carcass segments. The resulting $B value is not designed to be driven by one factor, such as quality, red meat yield, or weight. Instead, it is a dynamic result of the application of commercial market values to Angus genetics for both feedlot and carcass merit.

Charolais Indexes

- **Terminal Sire Profitability Index –** This index combines EPD information with an economic analysis to predict profit differences in dollar returns from progeny of bulls used in a terminal sire system with all of the offspring marketed based on carcass value. Thus, both growth and carcass merit EPD are considered as well as costs of production during finishing.

Hereford Indexes

- **Baldy Maternal Index (BMI$) –** This maternally focused index assumes a Hereford x Angus or other British cross herd with the progeny marketed through the Certified Hereford Program (CHB). Consequently, it considers performance in the cow herd and through finishing as well as carcass value. It places a positive emphasis on weaning weight and a slightly negative emphasis on yearling weight to moderate mature size.
- **Calving Ease Index (CEZ$) –** This index focuses on identifying bulls that can be used on heifers with the progeny marketed through the CHB program. Both direct and maternal calving ease carry significant weight in this index. There is little weight on growth or carcass traits.
- **Brahman Influenced Index (BII$) –** This is a maternally focused index assuming a Brahman X Hereford crossbred herd. The offspring will be sold in a commodity-based program. Consequently, there is considerable emphasis on growth through finishing, REA and IMF.
- **Certified Hereford Beef Index (CHB$) –** This is a terminal sire index with the goal of identifying Hereford bulls to be used in commercial herds to sire offspring to be marketed through CHB. There is some pressure on calving ease and considerable emphasis in growth, both before and after weaning. There is some negative pressure on fat and positive pressure on both IMF and REA. Since there is not any emphasis on reproductive traits, this should be considered a terminal index.

Limousin Indexes

- **Mainstream Terminal Index ($MTI) –** This index assumes that the Limousin bulls will be mated in a terminal system to British crossbred cows with the resulting progeny sold into the commodity beef market. It emphasizes post-weaning growth, yield grade, and quality grade. Since it is a terminal index, there is not any emphasis on calving ease or maternal traits.

Simmental Indexes

- **All-Purpose Index (API) –** This index focuses on identifying bulls for use on the entire cow herd including heifers with replacements retained from the offspring. The remaining offspring will be finished and sold grade and yield. Consequently, this index includes growth, reproduction, and carcass EPD as well as an estimate of the associated costs.
- **Terminal Index (TI) –** This index estimates profit differences when Simmental bulls are used on British-breed females with all of the offspring finished and sold grade and yield. Consequently, postweaning performance and carcass characteristics are emphasized while reproductive traits are not considered.

Accuracy of $Value Index Estimates

While $Value estimates offer a simplified approach to considering both the positive and negative impacts of several traits on profitability, it should be noted that there are several factors that influence the accuracy of $Value estimates in predicting future differences in profitability. Firstly, as discussed previously, there is the inaccuracy in EPD in predicting genetic merit, especially in animals with low accuracy values. Secondly, the market value of cattle is constantly changing; and thirdly, the cost of inputs such as feed is changing. This variation in market value and costs means that the relative ranking of bulls for $Value may change. However, even given these sources of inaccuracy in the calculation of $Values, they are still extremely useful for commercial producers because they provide a multi-trait evaluation and have an economic component.

References

American Angus Association Website: *www.angus.org*

American Hereford Association Website: *Hereford.org*

American Simmental Association Website: *www.Simmental.org* Beef 2007-08. 2009. USDA.

Beef Sire Selection Manual. National Beef Cattle Evaluation Consortium.

Guidelines for Uniform Beef Improvement Programs, 9th Ed. 2010. Beef Improvement Federation.

Hammack, S. P. and J.C. Paschal. 2009. Texas Adapted Genetic Strategies VIII: Expected Progeny Difference (EPD). Texas A&M Bulletin E-164.

Marshall, D.M. 1994. Breed Differences and Genetic Parameters for Body Composition Traits in Beef Cattle. JAS 72:2745-2755.

Northcutt, S. 2011. By the Numbers. Angus Journal April 2011. Pg 92.

Red Angus Website: *www.RedAngus.org*

Sire Evaluation Report. 2011. American Angus Association.

Sire Evaluation Report. Spring 2012. American Hereford Association.

Chapter 12

Selection of Bulls and Replacement Heifers

In a typical commercial cow/calf operation, 80 to 90 percent of the genetic progress comes from bull selection, which makes the process of selecting herd bulls extremely important. The following discussion provides a step-by-step approach for selecting herd bulls.

Step 1 – Assess the Resources

As discussed in Chapter 9, matching the genetics of the herd to the resources is essential for maximum profitability. The key question is, how much growth rate and milking ability can the forage base support without requiring high levels of supplementation? If the genetic potential of the herd exceeds the nutrition provided by the forage base, reproductive performance will suffer, unless adequate supplementation is supplied, and supplementation tends to be expensive. Moreover, with the high levels of performance available in many breeds, it is easy to select cattle that exceed the optimal level of growth and milking ability. Generally, cows that fit the forage resource will be most profitable. Thus, selecting the best bull requires an in-depth consideration of what type of cattle fit the resources.

Step 2 – Consider the Future Marketing Program

There are numerous marketing opportunities available to commercial cattle producers, some of which offer significant premiums. An evaluation of these programs may offer insight into the best direction for the breeding program. For example, is there an opportunity to produce cattle for one of the high-quality beef programs and receive a premium? Or, does producing cattle that yield a high percentage of retail product have potential for increased profitability? Is there potential to produce high-quality replacement females that merit a premium? Cattle breeding is a long-term process, and the genetics of the bulls selected today will influence the herd for many years if replacements are retained from the bull's progeny. Establishing a marketing goal is an essential part of bull selection.

Step 3 – Assess the Needs of the Herd

An honest evaluation of the status of the herd is also essential. Recognizing both the strengths and weaknesses of the herd will aid in determining which traits should be emphasized in the selection process. Current weaning weights will give some indication of the current level of milk production (see Chapter 17, Table 17-7). Moreover, reproductive performance is a good indicator of the current match of genetics to the resources. For example, an increase in the percentage of late calving cows is one of the first indications that the genetic potential is exceeding the resource base. Another indicator of a poor match is difficulty in getting first-calf-heifers rebred with their second calf. Do the calves have adequate frame to receive the top price in the market? Also, do they have adequate muscle? Unfortunately, color has a significant bearing on the price received for calves in many areas. Thus, color of the calves should be a consideration. It may be valuable to have an outside source evaluate the herd to remove any bias.

Step 4 – Consider the Breeding System

As noted previously, commercial cattle producers should be crossbreeding to take advantage of heterosis and breed complementarity. However, it is important that the crossbreeding system fits the operation and is easy to manage. The crossbreeding system will dictate the number of breeds that will be used and to some extent the traits to be emphasized in each breed.

Step 5 – Select the Breeds

Once the crossbreeding system has been selected, the next step is to select the breeds making sure they complement each other and fit the production environment.

Step 6 – Establish the Relative Importance of Economically Important Traits

Based on the needs of the herd and the intended use of the bull, a priority list of traits to be emphasized in the selection process should be established for each breed. A key question is whether replacements will be retained from the bull's progeny. For example, if the bull is to be used as a terminal sire, growth and carcass traits should be high on the list, while maternal traits could be ignored. Conversely, if the plan is to retain a significant number of replacements from a bull, maternal traits should receive emphasis.

Step 7 – Establish a Set of EPD Specifications for the Major Traits in Each Breed

For each trait that will be emphasized in the selection process, a minimum and possibly a maximum EPD for each trait should be established. For example, a minimum of +15 and a maximum of +25 might be set for milk EPD for an Angus bull. In some cases, an upper limit might be set without a minimum. For example, one might put an upper limit of +90 pounds on yearling weight EPD in Angus with the goal of minimizing the mature size of the bull's daughters.

Step 8 – Identify Quality Breeding Programs in Each Breed as Sources

Once the EPD specifications have been developed, the next step is to identify breeding programs that have EPD fitting the specifications. Other considerations include:

1. Does the breeder produce his or her cattle in an environment similar to the one where the bull will be used? For example, do his or her cattle require higher levels of supplementation than are profitable in a commercial herd?
2. Is the breeder committed to producing cattle that fit the commercial industry? For example, some breeders focus on producing cattle for the show ring or club calves with little consideration for the traits important in the commercial industry.
3. What is the breeder's reputation for honesty and integrity?
4. Does the breeder participate fully in his or her breed association's performance program? Good breeders are committed to submitting as much data as possible to their breed association as well as using the performance information supplied by the association.
5. Does the breeder use artificial insemination? It is difficult to produce superior bulls without using AI.
6. Does the breeder tolerate problems in his/her herd? For example, are there cows in the herd with bad udders or that have bad dispositions?

Step 9 – Make a List of Bulls that Meet the EPD Specifications

Once a breeding program has been selected, the next step is to prepare a list of bulls that fit the EPD specifications. Then they should be evaluated for other traits such as frame, muscling, scrotal circumference, disposition, and soundness. Bulls not meeting minimum standards for any of these traits should be removed from the list. In addition, bulls that have EPD meeting the specifications but that have individual performance below average should be eliminated from the list. Correspondingly, the dams of the bulls being considered should be evaluated if replacements are to be retained from the bull. Any bull whose dam has produced at a level below average in the herd should be eliminated from the list.

Step 10 – Consider Price

Since all of the bulls on the final list meet the EPD criteria and other specifications, price is the next consideration. In many cases there will be significant differences in the cost of bulls on the final list, and since the performance of their offspring should be similar, purchasing the lower cost bulls reduces the cost per calf produced.

Selection of Replacement Heifers

While some genetic progress can be made through selection of replacement heifers, the goal should be to select a uniform group and let fertility be a major selection criterion. The following program has proven to be effective in numerous cattle operations.

Step 1 – Retain Considerably More Heifers than the Number of Replacements Required

In making the determination of which heifers to retain, remove the extremes with respect to weight and/or frame. This will result in a more uniform cow herd at least with respect to mature size. Select heifers from cows that have been productive, have good udders, and good dispositions. In addition, replacement heifers should be selected from cows that have calved early in the calving season each year. These cows probably fit the environment and production system, which should be passed on to their daughters. Selecting heifers from cows that calved early means that these heifers will be the oldest, increasing the odds that they will be cycling at the start of the breeding season. Selecting the oldest heifers is supported by the data in Table 12-1 from a study at the University of Nebraska where the heifers born in the first 21 days exhibited a higher percentage cycling at the start of the breeding season. These heifers had a higher percentage calving in the first 21 days and a tendency for a higher pregnancy rate for their second calf and heavier calf weaning weight.

Step 2 – Breed Replacement Heifers for a Maximum of Two Cycles

Replacement heifers should be bred for a maximum of 45 days. If considerably more heifers are retained than required for replacements, the heifers could be synchronized with only the heifers that conceive during the synchronized estrus kept as replacements. Alternatively, heifers can be bred during the synchronized estrus and those that fail to conceive could be bred during the next estrus. This results in a breeding season of approximately 30 days. The goal is to keep the breeding season relatively short with some selection for fertility. A short breeding season has the added benefit of keeping the calving season short reducing the effort required to observe and assist heifers during calving.

References

Beef Sire Selection Manual. National Beef Cattle Evaluation Consortium.

Funston, R.N., J.A. Musgrave, T.L. Meyer, and D.M. Larson. 2012. Effect of Calving Period on Heifer Progeny. 2012 Nebraska Beef Cattle Report.

Lalman, D. and D. Doye. Editors. Beef Cattle Manual. 6th Edition. Oklahoma Cooperative Extension Service Publication E-913.

Table 12-1. Results of a Nebraska Study Evaluating the Performance of Heifers Born in the First, Second, or Third 21 Days of the Calving Season.

	Calving Period[1]		
Item	1	2	3
Calf weaning weight, lbs	483[a]	470[b]	434[c]
Pre-breeding weight, lbs	653[a]	644[b]	608[c]
Cycling at beginning of breeding, %	70[a]	58[b]	39[c]
Pregnancy rate, %	90[a]	86[a]	78[b]
Calved in first 21 days, %	81[a]	69[b]	65[b]
Calf weaning weight, lbs	425	417	410
Pregnancy rate after first calf, %	93	90	84

[1]1 = Calved in first 21 days, 2 = Calved in second 21 days, 3 = Calved in third 21 days of the spring calving season.
[abc]Means with different superscripts differ statistically (P<.05).

Chapter 13

Reproduction in Beef Cows and Heifers

As has been mentioned previously, reproduction has a major influence on profitability in cow/calf operations. Put in simple terms, the most important thing a cow can do is to provide a live calf at weaning. An understanding of the reproductive processes in a cow is crucial to manage a cost-effective reproductive rate. Figure 13-1 shows a diagram of a cow's reproductive tract.

Figure 13-1. Diagrammatic Sketch of the Beef Cow Reproductive Tract.

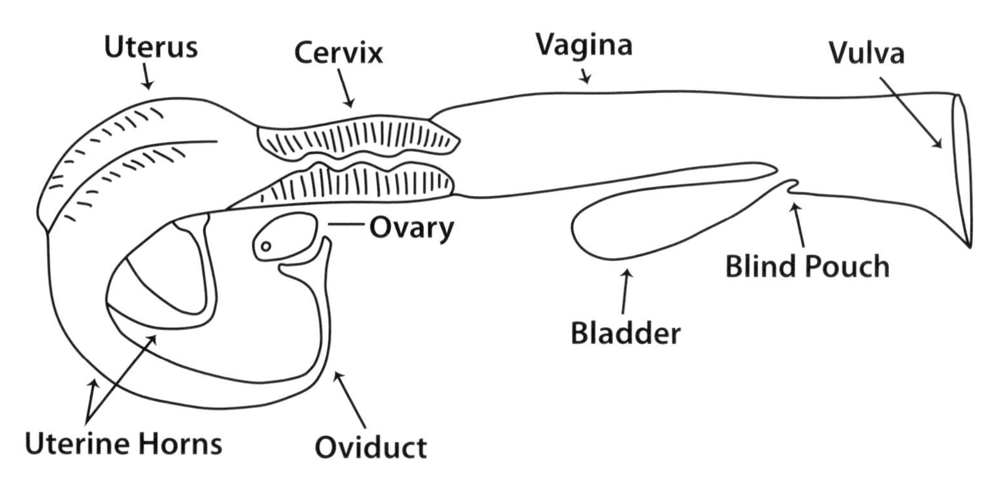

Each part plays a role in the process of producing a calf. The ovaries are the primary reproductive organs because they produce the egg or ovum, and they produce two hormones (estrogen and progesterone) required for pregnancy. The vagina serves as a receptacle for the penis and semen. The vagina, cervix, uterine horns, and oviducts assist in moving the sperm to the egg for fertilization. The uterus also serves as the site of attachment of the ovum and develops the tissues necessary to nourish the fetus.

Reproductive Hormones and Their Function

Table 13-1 shows the reproductive hormones and their function in the reproductive cycle of beef cattle. A number of hormones affect the reproductive cycle. Understanding how these hormones affect reproduction has allowed the development of systems to manage the reproductive cycle by stimulating cycling activity and/or synchronizing estrous.

The Beef Cow Estrous Cycle

The reproductive cycle of the cow consists of a series of events that occur in a sequence over a period of time. Following calving, a beef cow goes through a period, called the postpartum (anestrus) period, where she does not produce any eggs. The length of this period and the factors that determine its length will be discussed in Chapters 22 and 24. At the end of the postpartum period, the cow exhibits estrus, which is best characterized by her willingness to stand to be mounted by other females or a bull. The day the cow is in standing heat (or estrus) is termed day 0 of the estrus cycle. Near the end of standing heat, luteinizing hormone (LH) is released from the pituitary causing the mature follicle to release an egg in a process known as ovulation. The cells of the follicle change and form the corpus luteum (CL). By day 5 following ovulation, the corpus luteum has enlarged and is producing high

Table 13-1. Hormones that Influence Reproduction: Their Source and Function.

Hormone	Source	Function
Gonadotropin Releasing Hormone (GnRH)	Hypothalamus	Releases Follicular Stimulating Hormone (FSH) and Luteinizing Hormone LH from the Pituitary
Follicle Stimulating Hormone (FSH)	Pituitary	Promotes follicular Growth, Development and Estrogen Production
Luteinizing Hormone (LH)	Pituitary	Causes Ovulation, Development and the Function of the Corpus Luteum
Estrogens	Ovarian Follicles	Causes Growth and the Development of the Uterus. Estrus Behavior and Triggers Release of LH for Ovulation
Progesterone	Corpus Luteum	Maintains Pregnancy, Blocks Estrus
Prostaglandin ($PGF_{2\alpha}$)	Uterus	Causes Regression of the Corpus Luteum

levels of progesterone. During the next 7 days, the progesterone secreted by the CL inhibits LH release from the pituitary gland. On about day 16, if maternal recognition of pregnancy has not occurred, the CL regresses rapidly in response to prostaglandin (PG) produced by the uterus. By day 18, the CL no longer produces enough progesterone to prevent final maturation and ovulation of another follicle. Estrus will typically occur at about day 21 — day 0 of the new cycle.

At about 30 hours after the onset of estrus, the cow will ovulate an egg, which then may be fertilized by the sperm deposited by the bull during estrus or by artificial insemination. If the egg is not fertilized or maternal recognition of pregnancy does not occur, the cow will exhibit estrus again in approximately 21 days and ovulate another egg. Under normal conditions, this 21-day cycle is repeated until pregnancy. Figure 13-2 shows a schematic diagram of this cycle and the key structures and hormones involved. An important hormone not shown on Figure13-2 is Follicle Stimulating Hormone (FSH), which is released in waves throughout the cycle. As its name implies, it promotes development of follicles that contain eggs. The other key hormone not shown is Luteinizing

Figure 13-2. The Normal Estrous Cycle in a Beef Cow.

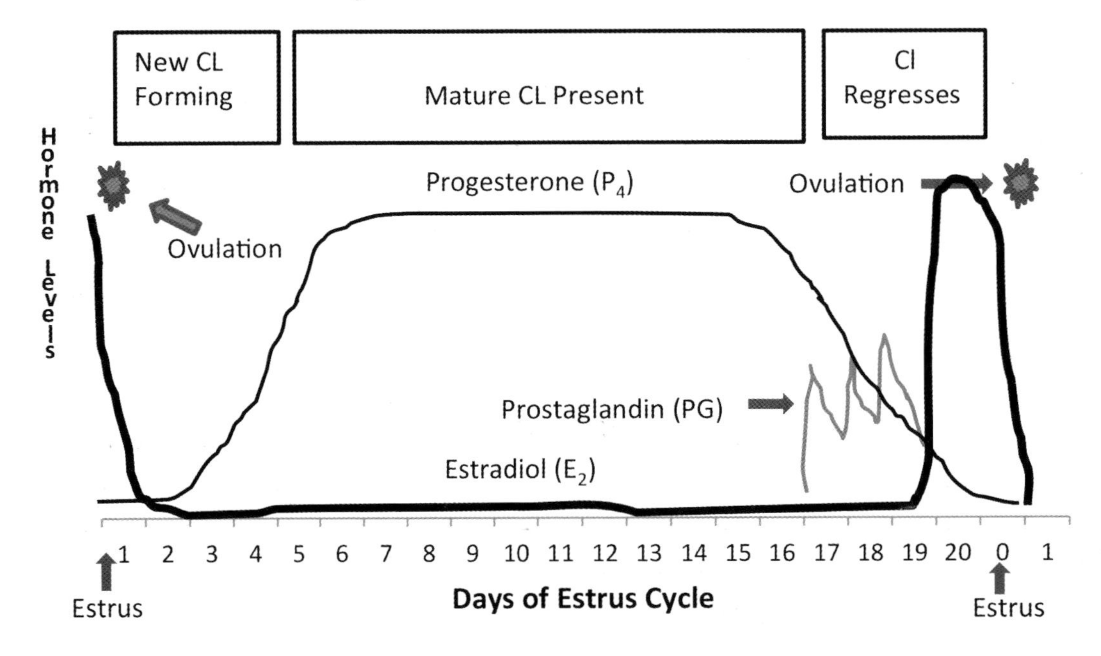

Hormone (LH), which is released by the pituitary towards the end of estrus. Its release results in the release of the egg from the follicle.

When pregnancy occurs, the progesterone produced by the CL limits follicle development on the ovaries and exhibition of heat (estrus). The increase in estrogen secretion at the end of a cycle is responsible for estrus and sexual receptivity.

Puberty in Heifers

Puberty in heifers is characterized by ovulation and development of a fully functional corpus luteum. Ovulation is controlled by the interaction of hormones produced by the ovary and those produced by the hypothalamus and pituitary located at the base of the brain. Prior to puberty, estrogen from the ovary prevents the hypothalamus from producing enough gonadotropin-releasing hormone (GnRH) to cause LH release from the pituitary and development of ovulatory follicles. As puberty nears, the hypothalamus becomes less sensitive to the negative effect of estrogen and starts producing significant pulses of GnRH, which acts on the pituitary to stimulate LH production. This LH causes follicles on the ovary to mature. These large follicles produce more estrogen, and as puberty nears, this estrogen actually stimulates releases of GnRH and subsequently LH. Ultimately one of these follicles releases an egg and ovulation occurs.

From a practical standpoint, puberty in beef heifers is primarily a function of age, weight, and breed. Table 13-2 shows the average age and weight at puberty for various breed crosses in the original U.S. Meat Animal Research Center data. Clearly, there were some major differences in these breeds for both the age and weight at which they reached puberty. Also, these data show only the sire breed effects, which should be roughly one-half of the actual breed differences.

A more recent breed comparison for the major breeds is shown in Table 13-3. The weight at which puberty was expressed increased significantly from the 1970s evaluation to the 2000s evaluation. However, the age at which puberty was expressed tended to

Table 13-2. Breed Group Averages for Age and Weight (1970s Evaluation).

Breed Type	Weight at Puberty, lbs	Age at Puberty, Days
Low to Moderate Growth, High Milk		
Jersey-X	518	308
Hereford-Angus-X	622	357
Red Poll-X	580	337
South Devon-X	639	350
Tarentaise-X	622	349
Pinzgauer-X	611	334
Brahman Influence		
Sahiwal-X	642	414
Brahman-X	712	429
High Growth, High Milk		
Brown Swiss-X	615	332
Gelbvieh-X	626	326
Simmental-X	666	358
Maine-Anjou	672	357
High Growth		
Limousin-X	679	384
Charolais-X	703	384
Chianina-X	699	384

Adapted from Beef Research Report No. 2, from U.S. Meat Animal Research Center

Table 13-3. Sire Breed Means for Growth and Puberty in Heifers (2000s Evaluation).

Sire Breed	N	18 mo. Wt., lbs	Puberty Expressed %	Puberty Wt., lbs	Age At Puberty Days	Preg. Rate %
Hereford	81	950	79.2	733	342	94
Angus	85	936	97.2	750	340	88
Red Angus	106	953	97.4	744	339	91
Simmental	103	961	91.6	757	335	90
Gelbvieh	111	922	91.5	711	322	83
Limousin	109	933	80.3	785	363	87
Charolais	103	950	88.5	758	349	91

Adapted from Germplasm Evaluation Program Progress Report No. 22 from the U.S. Meat Animal Research Center.

decrease, especially in some of the Continental breeds. The differences in pregnancy rate were not significant. It should be noted that the number of heifers evaluated was relatively small at the time of the publication of Progress Report 22. An evaluation involving more heifers should be available on the U.S. Meat Animal Research Center website.

References

Anderson, L. Managing Reproduction. University of Kentucky Beef Book.

Cundiff, L., T.L. Wheeler, K.E. Gregory, S.D. Shackelford, M. Koomaraie, R. M. Thallman, G.D. Snowder, and L.D Van Vleck. 2004. Germplasm Evaluation Program Progress Report No. 22.

Dyer, T.G. and W.M. Graves. 2009. Estrous Synchronization Procedures for Beef Cattle. University of Georgia Extension Bulletin 1232.

Lamb, C. Relationship Between Nutrition and Reproduction in Beef Cow. North Florida Research and Education Center.

Parish, J.A., J.E. Larson, and R.C. Vann. 2010. The Estrous Cycle of the Cow. Mississippi State University Extension Publication 2616.

Rich, T.D. and E.J. Tautm. Reproductive Tract Anatomy and Physiology of the Cow. Beef Cattle Handbook. GPE-8454.

Synchronizing the Estrous Cycle in Beef Cattle. 1980. Montana Agricultural Experiment Station Bulletin 730.

Waters, K. 2012. Replacement Heifer Development: The Attainment of Puberty. South Dakota State University Website: igrow.org/livestock/beef.

Wheeler, T.L., L.V. Cundiff, L.D. Van Vleck, G.D. Snowder, R. M. Thallman, S.D. Shackelford, and M. Koomaraie. 2006. Germplasm Evaluation Program, Progress Report No. 23 U.S. Meat Animal Research Center.

Williams, G.L. and M. Amstalden. 2010. Understanding Postpartum Anestrus and Puberty in the Beef Female. Proceedings, Applied Reproductive Strategies in Beef Cattle. January 28-29, San Antonio, TX.

Chapter 14

Estrus Synchronization

Artificial insemination is a powerful tool for beef cattle improvement. It allows the widespread use of proven bulls. Artificial insemination without estrus synchronization, however, is time consuming. Consequently, the approval of products that promote estrus synchronization has been a significant advancement in the beef cattle industry. These products are based on the use of naturally occurring hormones or their analogues. Currently, there are synchronization protocols that range from low-cost, simple systems to expensive, time-consuming protocols. Some of these protocols actually initiate cycling activity in heifers or cows that are close to cycling naturally. However, adequate nutrition and development are still important.

Selecting a Synchronization Protocol

As noted above, synchronization protocols differ greatly in complexity. Factors to consider in selecting a protocol include:

1. **Amount of Heat Detection Required** – The amount of heat detection required varies from zero in fixed-time AI programs to 12 days with other protocols.
2. **Level of Cycling Activity** – Some products, such as the products containing prostaglandins, are only effective in animals that are cycling. Thus, the percentage of animals cycling has a significant impact on the success of the program. Other products like a CIDR® or other progestin-containing products stimulate some cycling activity and may improve response in non-cycling animals.
3. **Management Considerations** – Protocols vary significantly in the number of times the cattle must be handled, making the quality of working facilities and labor availability considerations. Other protocols require feeding a product for several days, and it is crucial to the success of the program that every animal consumes an adequate level of the drug each day. Consequently, feed mixing capability and bunk space need to be considered in selecting a protocol.
4. **Cost** – Synchronization protocols vary in cost, which should be considered. The cost of semen and labor can be significant.

Products Used to Synchronize Estrus in Beef Females

Products used to synchronize estrus fall into three categories:

- **Prostaglandins (PG)** – The first commercially available prostaglandin product was approved in 1979, and these products are still used in many synchronization protocols.
- **Progestins** – The biologically active hormone in this group is the hormone progesterone or one of its analogues.
- **Gonadotropin Releasing Hormone (GnRH)** – The active ingredient in these products acts on the pituitary to stimulate the release of leutenizing hormone and follicle stimulating hormone, which both stimulate activity on the ovary as discussed in the previous chapter.

There are several products in each of these categories as shown in Table 14-1.

Table 14-1. Commercially Available Products Used to Synchronize Estrus in Beef Cattle.

Product	Dosage	Approved Label Use
	Prostaglandins	
EstroPLAN	2 ml, im	Beef heifers and cows
Estrumate	2 ml, im	Beef heifers and cows
IN-SYNCH	5 ml, im	Beef heifers and cows
Lutalyse	5 ml, im	Beef heifers and cows
ProstaMate	5 ml, im	Beef heifers and cows
	Progestins	
CIDR[a]	1/hd, intravaginally	Beef heifers and cows
MGA[b]	0.5 mg/hd/day, orally	Beef heifers
	GnRH	
Cystorelin	2 ml, im or iv	Bovine females
Factrel	2 ml, im	Dairy females
Fertagyl	2 ml, im or iv	Bovine females
OvaCyst	2 ml, im or iv	Dairy females

[a]Controlled internal drug release. [b]Melengestrol acetate – a synthetic progesterone. Modified from University of Nebraska Extension Bulletin EC283.

Synchronization Protocols

Synchronization protocols are usually grouped by the amount of heat detection required:

1. Those requiring heat detection.
2. Those requiring heat detection plus timed AI.
3. Those using only timed AI.

Heat Detection Protocols

In these systems, the females should be inseminated 6 to 12 hours after being observed in standing heat. Considerable time should be spent observing the cattle since some heifers and cows only show signs of estrus for a brief period.

Single Injection of Prostaglandin (PG)

This protocol involves heat detecting for 5 days and then injecting the heifers or cows that have not exhibited heat with PG. Heifers that aren't cycling will not respond to the injection of PG. Consequently, if a considerable number of heifers have not exhibited heat in the first 5 days of heat detection, it is best to not give the PG injection. If 100 percent of a group of heifers are cycling, expect approximately 5 percent of the heifers to be cycling each day.

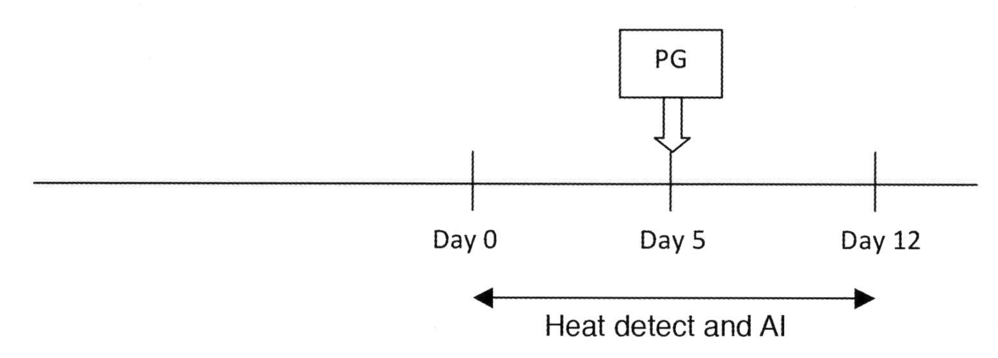

Feed MGA® - Inject PG

This protocol consists of feeding MGA for 14 days starting 33 days before the PG injection. Obviously, providing the MGA on a consistent basis is key to this protocol working effectively.

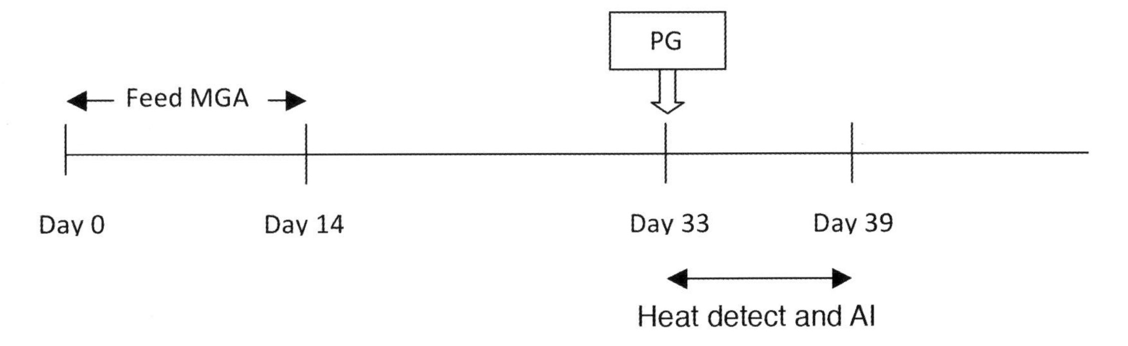

PG 6-day CIDR®

This protocol can be used in cows and heifers. It reduces the cost of synchronization products for those animals inseminated in the first 3 days. However, adequate synchronization products must be available to treat all animals. GnRH and CIDR® are administered to animals not detected in heat and inseminated by day 3.

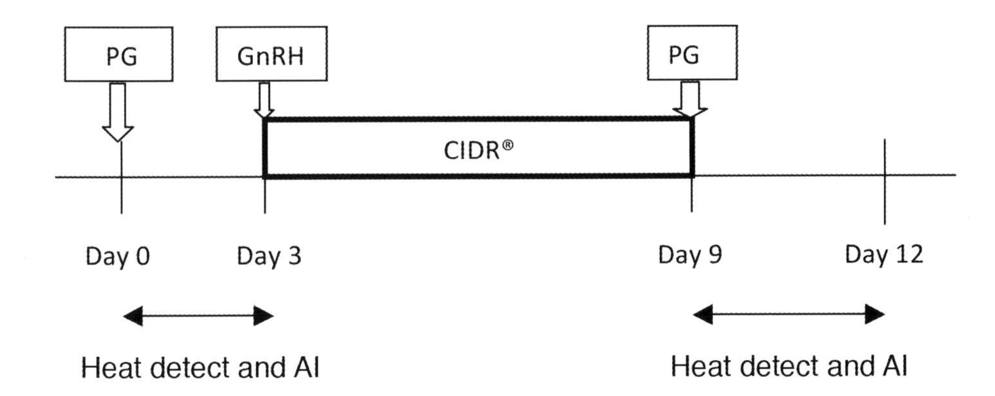

7-Day CIDR® - PG

This protocol is recommended in heifers instead of the Select Synch + CIDR®.

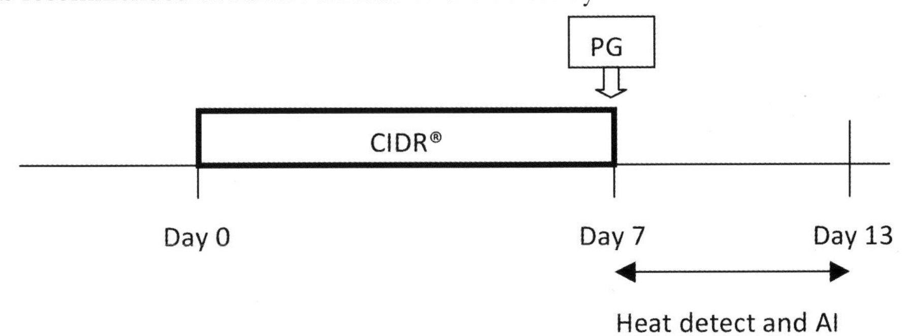

Select Synch and Select Synch + CIDR®

These protocols are for use in cows. The CIDR® should be used when a significant number of cows are likely to be anestrus. With Select Synch, 5 to 20 percent of the animals may show heat 1.5 to 2 days before the PG injection.

Select Synch

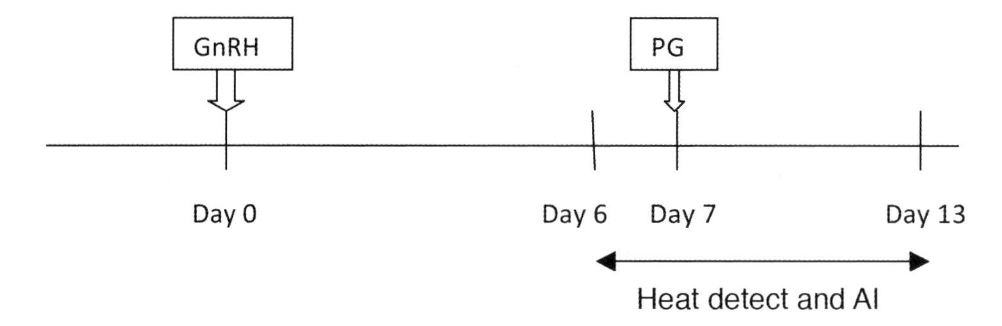

Select Synch + CIDR®

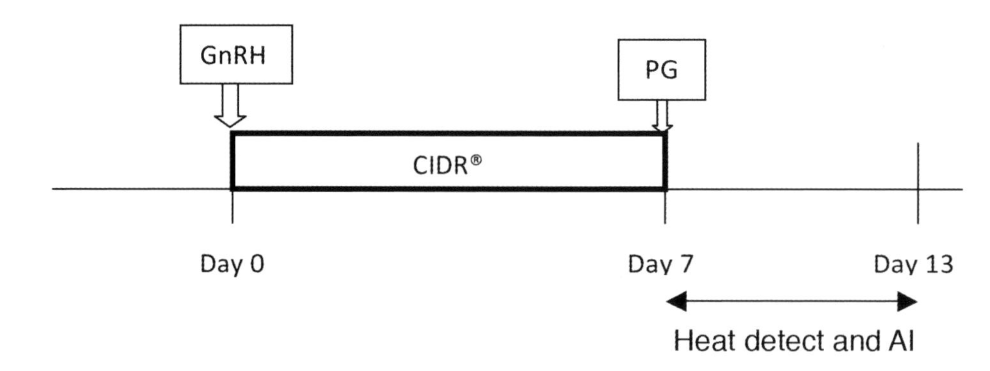

Heat Detection and Timed AI (TAI) Protocols

These protocols involve artificial insemination 6 to 12 hours after observed estrus for 3 days and then insemination of animals that have not exhibited heat 72 to 84 hours after PG. GnRH is injected at the timed insemination.

MGA®-PG & TAI

This protocol can be used in heifers. It combines heat detection and timed AI. Heat detect and AI day 33 to 36 and TAI all non-responders 72 to 84 hours after PG with GnRH at TAI.

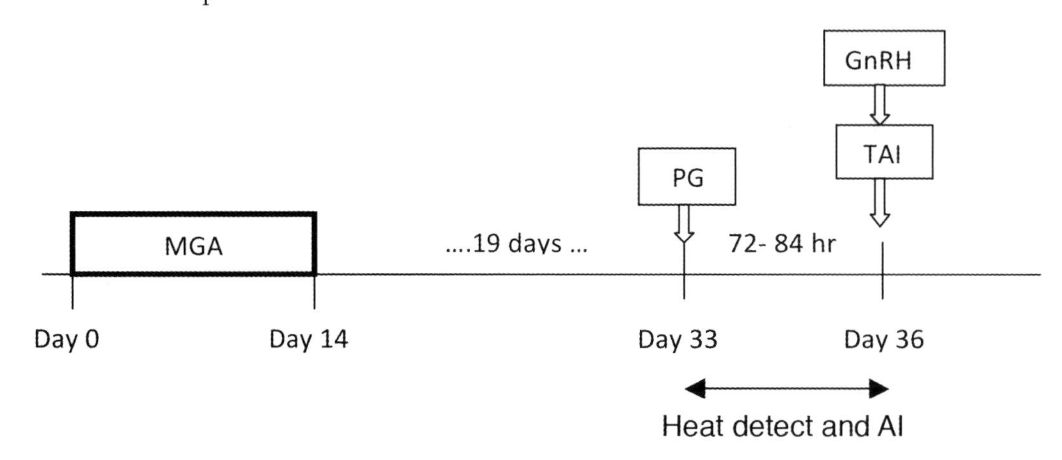

Select Synch + CIDR® & TAI

This is another protocol recommended for heifers. The GnRH injection is given when the CIDR® is administered because it reduces the variability of the response. Heat detect and AI day 7 to 10 and TAI all non-responders 72 to 84 hrs after PG with GnRH at TAI.

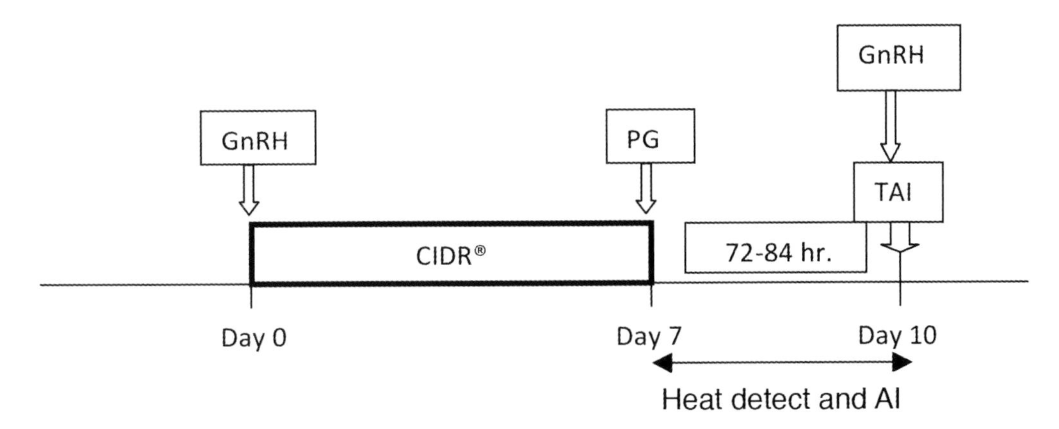

14-day CIDR® - PG

This protocol also works well with heifers. Heat detect and AI day 30 through 33 and TAI all non-responders 72 hrs after PG with GnRH at TAI.

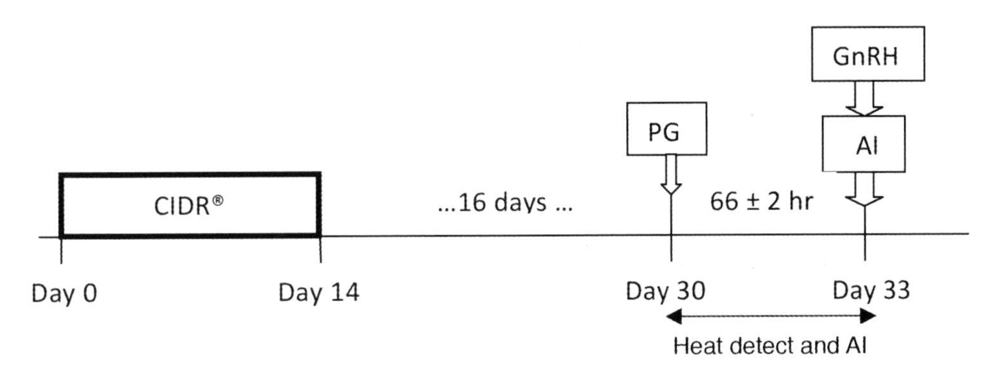

Select Synch & TAI

This protocol works best with cows. Heat detect and AI day 6 through 10 and TAI all non-responders 72 to 84 hours after PG with GnRH at TAI.

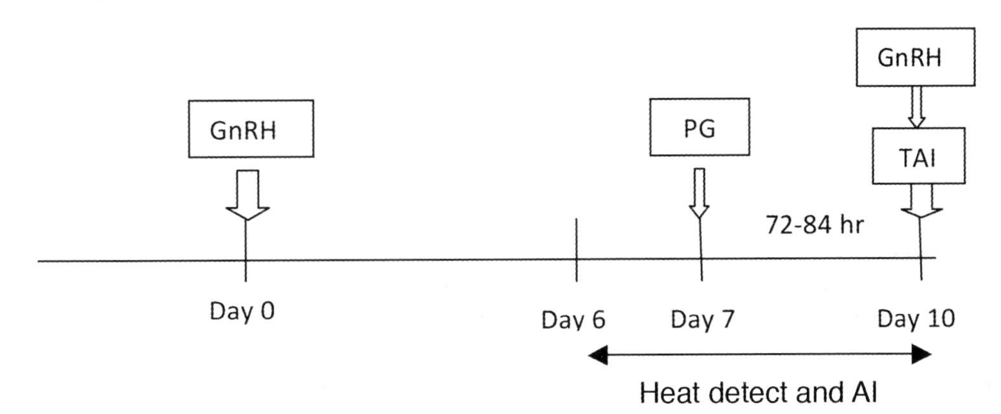

PG 6-day CIDR® & TAI

This protocol can be used with both heifers and cows. Heat detect and AI day 0 through 3. Administer CIDR to all non-responders and heat detect and AI days 9 through 12. TAI all non-responders 72 to 84 hrs after CIDR removal with GnRH at AI.

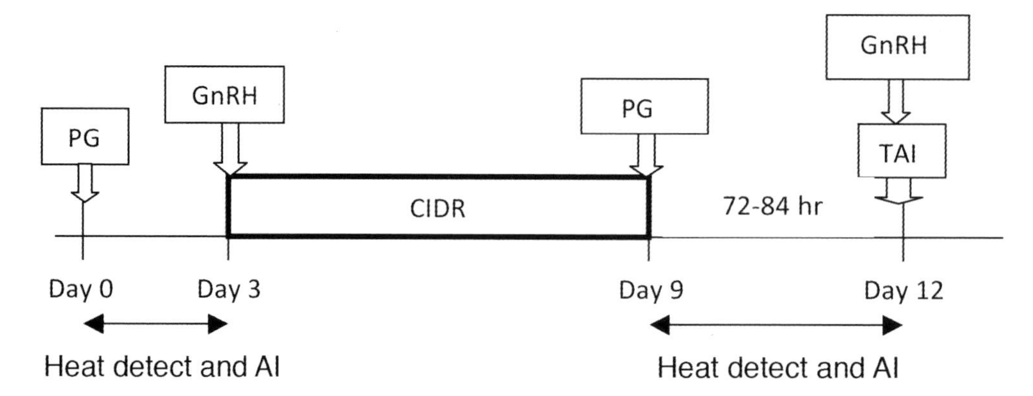

Fixed-Time AI Protocols

In fixed-time protocols, all animals are inseminated at a predetermined time. With cows, fixed-time protocols result in pregnancy rates similar to protocols that require 5 to 7 days of heat detection. With heifers, pregnancy rates from fixed-time protocols tend to be 5 to 10 percent lower than using heat detection alone. The time listed for insemination should be considered as the average. The maximum number of females synchronized should be the number that can be inseminated in 4 hours.

Fixed-Time Protocols for Heifers

MGA-PG

Pregnancy rates with this MGA protocol will probably be lower than with protocols that use CIDR®. TAI at 72±2 hrs after PG with GnRH at TAI.

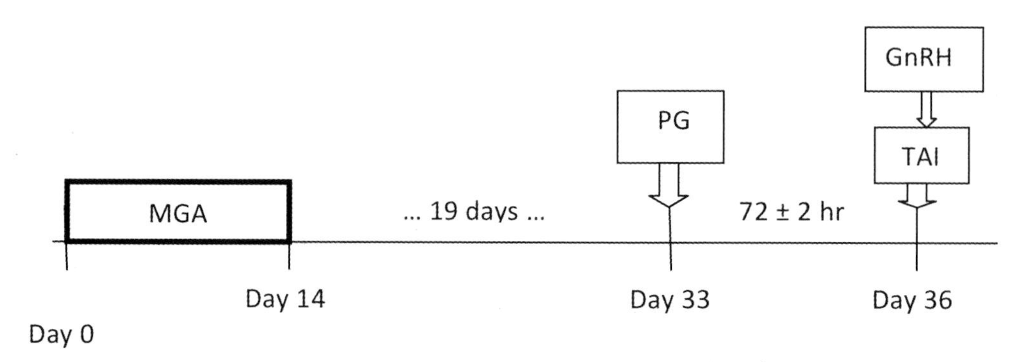

7-Day CO-Synch + CIDR® - Heifers

TAI at 54±2 hrs after PG with GnRH at TAI.

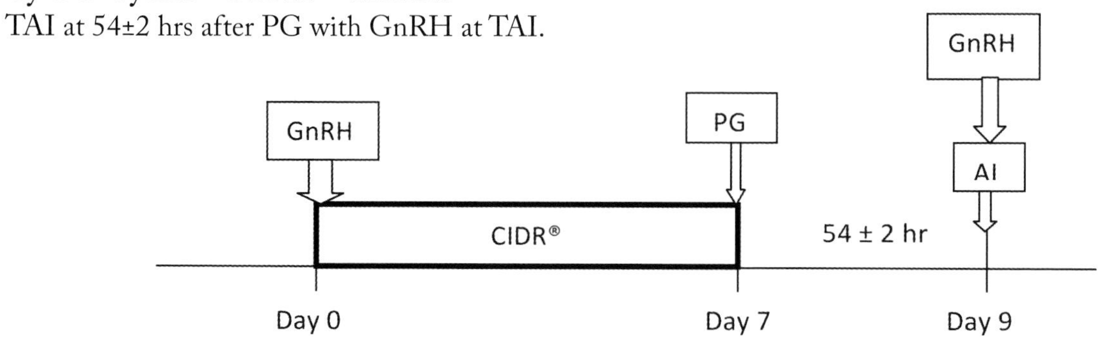

14-day CIDR® - PG

TAI at 66±2 hrs after PG with GnRH at TAI.

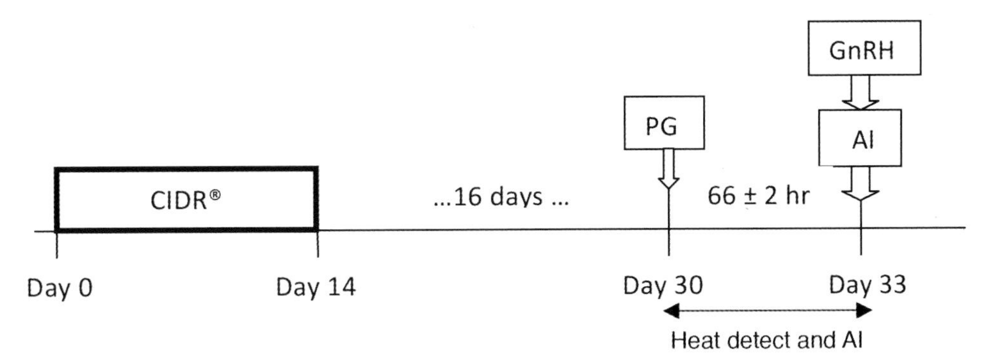

Fixed-Time Protocols for Cows

7-day CO-Synch + CIDR® - Cows

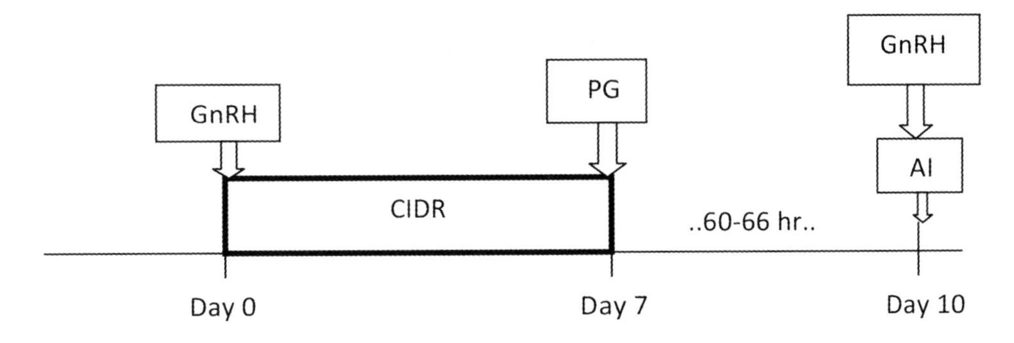

TAI at 60-66 hrs after PG with GnRH at TAI.

5-day CO-Synch + CIDR® - Cows

Two injections of PG 8±2 hrs apart are required for this protocol. TAI at 72±2 hrs after 1st PG injection with GnRH at TAI.

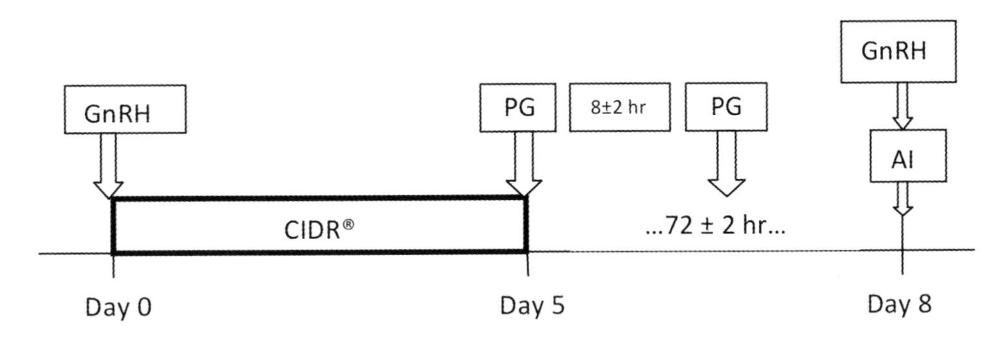

Comparison of Synchronization Protocols

Clearly, numerous synchronization protocols can be used to synchronize beef heifers. A comparison of these protocols based on the cost, labor requirements, and number of days required is shown in Table 14-2. Correspondingly, Table 14-3 shows a comparison of synchronization protocols for beef cows.

Table 14-2. Comparison of Beef Heifer Estrus Synchronization Protocols.

Protocol	Cost	Labor Required	Length (Days)
Heat Detection			
1 Shot PG	Low	High	12
MGA-PG	Low	Low/Medium	39
7-day CIDR®- PG	Medium	Medium	13
Heat Detection & Timed AI			
Select Synch	Medium	Medium	13
Select Synch + CIDR®	High	Medium	13
Timed Insemination			
MGA-PG	Medium	Medium	36
CO-Synch + CIDR®	High	Medium	9
14-day CIDR® - PG	High	High	33

Modified from University of Nebraska Extension Bulletin EC283.

Table 14-3. Comparison of Estrus Synchronization Protocols for Cows.

Protocol	Cost	Labor Required	Length (Days)
Heat Detection			
PG 6-day CIDR®	Medium	Medium	12
Select Synch	Low	Medium/High	13
Select Synch + CIDR®	High	Medium	13
Heat Detection & Timed AI			
Select Synch	Low	Medium/High	10
Select Synch + CIDR®	High	Medium	10
Timed Insemination			
7-day CO-Synch + CIDR®	High	High	10
5 day CO-Synch + CIDR®	High	High	8

Modified from University of Nebraska Extension Bulletin EC283.

References

Anderson, L. Managing Reproduction. University of Kentucky Beef Book.

Dyer, T.G. and W.M. Graves. 2009. Estrous Synchronization Procedures for Beef Cattle. University of Georgia Extension Bulletin 1232.

Johnson, S., R. Funston, J. Hall, G. Lamb, J. Lauderdale, D. Patterson, and G. Perry. Protocols for Synchronization of Estrus and Ovulation. Beef Reproduction Task Force.

Lamb, C. 2010. Estrus Synchronization Protocols for Cows. Proceedings of the Applied Reproductive Strategies Conference.

Patterson, D., D. Mallory, J. Nash, and M. Smith. 2010. Estrus Synchronization Protocols for Heifers. Proceedings of the Applied Reproductive Strategies Conference.

Rasby, R. and R. Funston. 2010. Synchronizing Estrus in Beef Cattle. University of Nebraska Extension Bulletin EC283.

Chapter 15

The Ruminant Digestive System

Beef cattle are ruminants with a four-compartment stomach as shown in Figure 15-1. The compartments include the rumen, reticulum, omasum, and abomasum.

Figure 15-1. Digestive Tract of Beef Cattle.

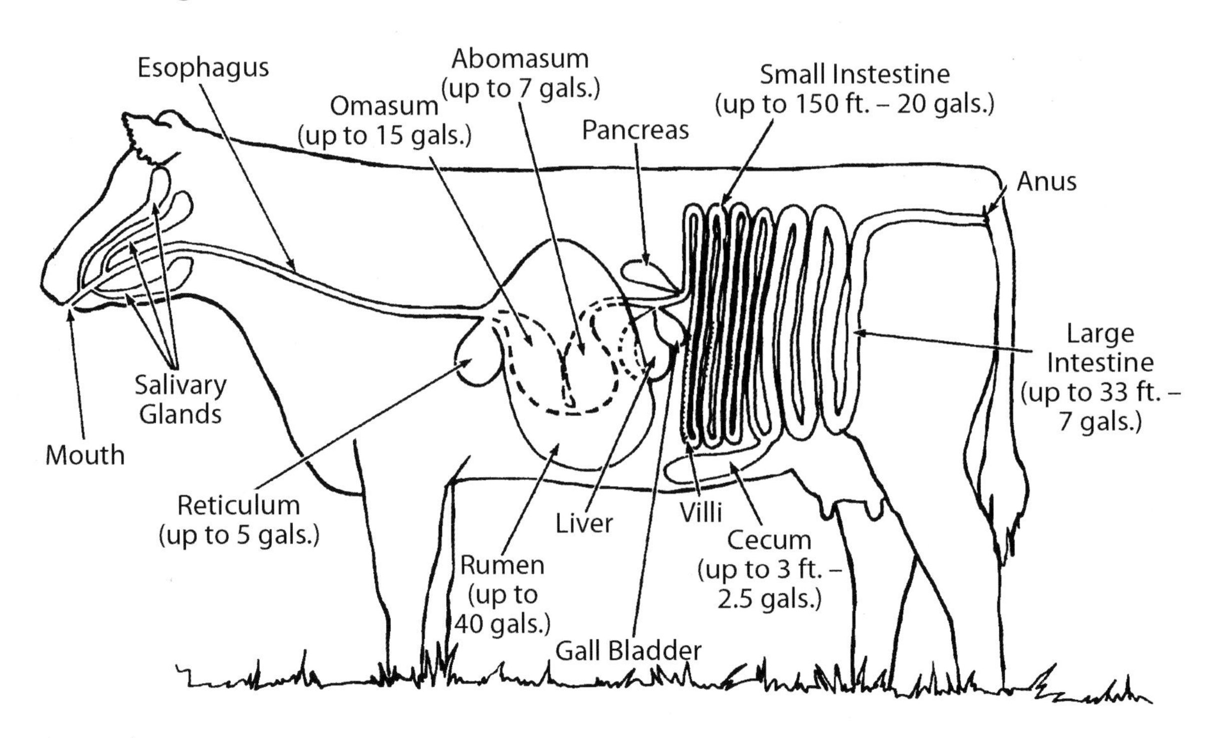

Flow of Feed through the Digestive System

Figure 15.2. Diagram of the Flow of Feed through the Digestive Tract.

Mouth ⟹ Esophagus ⟹ Rumen ⟹ Reticulum ⟹

Omasum ⟹ Abomasum ⟹ Small Intestine ⟹ Large Intestine

Parts of the Beef Cattle Digestive System and Their Function

Mouth

Obviously, the main function of the mouth is to acquire feed and chew it. In addition, the mouth adds saliva to the feed. Cows produce 20 to 35 gallons of saliva per day. This saliva moistens the feed, and it contains sodium bicarbonate, which buffers the rumen reducing the potential for a low pH.

Stomach

As noted, the stomach of a ruminant has four compartments:

Rumen

The rumen is the largest compartment (40 to 60 gallons in an adult cow) and serves as a large anaerobic fermentation vat. It contains billions of bacteria, protozoa, molds, and yeasts. The animal provides a warm, moist environment and substrate (feed) for their growth and multiplication. These microorganisms can digest feedstuffs high in fiber that the animal could not otherwise digest. After the microorganisms digest feed in the rumen, they are digested in the lower tract. Bacteria and protozoa do most of the digestion in the rumen. There are 25 to 50 billion bacteria and 200 to 500 thousand protozoa in every milliliter of rumen fluid. The primary product of the digestion of fiber by these organisms is volatile fatty acids, which are absorbed directly through the wall of the rumen into the blood stream. Approximately 60 to 80 percent of the energy needs of the cow are supplied in this manner.

The microorganisms use the protein and non-protein nitrogen (nitrogen not contained in amino acids) in the feeds to produce their own proteins, which become a major source of protein for the animal when the microorganisms are digested. Rumen microbes are 50 to 60 percent protein on a dry matter basis, which makes them an excellent protein source for the animal. Large quantities of gas (30 to 50 liters per hour in adult cattle) are produced in the rumen with the primary gases being carbon dioxide and methane. These gases must be released by belching otherwise bloat occurs. The rumen contracts at a rate of one to three times per minute, which mixes the feeds and aids in eructation of the gases.

Rumination – Ruminants spend a considerable portion of their day chewing their cud, which is essentially regurgitated feed. This process reduces particle size of the feedstuffs speeding up digestion. And as noted, considerable saliva is added to this material during the chewing process.

Advantages of Rumen Fermentation

1. **Forage Utilization** – The rumen allows beef cattle to digest high-fiber feedstuffs.
2. **Utilization of Non-protein Nitrogen** – Rumen microorganisms can use nitrogen-containing compounds of many types to manufacture amino acids and protein.
3. **Vitamin Synthesis** – Rumen microorganisms can synthesize the B-Complex vitamins and vitamin K, making supplementation of these nutrients unnecessary under normal conditions.

Disadvantages of Rumen Fermentation

1. **Excess Gas Production** – The production of carbon dioxide and methane during the digestion of carbohydrates results in lost energy.
2. **Inefficient Nitrogen Utilization** – Some of the ammonia produced during digestion is excreted in the urine, resulting in inefficient use of dietary protein.
3. **Heat of Fermentation** – Rumen microorganisms produce considerable heat in the process of digesting feedstuffs. This heat may be an advantage in cold weather, but otherwise it is wasted energy.
4. **Digestive Problems** – The large amount of gas produced during digestion can be a problem if something prevents it from being eructated. In addition, the low pH in the rumen on high concentrate diets can result in acidosis, which damages the rumen lining and causes liver abscesses.

The losses in energy noted in disadvantages 1 and 3 explain the differences in feed conversion between beef cattle and monogastric livestock like the pig and chickens. For example, it takes beef cattle on finishing diets about 6 pounds of feed for each pound of gain. Correspondingly, pigs need about 2.5 pounds and chickens about 1.5 pounds. In summary, the rumen has the advantage of providing the ability to digest high-fiber feedstuffs, but it is inefficient in terms of feed conversion relative to monogastric species.

Important Concepts Regarding the Rumen

Microorganisms Tend to Prefer Specific Substrates

There are many different species of bacteria in the rumen, and each species has a preference for a specific type of feed component. For example, some species specialize in breaking down the fibrous part of the feed, while other species specialize in breaking down starch, and still others specialize in breaking down protein. In fact, the bacteria in the rumen have been categorized based on their preferred substrate. From a practical standpoint, one can consider these organisms to be in competition. For example, if the diet is primarily composed of feedstuffs high in fiber, the strains of bacteria that prefer fiber as a substrate will predominate. Conversely, diets high in starch promote the growth of microorganisms that prefer starch.

The Diet Influences the Microbial Population

Another way to look at the concept discussed in the previous paragraph is that the composition of the diet alters the microbial population. It should be noted that it may require some time for this change to occur. For example, when biuret (a nitrogen source composed of two urea molecules bound together) is fed for the first time, it takes a period of adaptation before the microorganisms can synthesize the enzyme required to cleave the urea molecules. Thus, there is a lag period of a couple of weeks before the nitrogen in biuret can serve effectively as a source of nitrogen for the microbial population. This type of adaptation by the microbial population occurs to some extent any time the diet is changed dramatically.

Importance of Meeting the Nutrient Requirements of the Microbial Population

To do a good job of feeding beef cattle, consider both the requirements of the cattle and the requirements of the microbes. In many situations, giving the microbes what they need to grow is the best way to meet the nutritional needs of the cattle. For example, when cows are consuming low-quality roughages, such as corn stover or mature grass, the ability of the microbes to digest the forage is often limited by the level of protein in the diet. Supplying this protein is the best way to supply both energy and protein to the cow because with this protein, the microbes can extract the energy from the forage. *In summary, making the "bugs" happy is often the key to meeting the nutrient requirements of beef cattle.*

Reticulum

The reticulum, which has a honeycomb-like lining, is involved in rumination. It also traps foreign objects such as metal and rocks. Sharp metal objects may puncture the lining of the reticulum and enter the diaphragm, heart, or lungs — a condition known as hardware disease.

Omasum

Its primary function is to filter large particles back to the reticulorumen for further digestion. It also appears to produce a grinding action on foodstuffs.

Abomasum

Also known as the true stomach, it produces large quantities of acid and some enzymes to start protein digestion. For example, it produces lysozyme that breaks down bacterial cell walls.

Small Intestine

The microorganisms that are in the digesta are digested in the small intestine serving as a major source of protein. Also, a high percentage of the bypass protein in the feed (protein that is not broken down by the microbes) is digested in the small intestine. While the volatile fatty acids are mainly absorbed through the rumen wall, the major source of absorption of the other nutrients like protein, fats, and minerals occurs in the small intestine.

Large Intestine

Most of the absorption of water occurs in the large intestine. The other major role is the formation of feces.

Figure 15-3. Schematic Diagram of the Digestive Process in Ruminants.

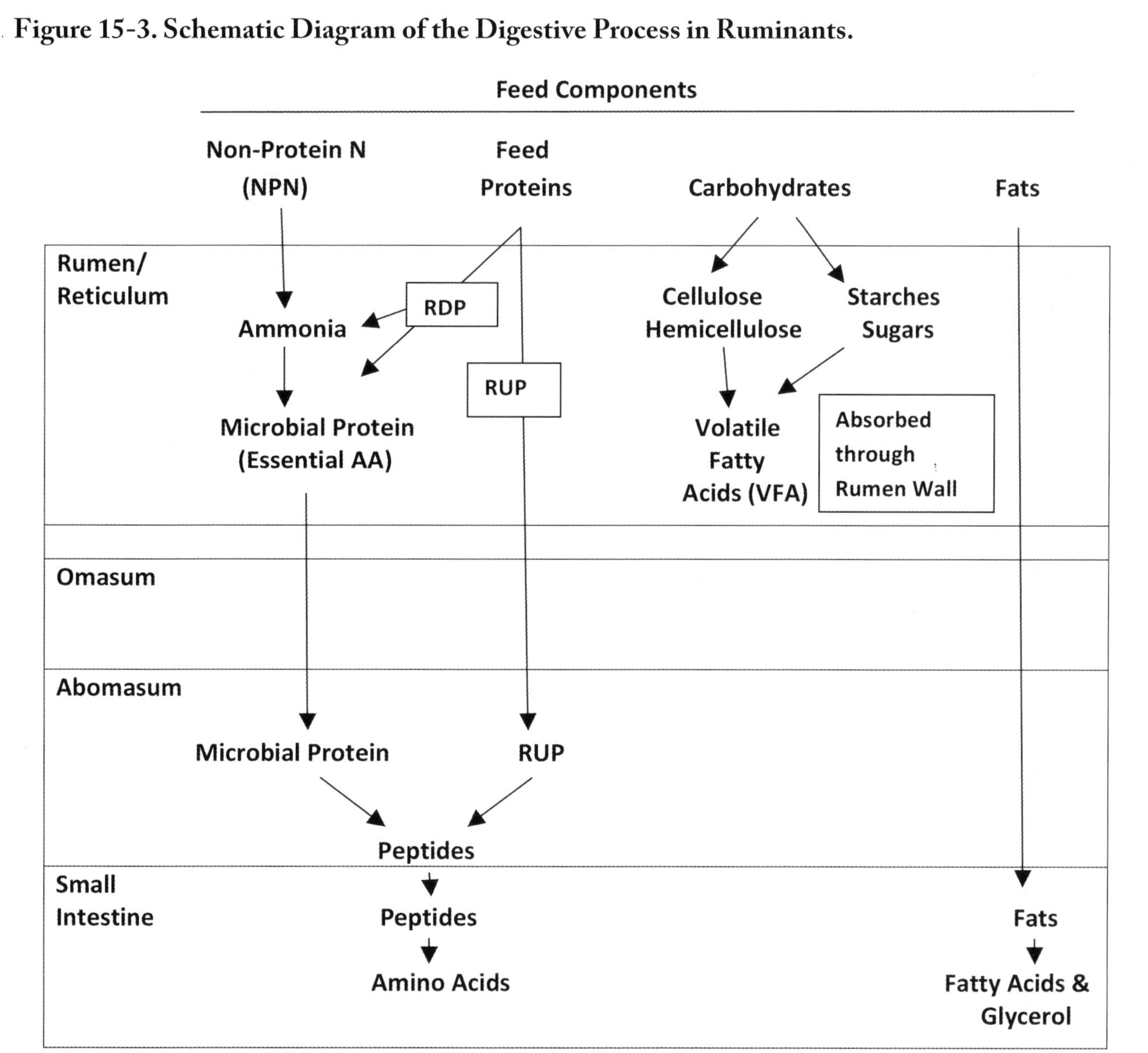

RDP = rumen degradable protein; RUP = rumen undegradable protein

Figure 15-3 shows schematic representation of the major sites of digestion of the primary nutrients in livestock feed:

1. Carbohydrates – Cellulose, Hemicellulose, Starches, and Sugars.
2. Protein – Rumen Degradable and Rumen Undegradable Proteins.
3. Non-Protein Nitrogen.
4. Fats.

Accessory Organs:

Liver – Center of metabolic activity in the body. Its major role in the digestive process is to provide the bile salts from the gall bladder, which are required for the digestion of fats.

Pancreas – Produces many of the enzymes required to digest the foodstuffs.

References

Coffey, R. Basic Functional Anatomy of the Digestive System – Ruminants. University of Kentucky.

Hall, J. and S. Silver. 2009. Nutrition and Feeding of the Cow-Calf Herd: Digestive System of the Cow. Virginia Cooperative Extension Publication 400-010.

Rounds, W. and D. Herd. The Cow's Digestive System. Texas Agricultural Extension Service Bulletin B-1575.

Chapter 16

Water

Water is the most abundant component in the body. It is needed for a number of functions including regulation of body temperature, digestion, metabolism, and excretion. Table 16-1 shows the approximate water consumption of various classes of beef cattle.

The information in Table 16-1 was based on some old research and should be used only as a general guideline. One concern is that there was no influence of temperature on water intake below 40 degrees F. Also, it may seem odd that the water intake of the smaller, wintering, pregnant cows was greater than the larger cows in this research. The interpretation of this was that the larger cows had more body fat and thus were consuming less dry matter and consequently required less water. In general, water intake would be expected to increase as body weight increases. The following factors influence the amount of water required:

1. **Stage of Production** – Production functions tend to increase water intake. For example, lactating cows consume more water than non-lactating cows. In addition, increased growth rate increases water intake.
2. **Ambient Temperature** – Water intake increases as the ambient temperature increases as one of the functions of water is to regulate body temperature.
3. **Dry Matter Intake** – As dry matter intake increases, water intake increases, because water is required to provide a suitable environment for microbial growth in the rumen.
4. **Salt Consumption** – As salt consumption increases, the amount of water required to "flush" the salt through the kidneys increases. This is especially relevant when salt is being used to limit intake as will be discussed in Chapter 36.
5. **Breed** – Zebu cattle consume less water than British and European breeds.
6. **Amount of Physical Activity** – Water intake increases because of physical activity. Water lost during respiration must be replaced and body temperature regulated.

In most production situations, water requires minimal consideration as a nutrient. However, there are some situations where special emphasis on water intake is required. For example, situations that restrict water intake below the level required by the animal will reduce feed intake and performance. A few of these situations include water that is high in salts, sulfate, and/or nitrate.

Water High in Salts

In some areas, water may be contaminated with salts including sodium chloride and various carbonates and nitrates. Salinity refers to the total amount of these salts dissolved in water and is measured as total dissolved solids. Table 16-2 presents a guide to using saline water for beef cattle.

Water High in Sulfate

There are a few areas where water available for livestock use is high in various sulfates. If the level of sulfate is high enough, water intake will be reduced with a corresponding reduction in feed intake, which will ultimately reduce performance. The generally recognized level where this occurs is more than 500 parts per million. Animals forced to drink water with very high levels of sulfates may exhibit polioencephalomalacia (PEM), which is also called Circling Disease or sulfur-induced polio. In the early stages, affected animals walk in circles. In the latter stages they lay on their sides and "swim." Animals with this condition are referred to as "brainers."

Water High in Nitrates

Numerous situations can lead to high nitrate levels in livestock water. Table 16-3 shows the nitrate levels and corresponding symptoms of toxicity.

Table 16-1. Approximate Total Water Intake by Beef Cattle in Gallons Per Day.

Weight, lbs	Environmental Temperature, Degrees Fahrenheit					
	40	50	60	70	80	90
Growing Heifers, Steers, and Bulls						
400	4.0	4.3	5.0	5.8	6.7	9.5
600	5.3	5.8	6.6	7.8	8.9	12.7
800	6.3	6.8	7.9	9.2	10.6	15.0
Wintering, Pregnant Cows						
900	6.7	7.2	8.3	9.7	---	---
1,100	6.0	6.5	7.4	8.7	---	---
Lactating Cows						
900	11.4	12.6	14.5	16.9	17.9	16.2
Mature Bulls						
1,400	8.0	8.6	9.9	11.7	13.4	19.0
1,600	8.7	9.4	10.8	12.6	14.5	20.6

Adapted from Nutrient Requirements of Beef Cattle, National Research Council, 1996.

Table 16-2. Guidelines for Using Saline Water.

Total Dissolved Solids, ppm	Production Recommendations
Less than 1,000	Presents no serious risk
1,000 – 2,999	Should not affect health or performance but may cause temporary mild diarrhea
3,000 – 4,999	Generally satisfactory, but may cause diarrhea, especially on initial consumption
5,000 – 6,999	Can be used with reasonable safety for adult ruminants but should be avoided for pregnant cattle and baby calves
7,000 – 10,000	If possible, avoid for all ages of cattle
More than 10,000	Unsafe, do not use

Adapted from Nutrients and Toxic Substances in Water for Livestock and Poultry, NRC, 1974.

Table 16-3. Nitrate Levels in Water and Potential Symptoms of Toxicity in Cattle.

Nitrate (NO_3) ppm	Nitrate Nitrogen (NO_3-N)	Symptoms
0 – 44	0 – 10	No harmful effects
45 – 132	10 – 20	Safe if diet is low in nitrates and nutritionally balanced
133 – 220	20 – 40	Could be harmful if consumed over long periods of time
221 – 660	40 – 100	Cattle at risk, possible death loss
661 – 800	100 – 200	Unsafe; high probability of death loss
More than 800	More than 200	Unsafe; do not use

Adapted from Nutrients and Toxic Substances in Water for Livestock and Poultry, NRC, 1974.

References

Burris, R. and J. Johns. Feeding the Cow Herd. University of Kentucky.

Nutrient and Toxic Substances in Water for Livestock and Poultry. 1974. National Research Council.

Nutrient Requirements of Beef Cattle. 1996. National Research Council.

Zimmerman, A.S., D.M. Veira, D.M. Weary, M.A.G. von Keyserlingk, and D. Fraser. Effect of Various Sulphate Salts on Water Consumption by Beef Cattle.

Chapter 17

Energy

Energy is defined as the ability to do work. While we do not think of cattle as working in a traditional sense, it does take energy to pump blood, breathe, and move. A certain amount of energy is required for maintenance of life without any production or change in body energy stores. Additionally, it takes energy for growth, lactation, and other production functions. The energy requirements of beef cattle and the energy composition of feedstuffs are expressed using two systems:

1. **Total Digestible Nutrients (TDN)** – This measure of energy expresses the requirement as a percentage of the diet assuming a certain level of intake, or alternatively as the total number of pounds required per day.
2. **Net Energy (NE) System** – This system partitions the energy requirements into that required for maintenance and that required for production. It recognizes that once the maintenance requirement has been met, the remaining energy in the diet is available for production functions. Energy requirements are expressed in Megacalories (Mcal), and each feedstuff has a Net Energy for Maintenance (NEm) and a Net Energy for Gain (NEg) value expressed as Mcal per pound or hundredweight.

Table 17-1. Energy Required Daily to Maintain Cows of Varying Weights During the Middle Third of Pregnancy.

Cow weight, lbs	TDN, lbs	NEm, Mcal
800	7.5	6.41
1,000	8.8	7.57
1,200	10.1	8.68
1,400	11.4	9.75

Adapted from Nutrient Requirements of Beef Cattle, NRC, 1984.

Either of these systems can be used to balance diets for beef cattle. The TDN and Net Energy systems both work well for beef cows, while the Net Energy System is a better approach for growing and finishing cattle. The NE system does a more accurate job of estimating the requirements for specific production functions like lactation, pregnancy, and growth.

Factors Influencing Energy Requirements

Body Weight

As body weight increases, the energy required for maintenance increases by a power of 0.75. This weight is referred to as the **"metabolic body weight."** The formula for the maintenance requirement in megacalories of NEm is: NEm requirement = Weight in kilograms $(kg)^{\wedge.75} \times 0.077$. Thus, the maintenance requirement is not directly proportional to size. In other words, a 1,600-pound cow does not require twice as much energy as an 800-pound cow. Table 17-1 shows the energy required for maintenance by cows of varying body weights.

Weather

Cold weather increases the amount of energy required to maintain an animal. Extremely warm weather may also negatively influence performance because of decreased intakes, but in most production situations there is little that can be done when the weather is too warm.

Table 17-2. Windchill Temperatures Based on Air Temperature and Wind Speed.

Wind Speed (mph)	Air Temperature, Degrees F												
	-10	-5	0	5	10	15	20	25	30	35	40	45	50
Calm	-10	-5	0	5	10	15	20	25	30	35	40	45	50
5	-16	-11	-6	-1	3	8	13	18	23	28	33	38	43
10	-21	-16	-11	-6	-1	3	8	13	18	23	28	33	38
15	-25	-20	-15	-10	-5	0	4	9	14	19	24	29	34
20	-30	-25	-20	-15	-10	-5	0	4	9	14	19	24	29
25	-38	-32	-27	-22	-17	-12	-7	-2	2	7	12	17	22
30	-46	-41	-36	-31	-27	-21	-16	-11	-6	-1	3	8	13
35	-60	-55	-50	-45	-40	-35	-30	-25	-20	-15	-10	-5	0
40	-78	-73	-68	-63	-58	-53	-48	-43	-38	-33	-28	-23	-18

Table 17-3. Estimated Lower Critical Temperature for Beef Cattle.

Coat Description	Critical Temperature °F
Wet or Summer Coat	59
Dry, fall coat	45
Dry, winter coat	32
Dry, heavy winter coat	18

Beef Cow Nutrition Guide, Kansas State University

Cattle perform optimally in the thermo-neutral zone where temperatures are neither too cold nor too hot. When the ***effective ambient temperature,*** an index of the heating or cooling power of the environment, is outside of the thermo-neutral zone, cattle performance is depressed. Effective ambient temperature considers wind, humidity, and solar radiation in addition to the actual air temperature. The lowest temperature in the thermo-neutral zone is termed the ***lower critical temperature.*** If the effective ambient temperature (also referred to as the wind chill index in cold weather) is below the lower critical temperature, additional energy is required for maintenance. Table 17-2 shows the ***wind chill index*** for varying combinations of wind and temperature. If cattle have shelter from the wind, the effective ambient temperature is the same as the air temperature.

The effect of cold weather on energy requirements of cattle is a function of the amount of insulation on the animal, which establishes the lower critical temperature. This relationship is shown in Table 17-3, which indicates that two factors establish the lower critical temperature:

- The thickness of the hair coat, and
- Whether it is wet or dry. Note that when the hair coat is wet, the hair provides little if any insulation, and the lower critical temperature is 59 degrees Fahrenheit.

Table 17-4. Increased Maintenance Energy Requirements for Cattle Per Degree of Coldness.

Coat Description	Cow Weight, lbs 1,000	1,100	1,200	1,300
	Percentage Increase Per Degree of Coldness			
Wet or Summer Coat	2.0	2.0	1.9	1.9
Dry, fall coat	1.4	1.3	1.3	1.3
Dry, winter coat	1.1	1.0	1.0	1.0
Dry, heavy winter coat	0.7	0.7	0.7	0.7

Beef Cow Nutrition Guide, Kansas State University

Table 17-4 shows the increase in maintenance energy required for cows of varying weights and coat condition expressed as a percentage of the maintenance requirement. It clearly shows the importance of coat condition on the additional energy required for maintenance. The ***degree of coldness*** is the difference between the effective ambient temperature (wind chill index) and the lower critical temperature. The information in Table 17-4 also shows why a cold rain with wind can stress cattle severely. This explains why late spring storms, after the cattle have shed their winter coats, and early fall storms, often have a greater impact on cattle than storms in the middle of the winter when the cattle have a dry, heavy, winter coat.

Since the most common situation requiring additional energy for maintenance in cold weather is in the winter when the cattle have a winter coat, a general rule-of-thumb is that the maintenance requirement is increased one per cent for each degree (F) that the effective ambient temperature is below the lower critical temperature. If the additional energy is not provided, the cow will use body stores as a source of energy resulting in a loss of condition.

Calculation of the Additional Energy Required because of Cold Stress

1. Assume a cow with a winter coat, an ambient temperature of 5 degrees F, and a wind speed of 20 mph.
2. Table 17-3 indicates that the lower critical temperature is 32 degrees F.
3. Table 17-2 indicates that the effective ambient temperature (wind chill) is –15 degrees F.
4. The lower critical temperature (32 degrees) minus the effective ambient temperature produces a difference of 47 degrees of coldness.
5. Using a 1 percent increase in energy per degree of coldness, the energy required for maintenance should be increased by 47 percent.

From a practical standpoint, a few days of cold weather will have minimal impact on performance. However, weather stress extending beyond a few days requires an increase in the energy level in the diet to prevent loss of body condition.

Another aspect of weather that can increase maintenance energy requirements is severe mud, primarily because of the additional energy required for movement. Also, when a significant portion of the animal is covered with mud, the insulating ability of the hair coat is reduced, raising the lower critical temperature. Muddy conditions also tend to reduce feed intake as will be discussed later.

Age

Until a cow reaches approximately 48 months of age, there is an additional energy requirement for growth. This requirement is greatest before the first calf and gradually declines as the cow approaches 4 years of age as shown in Table 17-5. The frame size of the cow also influences the requirement as is shown.

Table 17-5. Approximate Energy Requirements (Mcal of NEm Per Day,) for Growth in Young Cows.

	Frame Size		
Age, months	Small	Medium	Large
18-24	1.8	2.2	2.6
24-36	0.8	1.0	1.1
36-48	0.3	0.3	0.4

Adapted from the Nutrient Requirements of Beef Cattle, 1996.

Milk Production

Lactation greatly increases the cow's energy requirement as is shown in Table 17-6. The level of milk production increases from the birth of the calf until approximately 60 days post-calving at which time it typically peaks. Milk production then declines gradually until weaning. In addition to the values shown in Table 17-6, the maintenance energy requirement is increased by 20 percent during lactation because of larger organ mass, i.e., liver and visceral organs required to support lactation. Fortunately, voluntary intake also increases significantly during lactation.

Predicting Peak Milk Production

The level of milk production significantly impacts the total energy required by a beef cow. However, since

Table 17-6. Approximate Energy Requirements (Mcal of NEm Per Day) for Lactation.

Daily Milk Production at Peak Lactation	Early Lactation	Late Lactation
Low, < 15 lbs	4.1	2.5
Moderate, 15-20 lbs	5.8	3.5
High, 20-25 lbs	7.4	4.4
Very High, >25 lbs	9.1	5.5

Adapted from Nutrient Requirements of Beef Cattle, 1996.

we do not typically measure milk production in beef cows, we must rely on weaning weight as an indicator of milk production as shown in Table 17-7.

To illustrate how this table is used, assume that the average cow size is 1,200 pounds and the average male calf weighs 585 pounds at 7 months of age. Based on this table, this combination fits a cow that has a peak milk production of 25 pounds per day. This information can be used to select the appropriate nutritional requirements in Appendix Table 1 for a cow of this size and milk production level. Obviously, there will be a range in both cow sizes and calf weaning weights in any herd. Consequently, obtaining the most accurate nutritional requirements from the Appendix tables will require good estimates of the average cow body weight and male calf weight at 7 months of age.

Table 17-7. Estimated Peak Milk Production Based on Cow Size and Male Calf Weaning Weight at 7 Months of Age.

Mature Wt., lbs	Peak Milk, lbs/Day: 10	15	20	25	30
	Average 7-Month Male Calf Weight in Pounds				
900	440	465	495	--	--
1,000	460	485	515	545	570
1,100	480	510	540	565	590
1,200	500	530	560	585	615
1,300	520	550	580	605	635
1,400	540	570	600	625	655

Adapted from Nutrient Requirements of Beef Cattle, 1996.

Table 17-8. Maintenance Requirements of Cattle Breeds.

Breed	Maintenance Requirement
British and Continental Breeds Except as Listed Below	100%
Holstein, Jersey, Simmental, and Braunvieh	120%
Holstein, Jersey, and Simmental Crosses	110%
Bos indicus breeds (Brahman, Sahiwal and Santa Gertrudis)	90%
Bos indicus crosses with British and Continental	95%

Adapted from Nutrient Requirements of Beef Cattle, 1996.

Breed Differences in Maintenance Requirements

The 1996 Nutrient Requirements of Beef Cattle publication indicates that there are differences among breeds in maintenance requirements as shown in Table 17-8. The research that Table 17-8 was based on indicated that breeds with higher milk production potential had higher maintenance requirements even when they were not lactating. It should be noted that the British breeds have increased their milk production potential over the past 30 years to essentially the same level as Simmental and other Continental breeds (Chapter 8). Consequently, it may be that the differences in maintenance requirements shown in Table 17-8 no longer exist.

Energy Required for Pregnancy

There is a significant energy requirement for the developing fetus in a pregnant cow, especially during late pregnancy as shown in Table 17-9.

Table 17-9. Energy Required to Support Fetal Growth Based on the Expected Birth Weight.

Calf Birth Weight, lbs	Mcal of NEm/Day
Less than 70 lbs	1.6
70-90 lbs	2.2
Greater than 90 lb	2.7

Based on Calculations in Nutrient Requirements of Beef Cattle, NRC, 1996.

Effect of Body Condition on Energy Use

Cows that have been on a restricted energy intake resulting in poor body condition are more efficient in using energy. The 1996 Nutrient Requirements of Beef Cattle uses a 5 percent reduction in the energy required for maintenance for each body condition score below a score of 5.

References

Burris, R. and J. Johns. Feeding the Cow Herd. University of Kentucky

Corah, L.R. Cow Herd Nutrition. Kansas Formula Feeds Conference.

Kunkle, B., J. Fletcher, and D. Mayo. 2002. Florida Cow-Calf Management, 2nd Edition – Feeding the Cow Herd. Florida Cooperative Extension Service Publication AN117.

Marston, T.T., D.A. Blasi, F.K. Brazle, G.L. Kuhl. 1998. Beef Cow Nutrition Guide. Kansas State University Extension Bulletin C-735.

Morris, J.G. and M.D. Sanchez. 1987. Energy Expenditure of Grazing Ruminants. Grazing Livestock Nutrition Conference Proceedings.

Pate, F.M. and W.E. Kunkle. 2001. Guidelines to Selecting a Liquid Feed for Winter Supplementation of Producing Beef Cows in South Florida. Florida Cooperative Extension Service SS-ANS-14.

Nutrient Requirements of Beef Cattle. 1984. National Research Council.

Nutrient Requirements of Beef Cattle. 1996. National Research Council.

Rasby, R. and I.G. Rush. Feeding the Beef Cow Herd – Part I Factors Affecting the Cow Nutrition Program. Beef Cattle Handbook BCH-5402.

Chapter 18

Protein

Protein, composed of chains of amino acids, is the major building blocks in the body. They are the primary components of muscle, blood cells, and most soft tissue structures. Most of the protein that cattle ingest is converted to ammonia by the ruminal microbes. They, in turn, use this ammonia as a source of nitrogen to synthesize amino acids and produce their own proteins. As noted earlier, these microbes are then digested in the lower digestive tract as a source of protein for the animal.

Protein in Feed

The amount of protein in a feedstuff is calculated by multiplying the amount of nitrogen by 6.25 (this coefficient is based on the assumption that the proteins in feeds are 16 percent nitrogen). This calculation gives the **Crude Protein** content. Most of the nitrogen in a feedstuff is found in the amino acids of the proteins; however, some of the nitrogen may be in other compounds. The nitrogen in amino acids is termed "true protein." The nitrogen not in true protein is termed **Non-Protein Nitrogen (NPN).** On a feed analysis report, the non-protein nitrogen usually is the protein equivalent from urea, but in a broader sense, all nitrogen not contained in amino acids is non-protein nitrogen. The crude protein in a feedstuff that is from nitrogen in amino acids is also referred to as **"Natural Protein."**

Classification of Protein Based on Site of Digestion

The protein in feed is further differentiated based on whether it will be broken down by the ruminal microbes or pass out of the rumen undigested. In this classification system, protein is either 1) **Degradable Intake Protein (DIP)** or 2) **Undegradable Intake Protein (UIP).**

Degradable Intake Protein (DIP)

The DIP is the fraction of crude protein that will be digested by the ruminal microbes. It includes that portion of the true protein that will be digested in the rumen and essentially all of the non-protein nitrogen.

Undegradable Intake Protein (UIP)

The UIP is the fraction of crude protein that will bypass digestion in the rumen. This fraction is also referred to as **By-Pass Protein.** Most of this protein will be digested in the lower tract and used by the animal.

The relative proportion of these types of protein can have a significant influence on animal performance. In fact, a good supplementation program should focus on providing the right type of protein for a given situation.

Factors Influencing Protein Requirements

Body Weight

Protein requirements increase slightly with body weight, as shown in Table 18-1.

Table 18-1. Relationship between Body Weight and Protein Requirements of a Dry, Mature Cow in the Middle Third of Pregnancy.

Cow Weight, lbs	Protein Requirement, lbs
1,000	1.3
1,100	1.4
1,200	1.4
1,300	1.5
1,400	1.6

Nutrient Requirements of Beef Cattle, NRC, 1984.

Table 18-2. Protein Requirements for Milk Production.

Milk Production, lbs/day	Protein Requirement, lbs
10	2.0
15	2.3
20	2.6
25	2.9

Nutrient Requirements of Beef Cattle, NRC, 1984.

Milk Production

It requires considerable protein for a cow to produce milk as is shown in Table 18-2.

Growth

Since growth in young animals is primarily muscle, considerable protein is required to support this growth. As noted previously, young cows are still actively growing. Table 18-3 shows the approximate amounts of additional crude protein required to support growth in young cows. Calves and yearlings have significant requirements for protein to support growth as shown in the Appendix tables.

Table 18-3 Estimated Additional Protein Requirements above Maintenance to Support Growth in Yearling Heifers and Young Cows.

Frame Type	ADG, lbs	Protein, lbs
Heifers (Precalving)		
Small	0.73	0.12
Medium	0.88	0.15
Large	1.02	0.17
Cows at 24 to 36 months		
Small	0.33	0.06
Medium	0.39	0.07
Large	0.45	0.08
Cows at 37 to 48 months		
Small	0.11	0.02
Medium	0.13	0.02
Large	0.15	0.03

The data shown in Table 18-3 are based on calculations in the Nutrient Requirements of Beef Cattle, *1996. They should be considered as estimates and used only as general guidelines.*

Pregnancy

During late pregnancy, approximately 0.12 pound of additional protein per day is required to support the growth of the calf. As might be expected, birth weight will influence the actual requirement with heavier birth weights increasing the amount of protein required by the fetus.

Relationship between Energy in the Diet and Protein Requirements

Meeting the cattle's nutrient requirements is often best accomplished by meeting the needs of the microbes. *The Nutrient Requirements of Beef Cattle, 1996,* recognized the relationship between the amount of energy in the diet and the protein requirements of the microbes. Microbial growth is maximized when the energy and protein levels in the rumen are in balance. The general guideline is that the optimal amount of degradable intake protein (DIP) is 13 percent of the amount of TDN. This percentage, termed the **microbial efficiency,** appears to be accurate in diets with medium levels of energy. However, in very low-energy diets the microbial efficiency is probably closer to 8 percent. This relationship and its implications for supplementation programs will be discussed in more detail in chapter 32.

References

Burris, R. and J. Johns. Feeding the Cow Herd. University of Kentucky.

Kunkle, B., J. Fletcher, and D. Mayo. 2002. Florida Cow-Calf Management, 2nd Edition – Feeding the Cow Herd. Florida Cooperative Extension Service Publication AN117.

Nutrient Requirements of Beef Cattle. 1984. National Research Council.

Nutrient Requirements of Beef Cattle. 1996. National Research Council

Coffey, R. Basic Functional Anatomy of the Digestive System – Ruminants. University of Kentucky.

Hall, J. and S. Silver. 2009. Nutrition and Feeding of the Cow-Calf Herd: Digestive System of the Cow. Virginia Cooperative Extension Publication 400-010.

Rounds, W. and D. Herd. The Cow's Digestive System. Texas Agricultural Extension Service Bulletin B-1575.

Chapter 19

Minerals

As considered in beef cattle nutrition, minerals are divided into two groups: 1) Macro Minerals, which are needed in relatively large amounts and 2) Trace Minerals, which are needed in comparatively small amounts. Requirements for macro minerals are expressed as a percentage of the diet, while requirements for trace minerals are expressed as parts per million (ppm) of the diet. Table 19-1 shows the minerals for which requirements have been demonstrated for beef cattle.

Table 19-1. Minerals Required for Optimal Production in Beef Cattle.

Macro Minerals	Trace Minerals
Calcium	Copper
Phosphorus	Zinc
Magnesium	Manganese
Salt (Sodium and Chloride)	Cobalt
Potassium	Iodine
Sulfur	Iron
	Selenium

Macro Minerals – Functions and Symptoms of Deficiency

Calcium (Ca)

Functions:

- Bone formation
- Blood clotting
- Nervous system function
- Activation of enzymes

Deficiency Symptoms:

- Improper bone growth
- Poor growth
- Poor reproduction

Special Considerations:

- Since roughages tend to be adequate in calcium, a deficiency is rarely a problem in beef cows. However, grains are low in calcium. Consequently, anytime a high-grain diet is being fed, calcium will likely be deficient. For example, cows on a high-grain, limit-fed diet in dry lot might be deficient in calcium. Calcium supplementation also may be required when by-products high in phosphorus are being fed. Calcium may need to be supplemented to keep the Ca:P ratio in balance. Both growing and finishing cattle typically require calcium supplementation to meet their requirements.

Phosphorus (P)

Functions:

- Bone formation
- Component in the energy transfer molecules
- Component of DNA
- Necessary for proper use of vitamin A

Deficiency Symptoms:

- Poor growth
- Reduced appetite
- Reduced digestibility of feedstuffs

Special Considerations:
- Phosphorus is typically low in mature, winter pastures and crop residues.
- Phosphorus is often at marginal levels in harvested forages.

Magnesium (Mg)
Functions:
- Required for several enzyme systems
- Constituent of bones and teeth

Deficiency Symptoms:
- Hyperirritability and tetany (Grass Tetany), which is discussed in Chapter 44
- Poor calcium metabolism

Special Considerations:
- Magnesium is often inadequate on lush pastures in the spring. In addition, high levels of calcium and potassium in forages and cold conditions can reduce magnesium absorption.
- Some species of grasses, especially cool-season annual grasses, appear to have a reduced ability to pull magnesium from the soil, resulting in a tendency to produce tetany problems.
- Lactating cows have a higher requirement. Consequently, tetany problems will often occur shortly after calving.

Potassium (K)
Functions:
- Required for proper acid-base balance in the body.

Deficiency Symptoms:
- Reduced growth.

Special Considerations:
- Typical growing and finishing diets require supplemental potassium.
- Potassium is usually adequate in most forages; however, cows have shown a response to supplementation on dry, winter grass.

Sulfur (S)
Functions:
- Component of several amino acids

Deficiency Symptoms:
- Reduced growth

Special Considerations:
- Potential problems when feeding high levels of non-protein nitrogen since sulfur is required for the microbes to synthesize sulfur-containing amino acids.
- High levels in the diet may cause sulfur-induced polio.

Trace Minerals – Functions and Symptoms of Deficiency

Cobalt (Co)
Functions:
- Component of vitamin B_{12} – Required by microbes to synthesize B_{12} in the rumen.

Deficiency Symptoms:
- Anemia
- Loss of weight
- Loss of appetite
- Weakness and even death

Special Considerations:
- Cobalt deficiency is a regional problem based on soil type and cobalt level. Areas with deficiency problems tend to have sandy soils.

Copper (Cu)
Functions:
- Required for the formation of hemoglobin in the blood
- Required for several enzyme systems

Deficiency Symptoms:
- Reproductive problems including delayed estrus, decreased conception rates, delayed puberty, and decreased libido in bulls.
- Reduced immune function
- Anemia

- Reduced growth
- Bleached hair coat
- Rough hair coat
- Diarrhea

Special Considerations:

- Since the level in plants is directly related to the level in the soil, copper will probably be deficient if the level in the soil is low
- High forage Molybdenum results in copper deficiency in many areas. The molybdenum binds with copper in the rumen making the copper unavailable. The Cu:Mo ratio should probably be at 4:1 or higher to ensure adequacy of copper
- High sulfur levels in the water or feed can also tie-up copper resulting in copper-deficiency symptoms
- High iron in the feedstuffs and/or water will also tie up copper

Iodine (I)

Functions:

- Component of the hormone thyroxin, which controls the body's metabolic rate

Deficiency Symptoms:

- Goiter
- Reduced growth
- Weak or stillborn calves
- Calves born hairless

Special Considerations:

- A regional problem based on soil iodine levels

Iron (Fe)

Functions:

- Component of hemoglobin in blood cells

Deficiency Symptoms:

- Anemia
- Deficiency is rarely a problem in beef cattle

Special Considerations:

- Deficiencies in cattle are rare, in fact, levels that are too high are much more common because iron also can tie up copper

Manganese (Mn)

Functions:

- Serves as an enzyme activator

Deficiency Symptoms:

- Leg deformities, especially in calves
- Impaired reproductive performance
- Increased abortions

Special Considerations:

- Rarely a problem. However, there are a few areas in the country where manganese levels in feedstuffs are below those required by beef cattle

Salt (NaCl)

Functions:

- Both sodium (Na) and Chloride (Cl) are involved in maintaining osmotic pressure, controlling water balance, and regulating acid-base balance

Deficiency Symptoms:

- Reduced growth and performance

Selenium (Se)

Functions:

- Required for proper vitamin E function

Deficiency Symptoms:

- White muscle disease in calves
- Lameness
- Retained placentas
- Weak or dead calves

Special Considerations:

- Some plants such as locoweed, accumulate high levels of selenium, and when consumed may cause selenium toxicity
- Selenium deficiency occurs on an area basis when the level of selenium in the soil is low

Zinc (Zn)

Functions:

- Required for several enzymes

Deficiency Symptoms:

- Reduced immune function

- Decreased testicular size and libido in bulls
- Reduced growth
- Skin problems
- Poor hoof quality, which may lead to foot rot

Special Considerations:

- The level of zinc in many soils is inadequate, resulting in forages that have zinc levels below those required by cattle

Factors Influencing Mineral Requirements

Recommended mineral levels (NRC Recommendations) in cattle diets are shown in Appendix Table 11. A review of these recommendations indicates that the following factors influence requirements:

Growth

The amount of calcium and phosphorus required is significantly increased during periods of rapid growth because these minerals are a major component of bone. In general, the requirements for other minerals do not increase dramatically as growth rate increases, at least when the requirement is expressed as a concentration (percentage or ppm) in the diet.

Lactation

The requirement for both calcium and phosphorus increases significantly during lactation in direct proportion to the level of milk production. In addition, there is a slight increase in the level of magnesium and potassium required in the diet.

Pregnancy

The calcium and phosphorus requirements are both increased by pregnancy, especially during late gestation.

Presence of Antagonists

There are numerous interactions among minerals that have an influence on the amount of a mineral required in the diet. As noted previously, high levels of molybdenum and sulfur can tie up copper, resulting in a copper deficiency even though the dietary level of copper is adequate based on NRC guidelines. Unfortunately, there are numerous other examples of interactions among various minerals that negatively affect availability.

Stress

While the actual requirement for minerals is not increased by stress, the reduction in feed intake associated with stress means that the concentration of minerals in the diet should be increased. Recommendations for mineral supplementation of cattle diets will be covered in Chapter 33.

References

Bull, R.C. Trace Minerals and Immunology. Beef Cattle Handbook B5454.

Corah, L.R. 1995. Understanding Basic Mineral and Vitamin Nutrition. Range Beef Cow Symposium.

Engle, T.E. and D.S, Baker. Effects of Mineral Nutrition on Immune Function and Factors that Affect Trace Mineral Requirements of Beef Cattle. Proceedings of the 23rd Western Nutrition Conference.

Gadberry, S. 2003. Mineral and Vitamin Supplementation of Beef Cows in Arkansas. 2006. University of Arkansas Cooperative Extension Bulletin FSA3035.

Greene, L.W., A. Bruce Johnson, J. Paterson, and R. Ansotegui. 1998. NCBA Educational Seminar Proceedings.

Johns, J., R. Hemken, and Patty Scharko. Trace Mineral Supplementation for Kentucky Beef Cows. University of Kentucky Cooperative Extension Bulletin ASC-155.

Kerley, M.S., J.A. Paterson, and J.C. Whittier. Meeting the Mineral Needs of the Cow Herd.

Kunkle, B., J. Fletcher, and D. Mayo. 2002. Florida Cow-Calf Management, 2nd Edition – Feeding the Cow Herd. Florida Cooperative Extension Service Publication AN117.

Mc Dowell, L.R. 1997. Minerals for Grazing Ruminants in Tropical Regions. University of Florida Institute of Food and Agricultural Sciences.

Nutrient Requirements of Beef Cattle. 1984. National Research Council.

Nutrient Requirements of Beef Cattle. 1996. National Research Council.

Parish, J. and J. Rhinehart. 2009. Mineral and Vitamin Nutrition for Beef Cattle. Mississippi State University Cooperative Extension Service Publication 2484.

Spears, J.W. Overview of Mineral Nutrition in Cattle: The Dairy and Beef NRC.

Ward, M. and G. Lardy. 2005. Beef Cattle Mineral Nutrition. 2005. North Dakota State University Extension Bulletin AS-1287.

Whittier, J.C., M.S. Kerley, and J.A. Paterson. 1991. Meeting the Mineral Needs of the Cow Herd. KOMA Beef Cattle Conference Proceedings.

Chapter 20

Vitamins

Vitamins serve a variety of functions in the body and are required for the animal to use other nutrients. They are divided into two groups: 1) vitamins that are soluble in fat, and 2) vitamins that are soluble in water.

Fat-Soluble Vitamins

Vitamin A

This is the most important vitamin from a practical standpoint. It plays a key role in growth, reproduction, bone development, and maintenance of epithelial tissues. Although vitamin A doesn't occur in plants, its precursors, carotenes, occur in high levels in growing plants. Since carotenes are destroyed by heat and sunlight, carotene levels are often inadequate when cattle are grazed on dormant grasses or crop residues, or are being fed harvested forages that have been stored for long periods. Cattle have the ability to store a 2- to 4-month supply of vitamin A in the liver during times when levels in the feed are high. However, this may not be enough to supply adequate vitamin A through the winter, particularly in northern latitudes where winters are longer.

Signs of Vitamin A Deficiency

- Reduced feed intake
- Rough hair coat
- Edema of the joints.
- Blindness, especially night blindness
- Poor reproduction
- Abortions

Typical Situations Where Vitamin A Deficiencies Occur

- In late winter because much of the carotene that was in the harvested forages has been destroyed by heat and sunlight. In addition, the vitamin A stores in the liver have been depleted.
- Cattle are being fed feedstuffs that have been bleached by the sun and weathered for long periods. For example, cattle grazing crop residues or dormant range for long periods are susceptible to vitamin A deficiency.
- National Research Council guidelines for vitamin A requirements are 1,270 International Units (IU) per pound of dry feed for pregnant heifers and cows and 1,770 IU per pound of dry feed for lactating cows and bulls. Consequently, any animal characteristic like size, lactation, growth, that tends to increase intake also increases the total vitamin A requirement as shown in Table 20-1.

Table 20-1. Total Vitamin A Levels that Should Be Adequate to Meet the Needs of Various Classes of Beef Cattle.

Class of Cattle	IU Per Day
Calves	20,000
Pregnant Yearling Heifers	25,000
Dry, Pregnant Cows	30,000
Lactating 2-Year-Old Heifers	35,000
Lactating Cows	45,000
Bulls	55,000
Finishing, Yearlings	23,000

Extrapolated from Nutrient Requirements of Beef Cattle, *NRC, 1984 (Requirements weren't changed in the 1996 version).*

Vitamin D

Vitamin D is required for calcium and phosphorus absorption, normal mineralization of bone, and the mobilization of calcium from bone. Thus, a deficiency of vitamin D results in abnormal bone development and growth. The vitamin D requirement is 125 IU per pound of dry diet, and as with vitamin A, any factor that increases intake increases the vitamin D requirement. However, animals exposed to sunlight can synthesize vitamin D, and sun-cured forages typically have high levels of vitamin D. Consequently, vitamin D supplementation is rarely required other than in situations where animals are not exposed to sunlight for prolonged periods.

Vitamin E

Vitamin E functions in the formation of structural components of membranes in the body and as an antioxidant. Its function is linked closely to selenium in maintaining the functional integrity of the muscles of the body. The requirement for vitamin E has not been accurately determined. From a practical standpoint, vitamin E is of greatest importance in young calves, especially during times of stress, such as during weaning. Vitamin E activity in feedstuffs is reduced by heat, oxygen, and moisture.

Vitamin K

Vitamin K is essential for the production of prothrombin, which is required for blood clotting. Dicumarol, produced by mold on sweet clover hay, interferes with the activity of vitamin K. The affected animals may bleed to death internally. This condition is known as sweet clover poisoning. Under normal conditions, ruminal microorganisms synthesize adequate quantities of vitamin K.

Water-Soluble Vitamins

B Vitamins

The B-vitamin complex includes thiamine, biotin, riboflavin, niacin, pantothenic acid, pyridoxine, folic acid, B_{12}, and choline. These vitamins function in a number of metabolic pathways as coenzyme and components of enzymes. Symptoms of deficiency are not specific and include poor appetite, slow growth, and poor condition. However, there is one B vitamin (thiamine) deficiency that produces specific symptoms.

Thiamine – A deficiency of thiamine produces a central nervous system disorder known as polioencephalomalacia (PEM) characterized by convulsions. This type of polioencephalomalacia is most commonly found in cattle on high-energy, high-concentrate diets.

Unless rumen function is depressed, the ruminal microorganisms synthesize all of the B vitamins in quantities adequate to prevent deficiency symptoms.

References

Nutrient Requirements of Beef Cattle. 1984. National Research Council.

Nutrient Requirements of Beef Cattle. 1996. National Research Council.

Parish, J. and J. Rhinehart. 2008. Mineral and Vitamin Nutrition for Beef Cattle. Mississippi State University Extension Service.

Sewell, H.B. Vitamins for Beef Cattle. University of Missouri Extension Bulletin G2058.

Chapter 21

Feed Composition – Nutrient Content

Common feedstuffs used in feeding beef cattle are classified as either roughages (high fiber, low energy) or concentrates (low fiber, high energy). Examples of roughages include hays and silages, while examples of concentrates include grains. Another major group of feedstuffs consists of the by-products from grain or oilseed processing. Many of these by-products do not fit into the traditional categories because they can be high in fiber and relatively high in energy.

Nutrients in Feedstuffs

Dry Matter – Water

As a general statement, all feedstuffs used in cattle diets contain some water with the exception of mineral supplements, which contain little moisture. All of the components other than water are termed "**Dry Matter (DM).**" The relative amount of water in the feed is important for the following reasons:

- **Animals Tend to Eat to Meet a Dry Matter Intake** – Regardless of the moisture content in the feedstuffs; cattle consume feed until they reach a certain level of dry matter intake.
- **Nutrient Requirements are Stated on a Dry Matter Basis** – Animal requirements are stated on a dry matter basis because the moisture content in feedstuffs is variable.
- **Feedstuffs Should be Priced on a Dry Matter Basis** – Feeds should be compared on a cost per unit of dry matter rather than on an "as fed" basis.

Example:

1. Alfalfa haylage – 45 percent moisture, $140 per ton as fed
   - $140.00 ÷ 0.55 (percent DM) = $254.55 per ton of dry matter
2. Alfalfa hay – 12 percent moisture, $210 per ton as fed
   - $210 ÷ 0.88 (percent DM) = $238.64 per ton of dry matter

In this example, the alfalfa hay is a better buy if all other factors like nutrient levels, anticipated wastage, and cost of feeding are equal.

Energy

The primary sources of energy are carbohydrates, which consist of cellulose, sugars, starch, hemicellulose, and lignin. Sugars and starches are usually highly digestible; cellulose is variable in digestibility; hemicellulose is of lower digestibility; and the lignin fraction is essentially indigestible. Furthermore, lignin reduces the digestibility of the other fiber fractions. The percentage of lignin in a plant increases as it matures, which explains the lower energy levels in hays and silages harvested as mature plants.

Another major source of energy in some feedstuffs is fat. It has 2.25 times the energy of carbohydrates. In most plants, the level of fat is relatively low, but in some oil seeds, such as soybeans, cottonseed, and sunflowers, it is quite high. Some by-products fed to cattle, including distiller's grains, are relatively high in fat. Another source of energy is protein when it is fed in excess of the animal's requirement. An example of this situation includes cattle on lush pasture, such as wheat pasture, where the protein content can be 25 percent or higher on a DM basis. Once the animal's protein requirement has been fulfilled, the excess natural protein is used as an energy source.

Energy Values on a Feed Analysis

Energy values provided as part of a feed analysis are not based on a direct chemical or physical measurement of energy. They are typically based on some measurement of the fiber content. This assessment of energy content is based on the assumption that as fiber content goes up, energy content goes down and visa versa. While this relationship is generally true, the relationship between fiber level and energy is not extremely strong. Because of differences in the digestibility of fiber, the energy level in a feed may not be directly related to the fiber content. This means that the energy values supplied by analytical laboratories often are not accurate.

From a practical standpoint, the inaccuracy of the estimated energy content of feedstuffs means that the value supplied by the laboratory should be compared to the value in feed composition tables. And, if the energy value from the feed analysis is much different from the book value, it should be used with caution. It is recommended that the energy value used in ration formulation should be adjusted close to the book value if the value on the feed analysis is significantly higher or lower than the "book" value.

Consider this example:

- Bromegrass Hay analysis shows: 11% CP, 58 Mcal NEm/cwt, 26 Mcal NEg/cwt, and 39% ADF.
- Appendix Table 11 shows "book" values for Bromegrass Hay of 10% CP, 55 Mcal NEm/cwt, 21 Mcal NEg/cwt, and 41% ADF.

If this hay will be a significant portion of the ration, it would be wise to adjust the energy level of the hay closer to the "book" value before formulating a ration to ensure that the desired level of performance is obtained. For example, appropriate values might be: 56 Mcal NEm/cwt and 23 Mcal NEg/cwt. The protein level from the analysis should be used since it is probably accurate.

Protein

Protein content in feedstuffs is typically expressed as crude protein, which is based on the total nitrogen content. Digestible protein values are often provided as part of a feed analysis. Be aware, however, that they are based on calculations using a standard formula rather than a chemical analysis. This limits the accuracy of digestible protein estimates on feed analysis reports.

Recent trends in protein nutrition have focused on describing the relative percentage of the crude protein fraction that will be digested in the rumen (DIP) as described in Chapter 18. Since determining this fraction requires animal feeding trials, this information is not provided as part of a routine feed analysis, which means that averages from animal feeding trials presented in feed composition tables must be used.

Minerals

Minerals can be accurately measured in most feedstuffs. The major minerals are expressed on a percentage basis, and the trace minerals on a parts-per-million (ppm) basis. Feed analysis laboratories often use near infrared spectrophotometry (NIR) for minerals, which is accurate for some standard feedstuffs like corn and alfalfa hay. For most feedstuffs, however, especially those of mixed composition, analysis based on wet chemistry is more accurate for minerals.

Vitamins

The content of fat-soluble vitamins in feeds is expressed as international units (IU). Green forages typically contain high levels of Vitamin A and E, but as noted in Chapter 20, the activity of vitamin A is reduced rapidly by heat or sunlight. Unless very green material is being fed, the feed should not be viewed as the primary source of vitamins — they should be provided in a supplement. Having a feedstuff analyzed for vitamin A is not useful in most production situations.

References

Chandler, P. 1990. Energy Prediction of Feeds by Forage Testing Explored. Feedstuffs August 27, 1990.

Guyer, P.Q. Analyzing Grain and Roughages. Great Plains Beef Cow-Calf Handbook GPE-1802.

Hall, M.B. Feed and Forage Evaluation – The New Frontier. Department of Dairy and Poultry Science University of Florida.

Hall, M.B. 1997. Interpreting Feed Analyses: Uses, Abuses, and Artifacts. Feed Management Vol: 48, No. 6.

Nutrient Requirements of Beef Cattle. 1984. National Research Council.

Nutrient Requirements of Beef Cattle. 1996. National Research Council.

Preston, R. 2012. Feed Composition Tables. Beef. March 2012.

Chapter 22

Feeding for Reproduction

When planning a cow herd nutrition program to optimize reproduction, keep in mind the biological priority for nutrients shown in Table 22-1.

Table 22-1. The Biological Priority for Nutrients by Beef Cows.

Priority	Function
1	Maintenance
2	Growth
3	Milk Production
4	Reproduction

The first priority for any animal is to maintain its life. The second priority, if it is still trying to grow, is the allocation of nutrients for growth. The third priority for animals that are lactating is the allocation of nutrients to milk production, and lastly, if there are sufficient nutrients in excess of those required for the first three functions, the animal will reproduce. Keeping these priorities in mind is extremely important when considering the most important thing a beef female can do from an economic standpoint is produce a calf each year.

While this priority ranking is to some extent an oversimplification, it does recognize that animals typically fail to reproduce when nutrition is inadequate to support other body functions. This is Mother Nature's way of preventing an animal from getting pregnant when pregnancy would threaten the animal's survival.

The Beef Cow Production Cycle – Four Production Periods

Breaking the beef cow year into four periods has proven to be a valuable way of looking at the nutritional requirements because it recognizes distinct changes in nutrient requirements during the year.

Period 1 – Calving to Rebreeding – 82 Days – "The Postpartum" Period

Nutritional Requirements:

- **Maintenance**
- **Growth** – If the cow is still growing, which is the case with 2- and 3-year-olds, there are significant nutrient requirements for growth.
- **Lactation** – Substantial requirement with peak lactation at approximately 60 days postpartum.
- **Reproduction** – Nutrient availability should be adequate to allow the cow to recover after calving and prepare to rebreed by 82 days postpartum.

Period 2 – Rebreeding to Weaning of the Calf – 123 Days (Assuming Weaning at 205 Days)

Nutritional Requirements:

- **Maintenance**
- **Growth** – Again, young cows have a significant nutrient requirement for growth.
- **Lactation** – Although the requirements for lactation are lower than during the first couple of months following calving, they are still significant considering that in many cases the quality of the available forage has declined.
- **Pregnancy** – Nutrient requirements are relatively small during early pregnancy, but they can be significant if nutrient availability is limited.

Period 3 – Weaning to 60 Days before Calving – 100 Days Nutritional Requirements:

- **Maintenance**
- **Growth** – Young cows still need nutrients for growth.
- **Pregnancy** – Nutrient requirements for pregnancy become significant, especially late in this period.

Period 4 – 60 Days before Calving – The "Prepartum" Period

While there is nothing magical about 60 days before calving, it is an excellent time to evaluate body condition because there is still time to make an improvement if needed. Some producers wait until only a few weeks before calving to recognize that body condition is less than desirable. With only a short period of time before calving, it is difficult and expensive to improve body condition. The importance of making an evaluation of cow condition with enough time to change condition is clear considering that poor nutrition before calving delays estrus even if nutrition after calving is adequate.

Nutritional Requirements:

- **Maintenance**
- **Growth** – Young cows still need nutrients for growth.
- **Pregnancy** – Nutrient requirements for pregnancy are at their highest.
- **Reproduction** – Numerous research trials show that nutrition during this period has a significant influence on reproductive performance, especially the interval between calving and the start of cycling activity.

Evaluating Reproductive Performance

Measuring and evaluating reproductive performance should be key management functions of every cow/calf producer. The following measures should allow an evaluation of the adequacy of the nutrition program based on reproductive performance.

Calf Crop

Although the term "calf crop" means different things to different people, the most valuable measure of calf crop is the percentage of exposed cows that actually wean a calf. With this definition, both reproductive failure and death loss are considered. The actual percent calf crop that is optimal varies across different environments. For example, in an extensive production system involving large range pastures, the acceptable percentage could be lower than in an intensive operation because of the difficulty in providing supplementation under range conditions and the difficulty of providing assistance during calving. Developing a good nutrition program requires a comparison of the cost of supplementation to the expected impact on the calf crop.

Calving Sequence

Calving sequence is a measure of the percentage of calves born at 21-day intervals from the start of calving. The percentages are usually calculated for 21-day periods because this matches the length of the cow's reproductive cycle. A good calving sequence for a 60-day breeding/calving program appears in Table 22.2.

Table 22-2. Calving Sequence Example.

Calving Period	Percentage of Cows Calving
1st 21 Days	60
2nd 21Days	30
3rd 21Days	10

The Importance of Calving Sequence

1. **Weaning Weight Is a Function of Age** – If all calves are weaned on the same date, the older calves typically will be heavier. Thus, one way to increase the total saleable weight at weaning is to get a higher percentage of the calves born early in the calving season. Consider the example in Table 22-3 where an additional 2,400 pounds of calf was weaned because of a better calving sequence.

Table 22-3. Impact of Differing Calving Sequences on Total Weaning Weight in Two 100-Head Cow Herds.

Calving Period	Herd A				Herd B			
	%	No.	Weaning Wt., lbs	Total lbs	%	No.	Weaning Wt., lbs	Total lbs
1st 21 Days	60	60	600	36,000	40	40	600	24,000
2nd 21 Days	30	30	540	16,200	30	30	540	16,200
3rd 21 Days	10	10	480	4,800	30	30	480	14,400
Total				57,000				54,600

2. **Time of Calving Impacts the Chance for Pregnancy** – If the breeding season is a defined period (60 or 90 days for example), cows that calved early have a greater probability of getting pregnant than those that calved late. To illustrate why, assume a 60-day breeding season, which results in a 60-day calving season (the calving season actually would be closer to 75 days in length because of varying gestation lengths). Compare a cow that calves on the first day of the calving season with a cow that calves on the 60th day.

   Assuming an average gestation length of 283 days, breeding should begin approximately 82 days from when the first calf is born in order to start calving on the same day next year. At the start of breeding, the cow that calved on the first day of calving is 82 days postpartum and probably cycling. If she has started to cycle, she may have three opportunities to conceive. Conversely, the cow that calved on the last day of calving is only 22 days postpartum at the start of breeding. There is very little chance that she is cycling. She'll have at most two chances to conceive, and it is likely that she will only get one chance. A cow that calves late this year is a good candidate to be an open cow next year — she simply has fewer chances to get pregnant. Probabilities for pregnancy based on time of calving and body condition will be discussed in more detail in Chapter 24.

3. **Cow Herds with a Tight Calving Sequence Have a Marketing Advantage** – Many studies show that the larger the number of head in a group at sale time, the higher the price per pound. For example, with a calving sequence that has a high percentage of the calves born early, the calves might be sorted into four groups at sale: Heavy and Light Steers and Heavy and Light Heifers. Conversely, a spread out calving sequence or a long calving season results in three to four groups for each sex when they are sorted at sale. The net result is fewer gross dollars for the same number of calves and pounds.

4. **Delayed Rebreeding Can Have Long-Term Effects** – Even one year with inadequate nutrition can have long-term effects on reproduction and profitability. Inadequate nutrition that causes a cow to calve late can mean that she will probably calve late throughout the remainder of her life and wean a smaller calf each year. Furthermore, as discussed previously, she is a likely candidate to fail to breed in subsequent years.

5. **Long or Year-Long Breeding Seasons** – Many producers resort to long breeding seasons (120 to 180 days, for example) or even leaving the bulls out year-round rather than developing a nutrition program that provides high pregnancy rates and good calving sequences. Eventually, even a thin cow will breed in a very long breeding

season. However, inadequate nutrition still reduces the number of calves in a given period of time. In other words, over a 5-year period, does the cow produce five calves, four, or even three calves? In addition, this type of breeding program results in a wide spread in calf weights at any time, which reduces the value of the calves at marketing as discussed previously.

Impact of Nutrition Prior to Calving on Reproduction

Inadequate nutrition before calving greatly increases the postpartum period resulting in a lower percentage of cows cycling and conceiving during the first 21 days of the breeding season. The impact on the pregnancy percentage at the end of the breeding season may be minimal if the breeding season is relatively long. The

Table 22-4. Effect of Daily Feed Level before Calving on Pregnancy Rate and Pregnancy in the First 21 Days of Breeding.

Lbs of TDN Before Calving	Lbs of TDN After Calving	Gain Before Calving lbs	Gain Calving to 90 Days After Calving	Number of Cows	Percent Pregnant	Pct. Preg. First 21 Days of Breeding
			Older Cows			
9.0	16.0	67	-14	21	95	60
4.5	16.0	-118	22	20	95	46
			2-Year-Old Cows			
8.0	13.0	150	87	37	73	56
4.3	13.0	35	138	41	73	27
			2-Year-Old Cows			
8.0	13.0	120	19	24	79	54
4.3	13.0	13	64	23	83	48

J.N. Wiltbank, 9th Annual Wyoming Beef Cattle Short Course, 1968.

Table 22-5. Effect of Daily Feed Level before Calving on Time When Cows Show Heat.

Lbs of TDN Before Calving	Lbs of TDN After Calving	Cows Showing Heat After Calving, % Days					
		40	50	60	70	80	90
			Older Cows				
9.0	16.0	--	65	80	90	90	95
4.5	16.0	--	25	45	70	80	85
			2-Year-Old Cows				
8.0	13.0	18	68	82	90	92	97
4.3	13.0	7	27	49	66	73	83
			2-Year-Old Cows				
8.0	13.0	21	38	71	92	96	100
4.3	13.0	13	30	52	70	83	91

J.N. Wiltbank, 9th Annual Wyoming Beef Cattle Short Course, 1968.

impact of inadequate nutrition before calving on pregnancy in the first 21 days of breeding and on overall pregnancy rate is shown in Table 22-4. This research was conducted by Dr. J. N. Wiltbank in the late 1950s and early 1960s, and numerous research trials since this original research have confirmed this relationship. In this research, inadequate nutrition before calving did not reduce pregnancy, but the percentage of cows breeding early was reduced. Breeding was delayed because cycling activity was delayed as shown in Table 22-5.

Impact of Nutrition after Calving on Reproduction

Inadequate nutrition after calving typically reduces both the pregnancy rate and the percentage calving in the first 21 days in the following year as shown in Table 22-6.

The reduction in overall pregnancy rate and the percentage of cows breeding in the first 21 days is explained by the data in Table 22-7, which show that not only was the percentage of cows exhibiting heat

Table 22-6. Effect of Daily Feed Level after Calving on Pregnancy Rate and Pregnancy in the First 21 Days of Breeding.

Lbs of TDN Before Calving	Lbs of TDN After Calving	Gain Before Calving, lbs	Gain Calving to 90 Days After Calving	Number of Cows	Percent Pregnant	Pct. Preg. First 21 Days of Breeding
			Older Cows			
9.0	16.0	67	-14	21	95	60
9.0	8.0	89	97	22	77	34
			2-Year-Old Cows			
8.0	13.0	131	81	37	73	56
8.0	7.0	135	-79	42	64	31
			2-Year-Old Cows			
8.0	13.0	120	19	24	79	54
8.0	7.0	184	-136	13	76	23

J.N. Wiltbank, 9th Annual Wyoming Beef Cattle Short Course, 1968.

Table 22-7. Effect of Daily Feed Level after Calving on Conception Rate at First Service and Exhibition of Heat.

Lbs of TDN Before Calving	Lbs of TDN After Calving	No. of Cows	% Preg. From 1st Breeding	% In Heat During Exp.
		Older Cows		
9.0	16.0	21	67	100
9.0	8.0	19	42	86
		2-Year-Old Cows		
8.0	13.0	36	64	100
8.0	7.0	34	58	81
		2-Year-Old Cows		
8.0	13.0	24	50	100
8.0	7.0	12	33	92

J.N. Wiltbank, 9th Annual Wyoming Beef Cattle Short Course, 1968.

reduced, but the number conceiving out of those bred was also reduced.

References

Banta, J.P., D.L. Lalman and R.P. Wettemann. 2005. Post-Calving Nutrition and Management Programs for Two-Year-Old Beef Cows. The Professional Animal Scientist 21:151.

Hughes, J.H., D.F. Stephens, K.S. Lusby, L.S. Pope, J.V. Whiteman, L.J. Smithson and R. Totusek. 1978. Long-Term Effects of Winter Supplement on Growth and Development of Hereford Range Females. Journal of Animal Science. 47:805.

Hughes, J.H., D.F. Stephens, K.S. Lusby, L.S. Pope, J.V. Whiteman, L.J. Smithson and R. Totusek. 1978. Long-Term Effects of Winter Supplement on the Productivity of Range Cows. Journal of Animal Science. 47:816.

Lamb, C. Relationship Between Nutrition and Reproduction in Beef Cows. North Florida Research and Education Center.

Wiltbank, J.N. A Management System for Improving Reproduction. Department of Animal Science, Colorado State University.

Wiltbank, J.N. 1968. Managing Reproduction in the Beef Female. 9th Wyoming Beef Cattle Short Course Proceedings.

Chapter 23

Body Condition Scores

In many production situations, it is difficult to know with any certainty how much feed the cattle are actually consuming and the nutrient content of the diet. This makes it difficult to determine if the diet is meeting the animal's nutritional requirements. This lack of knowledge makes it essential to monitor body condition as a means of judging the adequacy of the nutritional program.

Table 23-1. Nine-Point Beef Cattle Body Condition Scoring (BCS) System.

Condition Score	Description
1	Bone structure of shoulder, ribs, back, hooks, and pins is sharp to the touch and easily visible. Little evidence of fat deposits or muscling.
2	Little evidence of fat deposition, but some muscling in the hindquarters. The spinous processes feel sharp to the touch and are easily seen with space between them.
3	Beginning of fat cover over the loin, back, and foreribs. The backbone is still highly visible. Processes of the spine can be identified individually by touch and may still be visible. Spaces between the processes are less pronounced.
4	Foreribs are not noticeable but the 12th and 13th ribs are still noticeable to the eye, particularly in cattle with a big spring of rib and width between ribs. The transverse spinous processes can be identified only by palpation (with slight pressure) and feel rounded rather than sharp. Full but straight muscling in the hindquarters.
5	The 12th and 13th ribs are not visible to the eye unless the animal has been shrunk. The transverse spinous processes can only be felt with firm pressure and feel rounded but are not noticeable to the eye. Spaces between the processes are not visible and are only distinguishable with firm pressure. Areas on each side of the tail head are well filled but not mounded.
6	Ribs are fully covered and are not noticeable to the eye. Hindquarters are plump and full. Noticeable sponginess over the foreribs and on each side of the tail head. Firm pressure is now required to feel the transverse processes.
7	Ends of the spinous processes can only be felt with firm pressure. Spaces between processes can barely be distinguished. Abundant fat cover on either side of the tail head with evident patchiness.
8	Animal takes on a smooth, blocky appearance. Bone structure disappears from sight. Fat cover is thick and spongy and patchiness is likely.
9	Bone structure is not seen or easily felt. The tail head is buried in fat. The animal's mobility may actually be impaired by excessive fat.

Texas Extension Service Bulletin B1526.

Table 23-2. Key Physical Signs Useful in Body Condition Scoring Beef Cattle

	Body Condition Score								
Description	**1**	**2**	**3**	**4**	**5**	**6**	**7**	**8**	**9**
Physically weak	Yes	No	No	No	No	No	No	No	No
Muscle atrophy	Yes	Yes	Slight	No	No	No	No	No	No
Outline of spine is visible	Yes	Yes	Yes	Slight	No	No	No	No	No
Outline of ribs is visible	All	All	All	3–5	1–2	0	0	0	0
Fat in brisket and flanks	No	No	No	No	No	Some	Full	Full	Extreme
Outline of hip bones is visible	Yes	Yes	Yes	Yes	Yes	Yes	Slight	No	No
Fat udder and patchy fat is visible around the tail head	No	No	No	No	No	No	No	Slight	Yes

South Dakota State University.

Using body condition as a guide in feeding beef cows certainly is not a new concept. The adage that "the eye of the master feeds the cattle" has been recognized for many years. While one can broadly assess body condition, using a recognized scoring system simply makes it easier to record differences for future reference, to communicate body condition differences, and to quantify requirements. The beef cattle industry has settled on the nine-point system outlined in Table 23-1.

Researchers at South Dakota State University defined this same nine-point system in slightly different terms as shown in Table 23-2, which may also be useful.

Figure 23-1 shows a graphical illustration of the body condition scoring system used in the beef cattle industry. These illustrations were generously provided by Elanco Animal Health.

Even with the guidelines shown in Tables 23-1 and 23-2, accurate body condition scoring is often difficult, especially when cattle have a heavy hair coat. In addition, body condition scores are more useful when each score is more detailed than just the whole number. For example, many producers use a plus/minus system where each score is broken down into three classes: below average (minus), average, and above average (plus). Scoring body condition using this system can be useful, especially for cows in body condition score 4 because in many environments there appears to be a significant difference in the reproductive performance of 4 minus cows compared to 4 plus cows.

Another system used by researchers and some producers is to break each body condition score into tenths in an attempt to provide an even more accurate assessment of body condition. While this system is useful in a research setting, in most commercial situations, the plus/minus system is probably adequate.

Figure 23-1. Illustrations of Body Condition Scores in Beef Cattle.

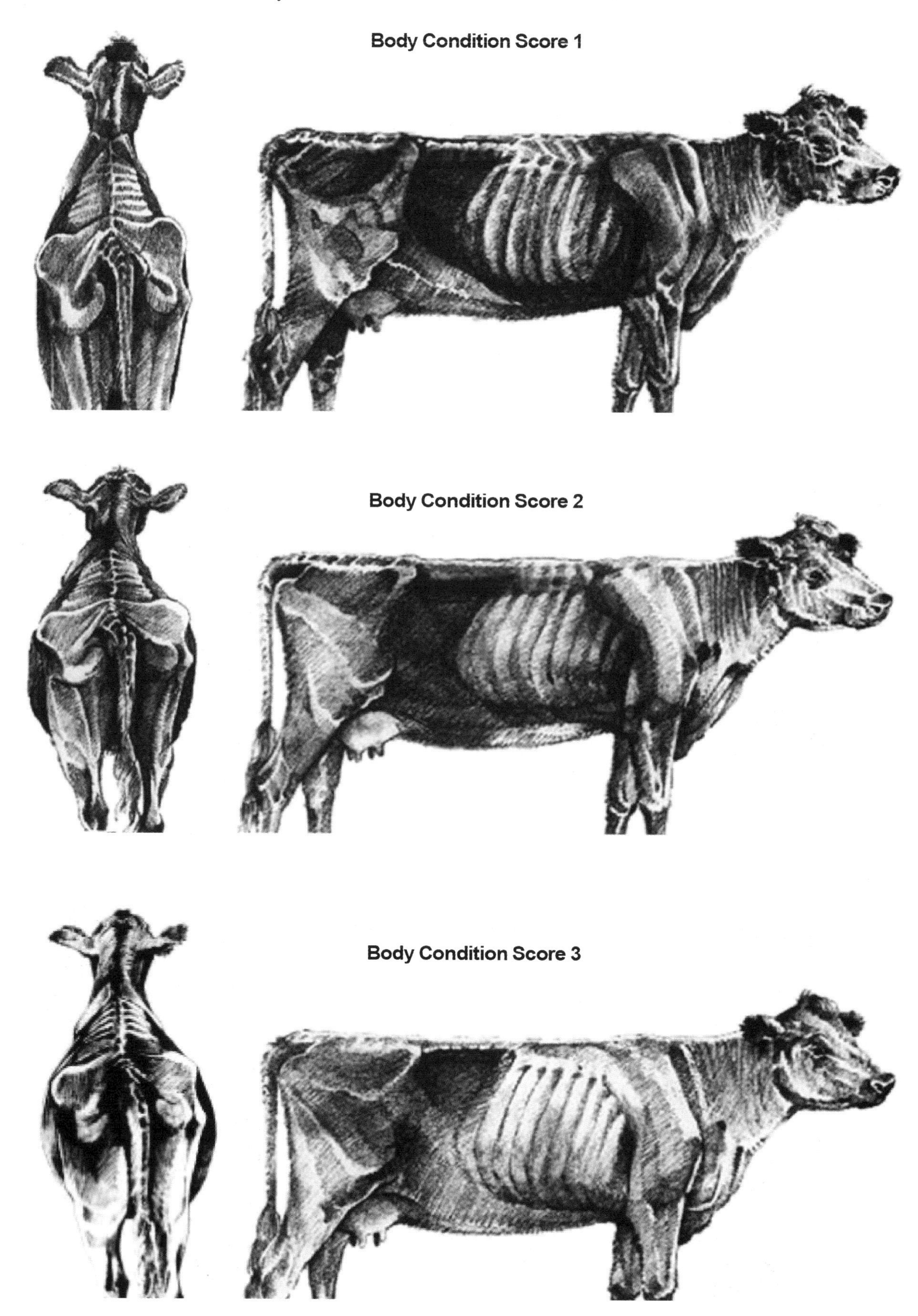

Body Condition Score 4
Body Condition Score 5
Body Condition Score 6

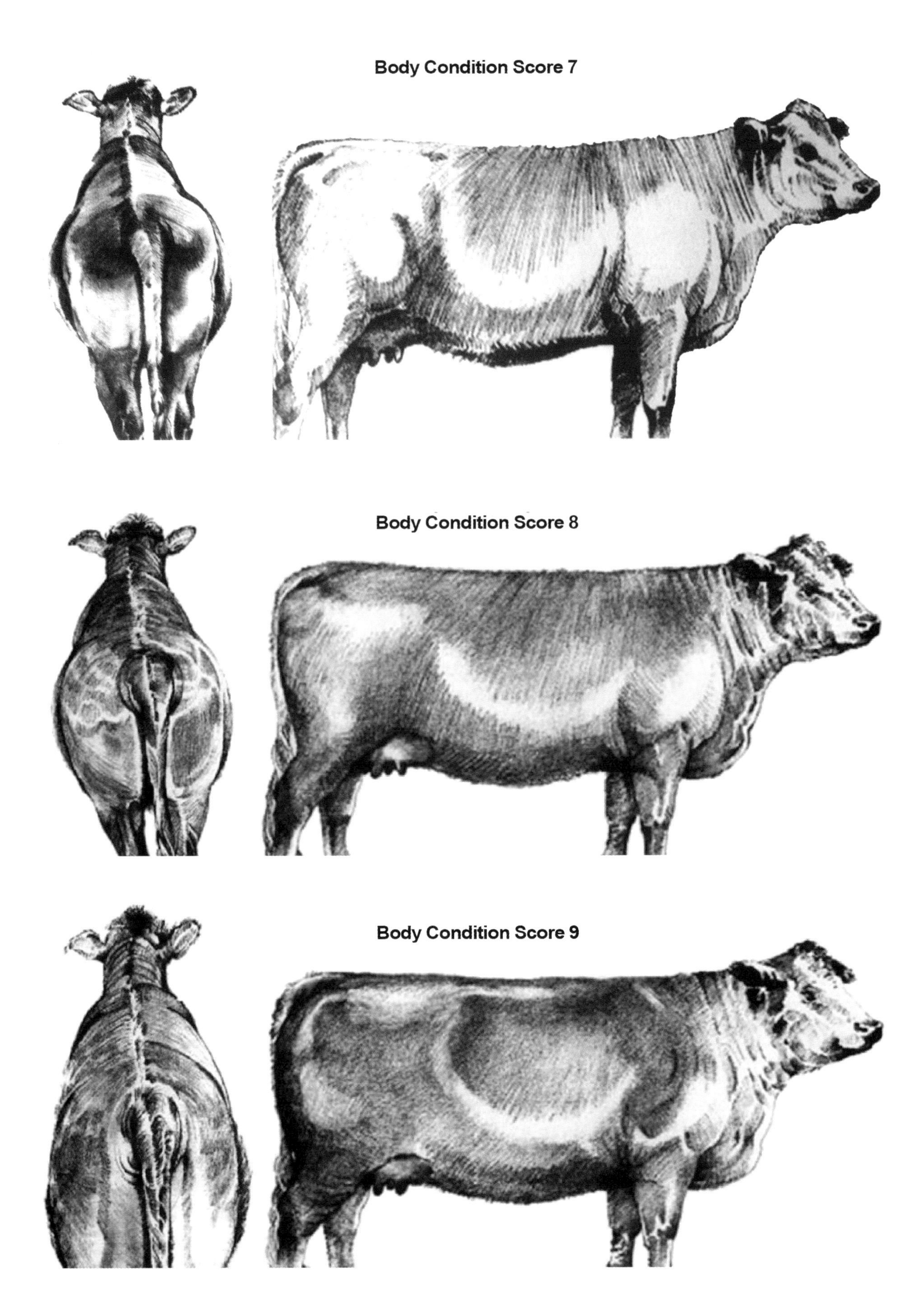
Body Condition Score 7
Body Condition Score 8
Body Condition Score 9

References

Blasi, D.A., R.J. Rasby, I.G. Rush, and C.R. Quinn. Cow Body Condition Scoring Management Tool for Monitoring Nutritional Status of Beef Cows. Beef Cattle Handbook BCH-5405.

Corah, L.R., P.L. Houghton, R.P. Lemenager, Dale A. Blasi. 1991. Feeding Your Cows by Body Condition. Kansas State University Cooperative Extension Bulletin C-842.

Herd, D.B. and L.R. Sprott. 1986. Body Condition, Nutrition and Reproduction of Beef Cows. Texas A&M University Cooperative Extension Bulletin B-1526.

Mathis, C.P., J.E. Sawyer, and R. Parker. 2002. Managing and Feeding Beef Cows Using Body Condition Scores. New Mexico State University Cooperative Extension Circular 575.

Pruitt, R.J. and P.A. Momont. 1988. Effects of Body Condition on Reproductive Performance of Beef Cows. South Dakota State University Beef Report.

Smith, T. 2008. Calving Condition – Do Cows have to be BCS 5 at Calving? Angus Journal, December 2008.

Chapter 24

Impact of Body Condition Score on Reproduction

The relationship between cow body condition and reproduction has been known for many years. Table 24-1 shows some of the early research that demonstrated this relationship.

In this research summary, it is clear that body condition at calving had a very significant influence on how many cows were cycling by either 60 or 90 days after calving. Remember, a cow must cycle before she can rebreed, and if she is not rebred by about 82 days post-calving, she will calve later next year resulting in a smaller calf. Furthermore, if she calves later next year, it is more likely that she will fail to breed in subsequent years.

Texas researchers conducted four trials looking at the influence of body condition at calving on subsequent pregnancy rates with a summary shown in Table 24-2. Clearly, body condition score at calving had a major influence on the percentage of cows that were bred in the first 60 days.

There have been a couple of large field studies looking at the relationship between body condition score at pregnancy check and pregnancy rates (Table 24-3).

Researchers at South Dakota State University expanded this concept to calculate the probability of a cow cycling by the start of the breeding season as shown in Table 24-4. Based on their research, as body condition increased, the probability of cycling by the start of breeding increased.

Table 24-4 also shows the probabilities of pregnancy during a 60-day breeding season and conceiving during the first 21 days of the breeding season. The data are also separated into early and late calvers. The probability of getting pregnant during the 60-day

Table 24-1. Body Condition at Calving and Heat after Calving.

Body Condition at Calving	No. Cows	% Cycling – Days Post-Calving 60 Days	90 Days
Thin (1-4)	272	46	66
Moderate (5-6)	364	61	92
Good (7-9)	50	91	100

Whitman, Thesis, Colorado State University.

Table 24-2. Combined Results of Four Trials Evaluating the Relationship between Body Condition Score at Calving and Subsequent Pregnancy Rates.

	Body Condition Score at Calving		
Item	4 or less	5	6 or more
No. of Cows	303	483	252
% Pregnant After 60 Days	60	76	91

Extrapolated from Texas Extension Service Bulletin B1526.

Table 24-3. Summary of Two Large Field Trials Examining the Relationship between Body Condition Score at Pregnancy Check and Pregnancy Rates.

Trial	Body Condition Score at Weaning or Pregnancy Check 3	4	5	6
Florida trials, >4,000 head, nine locations, % pregnant	33	65	89	92
Padlock Ranch, Wyoming, 7 years, more than 78,000 head, % pregnant	74	83	93	96

Table 24-4. Probability of Pregnancy and Conception as Influenced by Body Condition Score at Calving and at the Start of the Breeding Season.

	Probability of Pregnancy During a 60-Day Breeding Season		Probability of Conceiving in the First 21 Days of the Breeding Season	
Condition Score	Early Calvers	Late Calvers	Early Calvers	Late Calvers
Based on Condition Score at Calving				
3	.88	---	.51	---
4	.93	.88	.58	.41
5	.96	.93	.65	.56
6	.98	.96	.72	.70
7	.99	.97	.77	.81
8	.99	.99	.82	.89
Based on Condition Score at the Beginning of the Breeding Season				
2	.81	.60	.29	.23
3	.91	.80	.44	.36
4	.96	.91	.60	.50
5	.98	.97	.75	.65
6	.99	.99	.85	.77
7	1.00	.99	.92	.86

South Dakota State University.

breeding season was certainly influenced by body condition score at both calving and breeding.

These data also illustrate the concept of diminishing returns, since the **incremental increase** in the probability of getting pregnant decreased as the body score increased. For example, a body condition score of 4 compared to a 3 at calving increased the probability of pregnancy from 0.88 to 0.93 for an incremental increase of 0.05 points. Correspondingly, the difference between a 5 and a 6 was only 0.02 points (0.96 to 0.98). There is certainly a point where the incremental improvement probably will not justify the additional cost of feed. This relationship will be discussed in more detail in subsequent chapters.

The data in Table 24-4 also can be used to estimate the additional calves and the influence on calving sequence associated with raising body condition score. For example, assuming 100 head of early calving cows in condition score 4 at calving, feeding to increase their condition to a score of 5 would result in two more pregnant cows (98 vs. 96). Additionally, the percentage calving in the first 21 days next year would increase by 15 percent based on a probability of conceiving in the first 21 days for a body condition 4 of 0.60 and for a body condition score 5 of 0.75, respectively. Again, the cost to raise the condition score from 4 to 5 would depend on the cost of feed, but it should also be noted that there would be a long-term benefit from raising the score from a 4 to a 5 because of the higher probability of an early calving cow getting pregnant in the next year.

Unfortunately, assessing the economic effect of raising body condition score is not as simple as proposed in the previous paragraph because in a typical herd situation, there is a spectrum of body condition scores. In other words, there might be 30 percent, 60 percent, and 10 percent in body condition scores 4, 5, and 6, respectively. And in most cases, all of these cows will be fed together. Obviously, this makes the calculation of additional calves that might be expected much more difficult. Furthermore, it points out the value of sorting cows into feeding groups based on body condition, which maximizes the reproductive response to supplemental nutrition. Clearly, providing the additional feed to raise the condition of only the body condition score 4 cows instead of all of the cows will be more cost effective.

Key Times to Assess Body Condition Score

1. **60 to 80 Days Prior to Calving** – Assessing body condition scores 60 to 80 days prior to calving is important because there is still time to change the score prior to calving. It is often too late if one waits until 30 days pre-calving because of the high plane of nutrition required to improve condition in a short period of time. Furthermore, weather often has a significant influence on requirements, especially in spring calving herds.
2. **At Calving** – This is another key time because one can attempt to change condition before the breeding season starts (approximately 82 days from the start of calving). Changing body condition score between calving and breeding is somewhat more difficult than prior to calving because the cows are lactating.

   They tend to use the additional nutrients to produce more milk rather than improving body condition — the biological priority for nutrients at work. This reinforces the importance of having cows in good body condition before calving.
3. **At Breeding** – Obviously, it is too late to change condition and impact breeding significantly, but scoring at this time allows one to establish the relationship between body condition score and reproduction for a specific production system as will be discussed in a following section.
4. **At Weaning/Pregnancy Check** – The greatest value of scoring body condition at this time is to establish the nutritional requirements for the period from weaning to calving. If the cows are in good condition at pregnancy check, less supplementation may be required. Conversely, if they are in poor condition at pregnancy check, a higher plane of nutrition will be required to reach an acceptable body condition score at calving.

Optimal Body Condition Scores

There is a tendency to say that cows should calve in a body condition score of at least 5, and this may be a good general recommendation. However, if the goal is to maximize profitability, the optimal score may be lower. The two major considerations in establishing the optimal body condition score are the time of calving relative to the availability of green grass and the cost of providing supplemental feed.

1. **Time of Calving Relative to Availability of Green Pasture** – Nothing stimulates reproduction like lush, green grass. Consequently, if cows have access to high-quality forage before the initiation of breeding, they can be in a lower body condition score than will be required without access to lush, green grass. This means that the optimal body condition score might be a 4 at calving for cows that calve 30 days before the availability of green forage and a 5 for cows that calve 60 days before the availability of green pasture.
2. **Cost of Providing Supplemental Energy and Protein** – The cost of providing supplemental energy and protein varies widely across different environments. In many areas where by-products are produced in large quantities, it costs relatively little to keep cows in good body condition. Conversely, in other areas, the cost of

supplemental feed and/or the cost of delivery are high. In these areas, a lower body condition score, possibly a 4.5, might be optimal from a profitability standpoint.

Determining the Optimal Body Condition Scores for an Operation

As noted, a cow in a body condition score of 4 at calving does not reproduce at the same level in all environments. By evaluating body condition scores at key times, one can establish the average reduction in reproductive performance of a 4 versus a 5, for example, in a specific environment. This allows a much more accurate estimate of the amount of supplemental feed that will maximize profitability. There is tremendous variation in the relationship between body condition and reproduction from year to year. For example, if grass "green-up" is delayed because of cold weather, cows in a body condition 4 might rebreed at a lower level than in a normal year. Good records on body condition scores and reproductive performance can be useful in optimizing the supplementation program and maximizing profitability.

References

Bohnert, D.W., R. Mills, L.A. Stalker, A. Nyman, and S.J. Falck. 2010. Influence of Cow BCS and Late Gestation Supplementation: Effects on Cow and Calf Performance. Oregon State University Beef Research Report BEEF0027.

Blasi, D.A., R.J. Rasby, I.G. Rush, and C.R. Quinn. Cow Body Condition Scoring Management Tool for Monitoring Nutritional Status of Beef Cows. Beef Cattle Handbook BCH-5405.

Cherni, M. 1993. Using Body Condition to Score Beef Cattle. Montana Farmer-Stockman March 1993.

Corah, L.R., P.L. Houghton, R.P. Lemenager, Dale A. Blasi. 1991. Feeding Your Cows by Body Condition. Kansas State University Cooperative Extension Bulletin C-842.

Herd, D.B. and L.R. Sprott. 1986. Body Condition, Nutrition and Reproduction of Beef Cows. Texas A&M University Cooperative Extension Bulletin B-1526.

Pruitt, R.J. and P.A. Momont. 1988. Effects of Body Condition on Reproductive Performance of Beef Cows. South Dakota State University Beef Report.

Chapter 25

Changing Body Condition Score

There is a clear relationship between body condition score and reproduction. At times, cow body condition must be increased to achieve an acceptable level of reproductive performance. Consequently, the nutritional requirements needed to change body condition are extremely relevant. Table 25-1 shows the required weight gains of cows with different levels of body condition at weaning. In this table, the weight gain for pregnancy is 100 pounds.

A couple of key points are illustrated in Table 25-1: 1) the lower the body condition at weaning, the higher the gain required to get to moderate body condition, and 2) the number of days available to change condition dictates the rate of gain required. Table 25-2 illustrates this same concept except the initial body condition was evaluated at calving. The ADG illustrated would need to be above the gain for pregnancy.

It takes approximately 80 pounds of body weight for each condition score change on an average cow. Table 25-2, indicates that the desired rate of gain and energy level in the diet are functions of the pounds of gain required and the number of days available to raise body condition.

Appendix Table 2 shows the nutrient requirements to raise mature cows of different body weights from a BCS of 4 to a BCS of 5 during the last 90 days before calving. Correspondingly, Appendix Table 3 shows the

Table 25-1. Weight Gains Needed by Cows in Different Body Conditions at Weaning.

Body Condition		Weight Gain Required Prior to Calving, lbs				
Condition at Weaning	Desired at Calving	Weight for Pregnancy	Body Weight	Total	Days to Calving	Daily Gain, lbs
Thin	Moderate	100	160	260	120	2.2
Borderline	Moderate	100	80	180	120	1.5
Moderate	Moderate	100	0	100	120	0.8
Thin	Moderate	100	160	260	200	1.3
Thin	Moderate	100	160	260	100	2.6

Kansas State University Extension Bulletin C842

Table 25-2. Cow Weight Gains Required from Calving to Breeding to Achieve a Desired Body Condition.

Body Condition		Weight Gain Required by Breeding, lbs		
At Calving	Required at Breeding	Body Weight	Days to Breeding	ADG, lbs
Thin	Moderate	160	80	2.0
Borderline	Moderate	80	80	1.0
Moderate	Moderate	0	80	0.0

Kansas State University Extension Bulletin C842

Table 25-3. Energy Requirements to Change Body Condition Score of Beef Cows.

Current Body Condition Score	Desired Body Condition Score 2	3-4	5	6-7	8
	Mcal Per lb of Weight Gain (NEg)				
2	1.17	1.45	1.74	2.02	2.31
3-4		1.73	2.02	2.30	2.59
5			2.30	2.59	2.87
6-7				2.87	3.44
8					3.44

Purdue University

nutrient requirements to increase the BCS from a 3 to a 4 and a 4 to a 5 in either 30 or 60 days for cows of varying body weights. As shown in this table, increasing the BCS by one score in 30 days requires a significant energy level in the diet, which reinforces the need to evaluate body condition well ahead of calving to allow enough time to change body condition. Only the highest quality roughages like corn silage have enough energy to significantly change body condition in 30 days.

To illustrate the type of diets required to increase body condition, consider that a 1,200 pound cow in body condition score 5 requires a diet with 0.49 Mcal of NEm per pound to maintain condition. Most medium-quality hays have enough energy to meet this requirement. Now if this cow were a BCS of 4, 30 days before calving, increasing the BCS to 5 by calving would require a diet with 0.68 Mcal of NEm per pound (Appendix Table 3). As noted, there are few roughages with this much energy meaning that some type of concentrate will probably be required to provide adequate energy for this increase in BCS. For example, if the basal forage was bromegrass hay (NEm = 55 Mcal/cwt), it would take approximately 10 pounds of corn gluten feed (NEm = 86 Mcal/cwt) per day to provide a diet with a NEm of 68 Mcal/cwt.

Energy Required to Change Body Condition

Researchers at Purdue University fed cows of differing body condition to determine the energy required to change body condition score. Their results, expressed as NEg, are shown in Table 25-3, which shows that it takes less energy to put weight on a thin cow than a fat cow. This makes sense when considering that a pound of gain on a fat cow contains more energy than a pound of gain on a thin cow. Remember: Fat contains 2.25 times the energy of protein, and a thin cow is depositing mostly protein and some fat while a fat cow is depositing primarily fat. It is relatively easy to put weight on a thin cow, and it gets gradually more difficult and energetically expensive as the cow becomes fatter.

The data in Table 25-4, taken from Nutrient Requirements of Beef Cattle, 1996, show the Mcal of NEm that it will take to change body condition scores based on research at the U.S. Meat Animal Research Center. The energy required per condition score increases as the starting condition score increases and body weight increases. For example, changing a 1,300-pound cow from a 4 to a 5 requires 245 Mcal of NEm. Changing the same cow from a 5 to 6 requires 286 Mcal of NEm. Body weights for cow condition scores 1 through 9 are 76.5, 81.3, 86.7, 92.9, 100, 108.3, 118.1, 129.9, and 144.3 percent of condition score 5, respectively. For example, this 1,300 pound cow would weigh 1,127 pounds (1,300 × 0.867) at a body condition 3.

This table also shows the amount of energy that a cow can use from body stores when maintenance energy requirements are not met. The efficiency of this energy use is assumed to be 80 percent. Consequently, the 1,300-pound cow above at a condition score 5 can use 245 × 0.8 = 198 Mcal from body stores before reaching a condition score of 4. The calculations for balancing a ration to change body condition will be demonstrated in a subsequent chapter.

Table 25-4. Mcal of NEm Required to Change Body Condition Scores of Cows of Different Sizes.

Body Condition Score	Mature Weight at Body Condition Score 5, lbs						
	900	1,000	1,100	1,200	1,300	1,400	1,500
	Mcal of NEm Required or Provided for Each Condition Score						
2	114	126	139	151	164	177	189
3	129	143	157	172	186	200	214
4	147	163	180	196	212	229	245
5	170	188	207	226	245	264	283
6	198	220	242	264	286	308	330
7	234	260	285	311	337	363	389
8	280	311	342	373	405	436	467
9	342	380	418	456	494	532	570

Nutrient Requirements of Beef Cattle, *NRC, 1996.*

References

Blasi, D.A., R.J. Rasby, I.G. Rush, and C.R. Quinn. Cow Body Condition Scoring Management Tool for Monitoring Nutritional Status of Beef Cows. Beef Cattle Handbook BCH-5405.

Corah, L.R., P.L. Houghton, R.P. Lemenager, Dale A. Blasi. 1991. Feeding Your Cows by Body Condition. Kansas State University Cooperative Extension Bulletin C-842.

Herd, D.B. and L.R. Sprott. 1986. Body Condition, Nutrition and Reproduction of Beef Cows. Texas A&M University Cooperative Extension Bulletin B-1526.

Nutrient Requirements of Beef Cattle. 1996. National Research Council.

Pruitt, R.J. and P.A. Momont. 1988. Effects of Body Condition on Reproductive Performance of Beef Cows. South Dakota State University Beef Report.

Chapter 26

Impact of Nutrition on Calf Survival and Growth

Much of the focus in the last few chapters has been on feeding for reproduction, which is obviously critical to profitability; however, with respect to the number of calves that are available for sale at weaning, calf survival is also important. Nutrition influences calf survival and growth in several ways.

Effect of Prepartum Energy Level on Calf Survival

A study at the University of Wyoming in 1975 showed that the energy level before calving influenced calf survival as Table 26-1 illustrates. The energy level before calving had a significant influence on calf survival even though all of the cows were placed on adequate energy following calving. The major cause of death was calf scours. The authors of this study attributed the calf losses in the low energy group to weaker calves being slow to suckle and to the potential that the level of maternal antibodies in the colostrum was lower or poorly absorbed.

Effect of Prepartum Protein Level on Calf Survival

A severe deficiency of protein precalving has been implicated as the cause of **Weak Calf Syndrome** where calves are extremely weak and have difficulty nursing, which causes a high death loss. Research at Colorado State University (Table 26-2) showed that protein deficiency in 2-year-old beef heifers decreased colostrum production, reduced calf heat production, and increased the interval from calving to standing. While colostrum production was reduced, the concentration of antibodies in the colostrum was higher in the protein deficient treatment. However, the total amount of antibodies

Table 26-1. Effect of Late Gestation Energy Level on Calf Survival.

Item	Prepartum Energy Level	
100 to 30 days precalving	Low	Low
30 days precalving to calving	Low	High
Cow weight change (lbs) from 30 days precalving to calving	-119	-115
Cow weight change (lbs) from 30 days precalving to calving	-23	+93
Calves alive at birth, %	90.5	100
Calves alive at 2 weeks, %	80.9	100
Calves weaned, %	71.5	100

JAS 41:819.

tended to be higher in the adequate protein treatment. The amount of antibodies ingested by the calf from colostrum is extremely important because these antibodies are the major source of immunity in the calf for the first couple of months of life.

Effect of Body Condition Score on Calf Survival

Cow body condition score at calving can be an indicator of calf survival as shown in a study conducted at Colorado State University (Table 26-3) where low body condition increased the time to standing, and tended to reduce colostrum production and calf serum antibody levels.

Effect of Prepartum and Postpartum Nutrition on Calf Weaning Weights

It seems logical that cows in higher body condition at calving would produce more milk than cows in lower body condition resulting in heavier weaning weights; however, research trials have shown that the differences in milk production between cows of differing body condition scores are in fact minimal. To illustrate this point, consider the results of a research trial at Kansas State University shown in Table 26-4. In this study, cows were fed varying amounts of soybean meal before calving while consuming dormant tallgrass prairie hay. After calving, all of the cows were fed the same amount of supplement and alfalfa hay until sufficient new grass growth. The different levels of soybean meal resulted in significant differences in body weight change and body condition score before calving. However, weaning weights were essentially the same for all treatments. This result is explained by the fact that thin cows tend to compensate when forage quality is high by consuming more forage. This allows them to milk at a level similar to cows in better body condition. Moreover, calves nursing cows that give less milk tend to consume more forage. These factors combine to result in similar weaning weights. The additional

Table 26-2. Effects of Inadequate Protein Intake Prepartum in Two-Year-Old Beef Heifers.

Item	Adequate	Inadequate
Protein intake, lbs/day	1.36	0.84
Colostrum quantity (ml)	2,683	1,931
Colostral IgG (mg/dl)	5,810	7,157
Colostral IgM (mg/dl)	511	639
Total colostral IgG (g)	138	134
Total colostral IgM	11.7	11.5
Heat production (Kcal/MBS)	118	104
Interval from calving to standing (minutes)	66	97

Colorado State University 1987 Beef Report.

Table 26-3. The Relationship between Body Condition Scores of Beef Heifers at Calving and the Interval from Calving to Standing, Colostrum Production, and Immunoglobulin Concentration.

	Heifer Body Condition Score at Calving		
Item	3	4	5
Interval from calving to standing, minutes	60	64	43
Colostrum production, ml	1,525	1,112	1,411
Calf serum IgG (mg/dl)	1,998	2,179	2,310
Calf serum IgM (mg/dl)	146	157	193

Colorado State University Beef Report, 1986.

Table 26-4. Effects of Increasing Amounts of Supplemental Soybean Meal on the Performance of Beef Cows Grazing Dormant, Tallgrass Prairie.

Item	Supplemental SBM, lbs per head per day							
	1.0	1.5	2.0	2.5	3.0	4.0	5.0	6.0
Precalving Wt. change, lbs[a]	-255	-212	-193	-201	-185	-140	-162	---
BC Score change	-1.23	-0.74	-0.86	-0.53	-0.35	-0.07	-0.08	0.02
Cow body Wt. change, breeding to weaning, lbs	259	238	261	192	198	216	213	173
Calf weaning Wt., lbs	513	489	512	537	519	527	515	503

[a]The precalving weight loss reflects the weight change from the start of the trial to a weight taken within the first 48 hours after calving. Cattlemen's Day Report of Progress, 1998, Kansas State University.

Table 26-5. Effect of Level of Cow Winter Nutrition on Fall Calving System Productivity.

Item	Low Supplement	High Supplement
Cow weight change, lbs, Jan-Apr	-101	-97
Cow body condition score change	-0.69	-0.55
Calf weight change, lbs, Jan-Apr	99	128
Weaning weight, lbs	621	641

Oklahoma State University, Animal Science Report 2002.

forage consumption by the thin cows also allows them to add more body condition before weaning. It should also be remembered that thin cows are more efficient in putting on condition because it takes less energy per pound of gain for a thin cow compared to a fatter cow. The increase in intake and greater efficiency resulted in similar body condition scores for cows with all levels of supplementation when the calves were weaned in October.

The results in Table 26-4 are typical of what might be expected in a spring calving situation where high-quality forage after calving allows cows in low body condition to compensate. In fall calving situations, inadequate nutrition pre- or postpartum will typically have a greater influence on weaning weight because high-quality forage often is not available when nutrient requirements for lactation are high.

Table 26-5 shows the results of a study at Oklahoma State University looking at the effect of supplementation following calving on calf weaning weights in a fall calving herd. In this study, the cows were managed similarly through calving until the end of breeding in January. From the end of breeding until forage green-up in April, the cows in the low nutrition treatment received 2 pounds per head per day of a 40 percent protein cube while cows in the high nutrition treatment received 6 pounds per head per day of a 20 percent protein cube. Cows in both groups were grazing dormant tallgrass prairie.

In this research study, calf weaning weight was reduced by a lower level of supplementation from breeding until grass green-up. The cows in lower body condition gave less milk as reflected by lower weaning weights, and the forage quality did not allow the cows and calves to compensate as was the case in the Kansas State University study with spring calving cows. The decrease in weaning weight was relatively small, which means that the cost of the additional supplement probably would be higher than the value of the additional weaning weight. As a general statement, it is usually more profitable to increase calf performance by creep feeding rather than by feeding the beef cow to increase milk production and weaning weight.

References

Corah, L.R., T.G. Dunn, and C.C. Kaltenbach. 1975. Influence of Prepartum Nutrition on the Reproductive Performance of Beef Females and the Performance of Their Progeny. JAS 41:819.

Holland, M.D., K.G. Odde, and D.E. Johnson. 1987. Effects of Dietary Protein Intake in Two-Year Old Beef Heifers on Gestation Length, Birth Weight, and Colostral and Calf Serum Immunoglobulin Levels. Colorado State University Beef Program Report.

Mathis, C.P., R.C. Cochran, J.S. Heldt, B.C. Woods, K.C. Olson, and G.L. Stokka. 1998. Effects of Increasing Amounts of Supplemental Soybean Meal on Intake and Digestibility of Tallgrass-Prairie Hay. 1998 Cattlemen's Day Report of Progress, Kansas State University.

Mayo, S.J., D.L. Lalman, G.E. Selk, R.P. Wetteman, and D.S. Buchanan. 2002. Effect of Level of Cow Winter Nutrition and Calf Creep Feeding on Fall Calving System Productivity. 2002 Oklahoma Beef Cattle Research Report.

Odde, K.G., L.A. Abernathy, and G.A. Greathouse. 1986. Effect of Body Condition and Calving Difficulty on Calf Vigor and Calf Serum Immunoglobulin Concentrations in Two-Year-Old Beef Heifers. Colorado State University Beef Progress Report.

Odde, K. 1992. They are What They Eat – Impact of Cow/Calf Nutrition on Reproduction, Calf Development, and Disease Resistance. Topics in Veterinary Medicine, Volume 3.

Chapter 27

Feed Intake by Beef Cattle

The total nutrient intake by beef cattle is a function of the nutrient content of the diet and the amount consumed (intake). Unfortunately, predicting intake, especially in grazing situations, is difficult because there are many factors influencing intake. As a general statement, intake is a function of the energy content in the diet with intake increasing as the quality (energy density) of the diet increases. This generality is true with the exception of very high-energy diets. For example with a feedlot diet, dry matter intake will be lower than on a very high-quality forage diet even though the energy density is higher in the feedlot diet.

Factors Influencing Feed Intake in Beef Cows

Body Weight

Feed intake increases in direct proportion to the metabolic body weight (body weight to the 0.75 power). Again, this means that a 1,600-pound cow does not eat twice as much as an 800-pound cow.

Energy Content of the Diet

For the energy densities typical in cow diets, there is a fairly direct and positive relationship between the quality of the diet and dry matter intake. In other words, as the energy content of the diet increases, intake increases, and visa versa.

Breed Effects

Holstein and Holstein x beef crosses exhibit intakes 8 percent and 4 percent above the beef breeds, respectively.

Body Condition

Thin cows offered a good quality diet consume more than cows in good condition on the same diet. Also, as cows become relatively fat (body condition score 7, for example) they reduce intake to some extent.

Lactation

Intake increases in direct proportion to the level of milk production with approximately 0.25 pound of additional intake for each pound of milk produced.

Temperature

Intake is reduced at high temperatures, especially when there is not sufficient nighttime cooling. Intake increases at low temperatures as shown in Table 27-1.

Lot Conditions

When cattle are forced to walk through mud to get to feed or water, they tend to reduce intake. Mild mud conditions (4 to 8 inches deep) reduce intake by approximately 15 percent, while severe mud (12 to 14 inches deep) reduces intake by 30 percent.

Protein Level in the Diet

A protein deficiency reduces feed intake because the protein shortage reduces the microbes' ability to digest the diet. Passage of the feed slows, and intake declines.

Intake Guidelines for Beef Cows

Table 27-2 shows general guidelines for intake by beef cows based on Oklahoma State University research. Examples of low-quality forage would include dry, dormant winter range, crop residue in

Table 27-1. Effect of Temperature and Night Cooling on Feed Intake.

Temperature, Degrees F	Intake Percentage
>95, no night cooling	65
>95, with night cooling	90
78 – 95	90
59 – 78	100
41 – 59	103
23 – 41	105
5 – 23	116

Nutrient Requirements of Beef Cattle, NRC, 1996.

winter, and low-quality hay. Average quality forages include good grass hay and crop residue shortly after harvest, while examples of high-quality forages include alfalfa hay and corn silage.

To illustrate the usage of Table 27-2, consider a dry, gestating cow offered prairie hay, which is a low quality forage. A 1,200-pound cow would be expected to consume 1.8 percent of her body weight with adequate protein supplementation. Expected dry matter intake would be 1,200 × 0.018 = 21.6 pounds. Assuming a moisture content of 11 percent for this hay, this cow would be expected to consume 21.6 ÷ 0.89 (dry matter content) = 24.3 pounds of hay on an as fed basis. Wastage might be expected to be as low as 5 percent and as high as 30 percent depending on how the hay is fed. For example, assume the hay is being fed in a bale feeder with a wastage percentage of 10 percent. This means that 24.3 ÷ 0.90 (delivery minus wastage) = 27.0 pounds of hay should be fed to provide the expected dry matter intake.

The same 1,200-pound cow offered a high-quality roughage such as alfalfa hay would be expected to consume 2.5 percent of body weight. Thus, the expected consumption would be 1,200 × 0.025 = 30 pounds of dry matter. Assuming the same percentage of dry matter content (89 percent) in the hay, the cow would consume 30.0 ÷ 0.89 = 33.7 pounds of hay. If this hay was fed on the ground with an expected wastage of 25 percent, a total of 33.7 ÷ 0.75 (delivery minus wastage) = 44.9 pounds would need to be fed to provide the expected dry matter intake.

Feed Intake by Growing Cattle

Several factors influence feed intake by growing cattle including body weight, diet energy density, prior nutritional regime, and weather. There is little effect of sex on dry matter intake by growing cattle.

Body Weight

Body weight influences dry matter intake as shown in Table 27-3. This table also shows that intake as a percentage of body weight decreases as growing cattle increase in body weight.

Table 27-2. Estimated Forage Intake of Cows in Different Physiological States and Subjected to Different Supplementation Programs.

	Forage Dry Matter Intake, Percentage of Body weight	
Roughage Type/Conditions	Dry, Gestating	Lactating
Low Quality/Unsupplemented	1.5	2.0
Low Quality/Protein Supplemented	1.8	2.2
Low Quality/Energy Supplemented	1.5	1.5
Average Quality/Unsupplemented	2.0	2.3
Average Quality/Protein Supplemented	2.2	2.5
Average Quality/Energy Supplemented	2.0	2.3
High Quality/Unsupplemented	2.5	2.7
High Quality/Protein Supplemented	2.5	2.7
High Quality/Energy Supplemented	2.5	2.7

Hibberd and Thrift, 1992.

Table 27-3. Effect of Body Weight on Dry Matter Intake by Growing Cattle.

Body Weight, lbs	Dry Matter Intake, lbs	Dry Matter Intake, % of Body Weight
300	8.6	2.9
400	10.7	2.7
500	12.7	2.5
600	14.6	2.4
700	16.3	2.3

Adapted from Nutrient Requirements of Beef Cattle. 1996. National Research Council.

Table 27-4. Dry Matter Intake by a 500 lb Growing Animal Consuming Diets of Different Energy Densities.

Diet NEg, Mcal/lb	ADG, lbs	Dry Matter Intake, lbs
0.24	0.5	11.5
0.30	1.0	12.2
0.36	1.5	12.6
0.42	2.0	12.6
0.50	2.5	12.6
0.58	3.0	12.2

Adapted from Nutrient Requirements of Beef Cattle. 1996. National Research Council.

Diet Energy Density

In general, dry matter intake increases as diet energy increases. However, at very high energy levels intake decreases slightly as shown in Table 27-4. The data in this table are based on a 500 pound calf as shown in Appendix Table 6.

Previous Nutritional Regime

Growing cattle that were on a restricted energy level will consume more dry matter than cattle that previously were on a good-quality diet, especially early in the feeding period. Conversely, growing cattle that were on a very high quality diet and are fleshy will consume less dry matter than an animal in average condition.

Weather

Weather impacts the dry matter intake of growing cattle in the same way it influences intake by beef cows as discussed previously.

Feed Intake by Finishing Cattle

Several factors influence the dry matter intake by finishing cattle with age and previous nutritional regime being major factors. On average, finishing cattle consume about 2 percent of their body weight in dry matter daily. Yearling cattle tend to increase intake during the first 40 to 50 days of the feeding period. Intake plateaus during the middle of the feeding period and declines late in the feeding period. It appears that cattle placed on a finishing diet as calves do not exhibit the plateau and decline exhibited by cattle placed on feed as yearlings.

Age

The improvement in growth rate made by the beef cattle industry over the past 30 years has greatly increased the number of cattle being placed on feed as calves. Table 27-5 shows the results of placing similar cattle on feed as calves or as yearlings in a study conducted at Kansas State University. As shown, the calves were on feed much longer and gained less per day. However, the calves had much better feed conversion

Table 27-5. Comparison of Feeding Calves Verses Yearlings.

Item	Calves	Yearlings
Slaughter Weight, lbs	1,085	1,265
Age at Slaughter, days	452	578
Days Fed	224	133
Dry Matter Intake, lbs	14.6	20.5
ADG, lbs	2.45	2.67
Feed/Gain	5.92	7.69

Kansas State University Cattlemen's Day Report 1992.

primarily because of their lighter body weight resulting in lower maintenance requirements.

Prior Nutritional Regime

As might be expected, cattle that were on a low plane of nutrition before being placed on a finishing diet tend to consume more dry matter than cattle that were on a high plane of nutrition.

Gender

Steers will consume from 1 to 3 percent more dry matter than similar heifers in a finishing program.

References

DeHaan, K., M.T. Van Koevering, and M.L. Gibson. 1995. The Effect of Age, Background, and Gender on Feed Intake by Feedlot Cattle. In Symposium: Intake by Feedlot Cattle. Oklahoma State University Cooperative Extension Publication P-942.

Hibberd, C.A. and T.A. Thrift. 1992. Supplementation of Forage Based Diets: Are Results Predictable. J. Animal Science 70 (Suppl 1):181 (Abstract).

Hickok, D.T., R.R. Schalles, M.E. Dikeman, and D.E. Franke. 1992. Comparison of Feeding Calves vs. Yearlings. Kansas State University Cattlemen's Day 1992.

Johnson, C.R., D.L. Lalman, M.A. Brown, L.A. Appeddu, R.P. Wetteman, and D.S. Buchanan. Effect of Parity and Milk Production Potential on Forage Intake of Beef Cows During Lactation. 2002 Oklahoma State University Animal Science Report.

Johnson, C.R., D.L. Lalman, L.A. Appeddu, M.A. Brown, R.P. Wetteman, and D.S. Buchanan. Effect of Parity and Milk Production Potential on Forage Intake of Beef Cows During Late Lactation. 2002 Oklahoma State University Animal Science Report.

Nutrient Requirements of Beef Cattle. 1984. National Research Council.

Nutrient Requirements of Beef Cattle. 1996. National Research Council.

Ovenell, K.H. K.S. Lusby, G.W. Horn, and R.W. McNew. 1991. Effects of Lactational Status on Forage Intake, Digestibility, and Particulate Passage Rate of Beef Cows Supplemented with Soybean Meal, Wheat Middlings, and Corn and Soybean Meal. JAS 69:2617.

Chapter 28

Feed Analysis

Feedstuff analysis is a valuable tool in developing a sound nutritional program. However, too much money can be spent relative to the benefit. *The key to a cost effective feed analysis program is to only analyze for nutrients that might be deficient or nutrients that will, if deficient, influence the supplementation program.* For example, why pay for a trace mineral analysis on forages if a highly fortified free-choice mineral will be fed anyway? This chapter provides guidelines for using and interpreting feed analyses. First, recognize that there are short-term and long-term goals in having feedstuffs analyzed.

Short-Term Goals:

- **Determine the Composition of Forages Currently Available and Plan a Supplementation Program** – It is difficult to visually assess the nutrient content of forages. Moreover, the protein and energy content of forages varies greatly from year to year depending on growing conditions, time of harvest, and other factors. Thus, feed analyses for key nutrients should be routine for any forage that will constitute a significant portion of the feed supply.

Long-Term Goals:

- **Compile a Database on the Average Composition of Grasses and Other Forages** – Feed analyses, combined with some record of the weather during the year, can be useful in planning for supplemental forage and protein requirements. Since trace mineral levels are largely a function of soil content, a database of the trace mineral levels in forages for each pasture could eliminate the need for annual expenditures for trace mineral analyses.

Analyzing Roughages

The following general guidelines should be applied when deciding which analyses to request.

Moisture

Animal requirements and expected intakes are stated on a dry matter basis, while feed is delivered to the cattle on an "as-fed" basis. Consequently, the moisture content is an essential part of any feed analysis. The moisture content is especially useful for silages and haylages because of their high moisture contents and inherent variability. However, even a small difference in the moisture content of hay can be important when the hay makes up a large portion of the ration.

Protein

The crude protein content should be requested because of the importance of adequate protein in cattle diets. Remember that the crude protein level on a feed analysis is based on the level of nitrogen in the feed, not all of which will actually be natural or true protein. Some of the nitrogen will be contained in other compounds. Another type of protein analyses that might be useful includes:

- **Heat-Damaged Protein** – Laboratories use a variety of names for heat-damaged protein including acid insoluble nitrogen (ADIN) or insoluble crude protein (ICP). This analysis provides a measure of the amount of the nitrogen that has been tied up as a result of heating and isn't going to be digestible. This analysis should be requested for all roughages exhibiting heat damage. Common examples are hays that have been baled too wet and silages ensiled at low moisture levels. These feeds will usually have a dark color and tobacco smell. For some reason, cattle like the taste of these feeds and often eat the caramelized portions of a bale before eating the part that is not heat-damaged. This is further proof that cows have little nutritional wisdom. They eat feeds that taste good rather than feeds that have a higher nutritive content, not unlike humans.

If a significant portion of the protein is heat-damaged, it should be subtracted from the total crude protein before ration formulation. For example, if heat-damaged alfalfa hay has 18 percent crude protein and 3 percent heat-damaged protein, the protein level should be reduced to 15 percent before formulating a ration.

Energy

As noted in a previous chapter, the energy content of a feedstuff cannot be measured directly in the laboratory. The energy content shown on a feed analysis is typically based on the acid detergent fiber (ADF) content. As a general rule, as the ADF level goes up, the energy level goes down. While there is an inverse relationship between ADF and energy, it is a weak relationship, which means that energy values based on ADF are often inaccurate. Consequently, the energy values provided as part of a feed analysis should be interpreted with caution. The energy value on a feed analysis should be compared to "book" or tabular values (Appendix Tables 11), and if the laboratory value is either much higher or lower than the book value, it should be adjusted closer to the book value before being used in ration formulation. Typically, the energy content will be expressed as NEm, NEg, and TDN. As noted previously, each of these measures can be used to formulate rations.

Calcium

Since almost all roughages contain adequate levels of calcium to meet beef cow requirements, calcium analysis is of little value, even though it is typically part of the roughage analysis package available from most laboratories. However, analyzing roughage that will be fed in a growing program for calcium can be quite useful since typical growing and finishing diets are low in calcium.

Phosphorus

The level of phosphorus in most roughages is often marginal or inadequate compared to beef cattle requirements. Consequently, phosphorus should be requested as part of a feed analysis. This analysis will be useful in establishing the optimum level of phosphorus in a protein supplement or free-choice mineral. For example, if the level in the roughage is marginal, a low level of phosphorus (4 percent, for example) in the mineral should be adequate. Conversely, if the level in the roughage is well below the requirement, the mineral should contain a higher level (8 to 12 percent, for example). Chapter 33 provides more detailed guidelines for the phosphorus level in a free-choice mineral based on the phosphorus level in the basal forage. Formulating the diet to meet the phosphorus requirement without overfeeding phosphorus minimizes the cost of supplementation.

Nitrate

There are several situations where roughages may be high in nitrate. For example, drought-stressed forages are often high in nitrate. In addition, some forages tend to accumulate nitrate. Examples include summer annuals such as forage sorghums, sudans, and oat hay. The nitrate level should be requested when dealing with any of these species or when weather conditions have been favorable for nitrate accumulation. The problem of high nitrate levels will be covered in Chapter 44.

Trace Minerals

If excessive health problems in calves or if lower than expected reproduction considering cow body condition occurs, forage trace mineral levels should be analyzed. Samples should be submitted for all major pasture units since there can be significant variation among pastures, especially if soil types differ. Since trace mineral analyses are costly and trace mineral levels are largely a function of soil content, it is not necessary to analyze roughage samples annually.

Vitamins

Since the color (the greener the color, the more Vitamin A) of a roughage is a good indicator of the vitamin A content, and vitamin A is relatively inexpensive to supply, it usually isn't cost effective to analyze for vitamin A. A good recommendation is to supplement vitamin A any time it might be deficient as discussed previously.

Mycotoxins

Stored roughages will often exhibit some mold growth. Generally, molds on roughages do not produce toxins that cause health problems or reduce performance. Thus, testing for mycotoxins on roughages is of little value unless health problems occur that cannot be explained by other factors. However, excessive mold growth on roughages will often reduce intake, which may significantly influence performance.

Acid Detergent Fiber (ADF)

ADF consists of largely indigestible parts of the plant, which explains the inverse relationship with energy. If the energy content of the forage is requested as part of the analysis, the ADF value will normally be provided.

Neutral Detergent Fiber (NDF)

NDF consists primarily of the cell walls of the plant. It is a fairly good indicator of the expected intake of the forage with an inverse relationship where intake is expected to be higher with low NDF and lower with high NDF.

Relative Feed Value

RFV was originally developed to measure the feed value of alfalfa hay for dairy producers. It is calculated from ADF and NDF to provide a single measure of feeding value. It is widely used in the marketing of alfalfa hay. It has some value with other forages, but since crude protein is not included in the formula, its value is somewhat limited. Relative feed value should not be used for feedstuffs other than roughages because with grains and by-product feeds the values are meaningless. Moreover, RFV is not appropriate for comparing relative nutritional value across different types or species of forage. The protein content should also be considered along with the RFV.

Analyzing Grains and Grain By-Products

Since grain usually constitutes a small fraction of a beef cow diet, it usually is not worth spending the money for an analysis. However, grain often is a major ingredient in growing rations for heifers and bulls and in finishing diets. Thus, an analysis in these situations is warranted. And, given the typical variability in the nutrient content of by-products, they should be routinely analyzed if they will constitute a significant portion of the diet. The following general guidelines should be used when requesting analysis of grains and/or by-products.

Moisture

As with roughages, the moisture content should be analyzed. This is especially important with wet by-products since they tend to be especially variable with respect to moisture content.

Protein

While grains exhibit variability with respect to protein content, their typically low inclusion level in the diet in beef cow herds makes analysis of limited value. On the other hand, by-products can be quite variable in protein content and should be analyzed any time they make up a significant portion of the diet. Grains or grain by-products showing evidence of heating should be analyzed for heat-damaged protein. For example, pelleted corn gluten feed comes in a range of colors largely created by the amount of heat applied during the drying and pelleting process and by the amount of steep liquor added. Dark corn gluten feed should be analyzed for heat-damaged protein.

Energy

Again, laboratory tests for energy can be inaccurate, and as noted with roughages, energy values should be compared to book values. Analysis values much different than those listed in feed composition tables should be used with caution as discussed previously.

Calcium

Grains tend to be low in calcium relative to beef cattle requirements. In fact, the calcium levels in grains are so low that it probably is not worth the money for an analysis. By-products are quite variable with respect to calcium levels. Thus, if by-products are to be a significant portion of the diet, a calcium analysis should be requested.

Phosphorus

Since grains tend to be high in phosphorus content, an analysis for phosphorus is probably of little value. However, the phosphorus contents of grain by-products are variable making an analysis useful.

Mycotoxins

Any mold on grain should be considered a potential problem. There are a wide variety of molds that grow on grains that produce toxic substances. The best course of action is probably to simply not feed moldy grains to cattle. If, however, they must be fed, they should be tested for mycotoxins.

Analyzing Supplements

Feed companies typically make a serious effort to meet the specifications on feed tags, and they are monitored by the feed regulatory agency in each state. However, it is still a good idea to have a sample of supplements analyzed periodically, especially for key nutrients.

Sampling Roughages and Grains

Unless a thoroughly representative sample of each feedstuff is submitted for analysis, it may be better to simply use book values. The following guidelines are provided to assist in sample collection:

Grain and By-Product Sampling

Samples should be collected at various times during unloading to get a truly representative sample of the entire load. A grain probe is also useful in getting a more representative sample.

Hay Sampling

There are several types of coring tools available to assist in getting a representative sample of baled hay. Numerous bales should be sampled to get a good representation of the roughage. A good rule-of-thumb is to sample a minimum of 10 bales and mix the samples thoroughly before submitting for analysis. If the quantities are large enough, it is advisable to sample different cuttings and fields separately. This allows the use of the best quality hays during the periods of highest nutritional requirements. It also allows feeding the highest quality hay to the animals with the highest requirements such as first- and second-calf-heifers.

Silage Sampling

Silage should be sampled after ensiling. Unfortunately, it is essentially impossible to get a representative sample of an entire bunker or pile of silage. Thus, it might be advisable to sample the silage periodically as it is fed to determine its composition and adjust the rations accordingly. Because silage often makes up a significant portion of growing rations, it should be analyzed periodically to ensure that rations provide the desired performance.

Chemical Analysis vs. Near Infrared Reflectance Spectroscopy (NIR)

Recently, more feed analysis laboratories have been using NIR to estimate nutrient content rather than traditional chemical analyses because it is less expensive. Near Infrared Reflectance Spectroscopy is fairly accurate with some common feedstuffs and yet inaccurate with others. The following guidelines are provided to assist in knowing when chemical analyses, also referred to as "wet chemistry," should be requested.

1. NIR is accurate when the roughage is composed of one species, such as alfalfa hay. Conversely, it is often inaccurate when the roughage is a mixture of a number of species such as plants from a range forage sample or a mixed ration.
2. The reliability of NIR is better for common roughages and grasses because NIR accuracy is to some extent a reflection of the size of the database for the feedstuffs. For example, NIR is fairly accurate for alfalfa hay and grains because large comparative databases have been established.
3. Other than for alfalfa and grains, wet chemistry methods of analysis should probably be requested for all minerals.

References

Belyea, R.L. and R.E. Rickets. 1982. Forages for Cattle. University of Missouri Cooperative Extension Guide 3150.

Chandler, P. 1990. Energy Prediction of Feeds by Forage Testing Explored. Feedstuffs August 27, 1990.

Guyer, P.Q. Analyzing Grain and Roughages. Great Plains Beef Cow-Calf Handbook GPE-1802.

Hall, M.B. Feed and Forage Evaluation – The New Frontier. Department of Dairy and Poultry Science University of Florida.

Hall, M.B. 1997. Interpreting Feed Analyses: Uses, Abuses, and Artifacts. Feed Management Vol: 48, No. 6.

Owen, F. G., B. Anderson, R. Rasby, and T. Mader. 1989. Testing Livestock Feeds. University of Nebraska NebGuide G89-915.

Thiex, N. and G. Kuhl. 1980. Feed Analysis for the Livestock Producer. South Dakota State University Cooperative Extension Service Bulletin FS 754.

Wilcox, R.A. 1972. How to Sample Feedstuffs. Kansas State University Cooperative Extension Service Bulletin L-316.

Chapter 29

Evaluating Diet Adequacy

In the *Nutrient Requirements of Beef Cattle,* 1996, the approach to expressing nutrient requirements was changed from nutrient requirement tables to providing formulas to calculate requirements. While this approach is more accurate, it makes checking diet adequacy by hand laborious. Fortunately, there are computer programs that can calculate the requirements and check for deficiencies rather easily. However, not everyone has access to these programs or wants to use them. Appendix Tables 1-10 provide general guides for animal requirements. Before illustrating the use of these tables, it should be noted that the requirements listed in the tables do not consider the following factors that influence actual requirements:

- Requirements to change body condition
- Activity
- Weather stress
- Breed differences in maintenance requirements

Consequently, when using Appendix Tables 1-10, the nutrient requirements should be adjusted when any of the above factors is considered to be significant. Guidelines for making these adjustments are provided at the end of this chapter.

Evaluating Diet Adequacy – Feeding Harvested Forage

Example One:

Assume we are feeding 32 pounds of brome hay to a 1,200-pound cow giving 20 pounds of milk in her second month of lactation. The nutrient requirements for this cow are taken from Appendix Table 1. The nutrients supplied are calculated using the nutrient analysis data for bromegrass hay taken from Appendix Table 11.

Steps in Determining the Nutrients Supplied

Step 1 – Select the correct feedstuff from the Appendix Table 11. This is especially important when several quality grades are listed for the same feedstuff. For example, several quality levels of alfalfa and other roughages are listed, and selecting the one that most accurately describes the feedstuff is important. Obviously, a feed analysis is better than using values from the table. Table 29-1 shows the nutrient composition of bromegrass hay.

Step 2 – Calculate dry matter intake. Since nutrient content is expressed on a dry matter basis, first determine the cow's dry matter intake (DMI). The DMI in this example is 32 pounds × 0.89 (percent DM) = 28.5 pounds. Note: This calculation does not include an allowance for waste, and in most situations waste is significant and must be considered.

Table 29-1. Nutrient Composition of Bromegrass Hay from Appendix Table 11.

DM, %	TDN, %	NEm, Mcal/ cwt	CP, %	Ca, %	P, %
89	55	55	10	0.40	0.23

Table 29-2. Diet Adequacy for a 1,200-Pound Cow Being Fed 32 Lbs of Bromegrass Hay.

Nutrient	DM, lbs	TDN, lbs	NEm, Mcal	CP, lbs	Ca, lbs	P, % lbs
Required	27.8	16.7	16.9	2.97	0.09	0.06
Supplied	28.5	15.7	15.7	2.85	0.11	0.07

Step 3 – Calculate the intake of each nutrient as follows:

a. TDN = 28.5 × 55 percent (0.55) = 15.7 pounds

b. NEm = 28.5 × 55 Mcal per cwt (0.55 per pound) = 15.7 Mcal

c. CP = 28.5 × 10 percent (0.10) = 2.85 pounds

d. Ca = 28.5 × 0.40 percent (0.004) = 0.11 pounds

e. P = 28.5 × 0.23 percent (0.0023) = 0.07 pounds

Step 4 – Compare the cow's nutrient requirements to the nutrients supplied as shown in Table 29-2. Assuming that this cow would consume the additional dry matter over that predicted by the intake equations in the Nutrient Requirements of Beef Cattle, 1996, she would still be slightly deficient in energy and protein. If she was in good body condition, this level of deficiency might not warrant supplementation. However, if she was in borderline condition, supplementation would be advisable. Furthermore, if any of the factors such as weather stress, activity, and other factors not considered in the tabular values were significant, this diet would be inadequate, which could result in a significant loss in body condition.

Example Two:

Feed the same cow 32 pounds of prairie hay with the nutrient composition shown in Table 29-3.

Table 29-3. Nutrient Composition of Prairie Hay from Appendix Table 11.

DM, %	TDN, %	NEm, Mcal/cwt	CP, %	Ca, %	P, %
91	50	50	7	0.40	0.15

Calculation of Nutrients Supplied:

1. Dry Matter = 32 × 0.91 = 29.1 pounds
2. TDN = 29.1 × 50 percent (0.50) = 14.6 pounds
3. NEm = 29.1 × 50 Mcal/cwt (0.50 per pound) = 14.6 Mcal
4. CP = 29.1 × 7 percent (0.07) = 2.04 pounds
5. Ca = 29.1 × 0.4 percent (0.004) = 0.12 pounds
6. P = 29.1 × 0.15 percent (0.0015) = 0.04 pounds

In this example, there is a significant deficiency of energy and protein and a slight deficiency of phosphorus (Table 29-4). The protein shortage could reduce the intake and digestibility of the hay resulting in a reduction in available energy. In other words, the lack of protein will reduce microbial growth and thus the energy that the microbes can extract from the hay. Consequently, a protein supplement is definitely required or the cows will lose condition, produce less milk, etc. Again, this evaluation of the diet assumes no weather stress and does not consider the other factors not included in the Appendix Tables.

Table 29-4. Diet Adequacy When Feeding 32 Pounds of Prairie Hay to a 1,200 Pound Lactating Beef Cow.

Nutrient	DM, lbs	TDN, lbs	NEm, Mcal	CP, lbs	Ca, lbs	P, lbs
Required	27.8	16.7	16.9	2.97	0.09	0.06
Supplied	29.1	14.6	14.6	2.04	0.12	0.04

Adjusting Energy Requirements for Factors Not Included in the Appendix Tables

In this section, several general guidelines are provided to assist in adjusting nutrient requirements. In most production situations, these guidelines provide good estimates of the additional energy required. If more accurate adjustments are warranted, one of the ration programs based on the *Nutrient Requirements of Beef Cattle,* 1996, should be used.

Basal Net Energy Requirements

Adjusting for the factors not included in the Appendix Tables requires calculation of the basal net energy of maintenance requirement of the animal, which is based on the formula:

NEm requirement = Weight in kilograms (kg)$^{\wedge.75}$ × 0.077

In this formula, the weight is converted to kilograms (kg), raised to the 0.75 power and multiplied by a constant of 0.077. Net energy requirements for cows of typical weights are shown in Table 29-5.

Table 29-5. Net Energy for Maintenance Requirements of Mature Beef Cows (Mcal/day).

Cow Weight, lbs	NEm Required Mcal per Day	Cow Weight, lbs	NEm Required Mcal per Day
1,000	7.57	1,250	8.95
1,050	7.86	1,300	9.22
1,100	8.13	1,350	9.48
1,150	8.41	1,400	9.75
1,200	8.68	1,450	10.02

Adjusting for Weather Stress

Guideline:

Under most cold weather conditions, NEm should be increased by 1 percent for each degree of coldness below the thermoneutral zone.

Example:

a. Heavy winter hair coat

b. Temperature = 5 degrees F

c. Wind speed = 15 mph

d. Body weight = 1,200 lbs

Using the Tables in Chapter 17:

a. Lower critical temperature = 18 degrees F (Table 17-3)

b. Effective ambient temperature = -15 degrees F (Table 17-2)

c. Degrees of coldness = (18 - (-15)) = 33 degrees

d. Adjustment of 0.7 percent per degree of coldness (Table 17-4)

e. Adjust NEm by 33 × 0.007 = 23 percent

f. Basal NEm = 8.68 × 0.23 = 2.0 Mcal additional NEm required to maintain body weight and avoid loss of condition

g. Total NEm required to maintain body condition = 8.68 + 2.0 = 10.68 Mcal

Ration Adjustment for Weather Stress

To illustrate adjusting a ration for weather stress, consider the 1,200-pound cow being fed brome hay with the nutrient requirements shown in Table 29-2. As shown, the cow was slightly deficient in energy. Her energy requirement with weather stress will be 16.9 + 2.0 = 18.9 Mcal of NEm. She will increase her consumption of hay slightly in response to the cold weather, but it is doubtful that she will be able to consume enough additional hay to avoid losing weight. Without considering the additional intake, this cow will be deficient by about 3.2 Mcal of NEm (18.9 required and 15.7 supplied). The question is, how should the additional energy be supplied? There are many options, but in this case feed her corn gluten feed (CGF) with 86 Mcal NEm/cwt to provide the additional energy. Approximately 3.7 pounds of CGF (3.2 Mcal deficiency ÷ 0.86 Mcal/lb) will provide the additional energy required.

Adjusting for Cow Body Condition Change

Guideline:

Each body condition score is approximately 7.5 percent of the cow's body weight at a condition score 5.

Example:

A beef cow would weigh 1,200 pounds at a body condition score 5.

- 7.5 percent of 1,200 = 90 pounds
- at BCS 4, she would weigh 1,200 - 90 = 1,110 pounds

Steps to Calculate Additional NEm Required to Change Body Condition Score

Step 1 – Determine the number of days available to change the body condition score.

Step 2 – Determine the Mcal of NEm required to raise the condition score (Table 25-4).

Step 3 – Calculate the Mcal of NEm above maintenance required to change body condition score on a daily basis.

Example:

a. At 60 days before calving, the cow is a condition score 4 with a goal of BCS of 5 at calving

b. Table 25-4 indicates that 226 Mcal of NEm will be required to raise her BCS to a 5

c. Mcal per day = 226 ÷ 60 = 3.77 Mcal of additional NEm per day above maintenance

Adjusting Energy Requirements for Activity

Guidelines:

- **Grazing relatively small pastures** – Increase NEm for maintenance by 10 to 20 percent
- **Grazing large, range pastures** – Increase NEm for maintenance by up to 50 percent

Example:

a. Basal NEm requirement for a 1,200-pound cow is 8.68 Mcal (Table 29-5)

b. Grazing a small pasture

c. Activity requirement = 8.68 × 0.20 = 1.7 Mcal

d. Total energy required for maintenance = 8.68 + 1.7 = 10.4 Mcal

Adjusting for Breed Differences

Guideline:

Table 17-8 shows adjustments required for breed differences.

Example:

a. Brahman-Cross cow

b. Using Table 17-8, the NEm requirement should be reduced by 5 percent

c. Basal NEm = 8.68 × 0.05 = 0.4 Mcal reduction in NEm required

d. Total NEM required to maintain body condition = 8.68 - 0.4 = 8.28 Mcal

Evaluating Protein Adequacy

In the previous section, protein adequacy was based on crude protein. As noted earlier, the *Nutrient Requirements of Beef Cattle,* 1996, used the metabolizable protein system to express protein requirements. This system recognizes that some of the protein ingested is available to the microbes and some "bypasses" digestion in the rumen.

For each feedstuff, there is a percentage of the protein that is degradable (available to the microbes) and a percentage that is undegradable (bypasses digestion in the rumen). Feedstuff compositional tables typically give a value for **Degradable Intake Protein (DIP)** and/or **Undegradable Intake Protein (UIP).** As noted in earlier chapters, doing a good job of feeding beef cows often means focusing on meeting the needs of the ruminal microbes. In other words, the crude protein requirement might be satisfied by a particular ration, but the microbes need for nitrogen might still be unsatisfied.

Evaluating the Adequacy of Degradable Intake Protein

As noted in earlier chapters, the amount of protein required for the microbes is a function of the amount of energy available in the diet. This relationship is numerically expressed as the **microbial yield.** *The Nutrient Requirements of Beef Cattle,* 1996, proposed a standard requirement for DIP of 13 percent of the available energy (expressed as TDN). For example, if 10 pounds of TDN were contained in the diet, 1.3 pounds (10 × 0.13) of degradable protein would be required. However, in low-quality diets typical of those fed to beef cows during mid-gestation, the requirement for DIP was estimated to be closer to 8.0 percent of the energy content. Table 29-6 provides estimates of the appropriate microbial efficiency value for diets low in energy.

Table 29-6. Estimates of Microbial Yield for Diets Low in Energy.

Diet TDN, %	Microbial Yield	Diet TDN, %	Microbial Yield
62	13.0	53	9.4
61	12.6	52	9.1
60	12.2	51	8.8
59	11.8	50	8.5
58	11.4	49	8.2
57	11.0	48	8.0
56	10.6	47	7.8
55	10.2	46	7.6
54	9.8	45	7.5

Based on a formula developed at the University of Nebraska.

Steps to Determine the Degradable Protein Requirement

Step 1 – Determine the energy content of the diet expressed as TDN. As an example, the bromegrass hay diet in Table 29-1 had 15.7 pounds of TDN

Step 2 – Select a microbial yield level. The bromegrass hay was 55 percent TDN. Table 29-6 suggests a microbial yield of 10.2 percent

Step 3 – 15.7 × 0.102 = 1.60 pounds of DIP required

Steps to Determine the Amount of Degradable Protein Supplied

Step 1 – Determine the degradability of the protein in the feedstuff. Appendix Table 11 shows that the protein in bromegrass hay is 67 percent degradable (UIP percent = 33)

Step 2 – Calculate the amount of total protein supplied. Bromegrass hay averages 10 percent CP, and 28.5 pounds of dry matter is being supplied. (28.5 × 0.10 = 2.85 pounds of CP)

Step 3 – Degradable intake protein supplied = 2.85 × 0.67 = 1.91 pounds

Step 4 – The DIP supplied exceeds the requirement (1.91 > 1.60)

Factors to Consider in Evaluating DIP Adequacy

- As a general rule, most forage-based diets will be adequate in DIP because the degradability of protein in forages tends to be high.
- Conversely, the protein in some by-products such as distillers grains, tends to be low in degradability, which can result in a deficiency of DIP even though crude protein may be adequate.
- Since the amount of DIP required per unit of energy (microbial yield) is not a constant, selecting the correct value for microbial yield is key to determining the accuracy of the calculated DIP

requirement. Given the importance of meeting the needs of the microbes for nitrogen, it is wise to be conservative and use a value close to the 13 percent standard whenever there is a question regarding the correct value. Table 29-6 should provide a good estimate of the microbial yield for diets varying in energy.

- The degradability of the crude protein in different sources of the same feedstuff may be variable. For example, heating reduces the degradability while processing — grinding grains for example — increases the degradability. The values presented in feed composition tables are good estimates of the degradability of the protein, but these estimates may need to be adjusted if heating or processing has occurred.

Importance of Meeting the DIP Requirement

Given all of the assumptions that must be made regarding the microbial yield and degradability of protein to determine the adequacy of DIP, one might decide to use only crude protein to assess protein adequacy and not worry about the DIP requirement. As noted previously, however, unless the microbes are provided with adequate nitrogen, they cannot extract the maximum energy from the diet.

References

Burris, R. and J. Johns. Feeding the Cow Herd. University of Kentucky.

Corah, L.R. Cow Herd Nutrition. Kansas Formula Feeds Conference.

Kunkle, B., J. Fletcher, and D. Mayo. 2002. Florida Cow-Calf Management, 2nd Edition – Feeding the Cow Herd. Florida Cooperative Extension Service Publication AN117.

Nutrient Requirements of Beef Cattle. 1984. National Research Council.

Nutrient Requirements of Beef Cattle. 1996. National Research Council.

Rasby, R. and I.G. Rush. Feeding the Beef Cow Herd – Part I Factors Affecting the Cow Nutrition Program. Beef Cattle Handbook BCH-5402.

Chapter 30

Evaluating Diet Adequacy in Grazing Situations

In typical cow/calf production systems, beef cattle obtain most of their nutrients from grazed forages. In these situations, there are two problems that limit the ability to adequately evaluate their nutritional status:

1. **We Don't Know How Much They Are Eating** – While intake can be predicted to some extent, in most production situations, intake estimates are guesses at best.
2. **We Don't Know the Nutrient Content of the Diet** – In typical grazing situations, only an educated guess of the composition of the diet can be made because cows selectively graze the available forage.

The inability to answer these two questions with a high level of accuracy makes the use of body condition scoring essential. Producers can only make educated estimates about the amount and composition of the diet, provide supplemental nutrients when they appear to be warranted, and use body condition changes over time to actually monitor the adequacy of the diet.

Use of Clipped Forage Samples

It seems logical that producers could take samples of the forage available in a pasture, have the samples analyzed, and use these data to balance the ration. The problem with this approach is that cattle selectively graze, which means that the diet they actually consume will be different than the clipped sample. Table 30-1 shows the results of a Kansas State University study comparing clipped samples with those collected by esophageal cannulae — what the animals actually ate. These data are from steers grazing tallgrass prairie,

Table 30-1. Selective Feeding Pattern of Steers on Bluestem Pasture.

	Protein		Digestible DM	
Month	Clipped	Esophageal	Clipped	Esophageal
January	3	—	—	—
February	3	4	36-37	41-49
March	3	5	35-44	39
May	14	16	60-61	54-55
June	9	11	50	49-50
July	7	9	40-41	39-40
August	5	7	33-34	39-40
September	4	6	35-36	48-56
October	4	6	36-49	44-49
November	3	5	33	44-51
December	3	5	38	43-44

Kansas State University

which is a mix of numerous grasses and forbs. This diverse mix of species tends to promote a difference between clipped and grazed samples.

The digestible dry matter is a fairly good indicator of the energy content of the forage. Several important points can be made from these data:

1. The differences in protein between clipped and esophageal samples were fairly consistent throughout the year, about 2 percent. This demonstrates that cattle selectively graze for a higher quality diet. This value (2 percent for protein) is probably a good rule-of-thumb to use in most range situations provided that adequate forage is available to allow selectivity. Remember: Clipped samples represent what the cattle did not eat, at least not yet.
2. The differences in digestible dry matter (energy), between clipped and esophageal samples varied depending on the time of year.
   a. During the period of highest quality (May through June) there was little difference.
   b. During the fall, when there would have been considerable forage available, the steers selected a diet much higher in energy than the clipped samples.
   c. During the late winter (February through March), when forage availability was lower, the ability of the steers to select a higher-quality diet was reduced.

Recent research in Florida comparing hand-clipped samples to those collected from rumen fistulated animals grazing Bahiagrass pastures showed a pattern similar to the work in Kansas (Table 30-1). The cattle selected a diet consistently higher in both energy and protein. In this research, the masticated samples averaged 1.75 percent higher in CP than the hand-clipped samples and the difference was fairly consistent through the year. Correspondingly, in vitro digestible organic matter (IVDOM) percentage averaged 11.3 percent higher in the masticated samples. The difference in energy (as indicated by IVDOM) between the hand-clipped and masticated samples was also consistent through the year.

Example of Using a Clipped Forage Sample to Evaluate Diet Adequacy

Assumptions:

1. 1,400 lb cow grazing native range in February.
2. Based on Table 27-2, intake will be 1.5 percent of body weight without supplementation.
3. Analysis of the clipped sample as shown in Table 30-2.
4. Adequate forage available for selective grazing.
5. Protein in the diet will be 2 percent higher than in the clipped sample.
6. Based on Table 30-1, the energy in the diet will be 20 percent higher than in the clipped sample.
7. Calculation of nutrient intake:
   a. 1,400 lbs × 0.015 = 21 lbs of DM intake
   b. Protein intake = 21 × 0.07 (5% + 2%) = 1.47 lbs
   c. Energy intake TDN basis = 21 × (0.46 + (0.46 × 0.20)) = 11.6 lbs TDN
   d. Energy intake net energy basis = 21 × (0.37 + (0.37 × 0.20)) = 9.3 Mcal of NEm
   e. Calcium intake = 21 × 0.0034 = 0.07 pounds
   f. Phosphorus intake = 21 × 0.0012 = 0.025 pounds

Table 30-2. Analysis of Clipped Forage Sample from Native Range.

DM %	TDN %	NEm, Mcal/cwt	Protein, %	Ca %	P %
80	46	37	5	0.34	0.12

Comparison of Nutrient Intake to Requirements (Appendix Table 1)

Table 30-3 shows a comparison of the requirements of this cow to the calculated nutrient intake. As shown, this cow will be deficient in energy, protein, and phosphorus.

Table 30-3. A Comparison of the Calculated Nutrient Intake of a 1,400-pound, Gestating Cow to the Requirements.

	DM lbs	TDN lbs	NEm Mcal	CP lbs	Ca lbs	P lbs
Req.	27.3	13.4	12.0	1.89	0.07	0.04
Diet	21.0	11.6	9.3	1.47	0.07	0.025

Consider feeding one pound of a 20 percent protein range cube per day. In practice, the best way to feed this cube would be 3.5 pounds twice weekly. Based on Table 27-1, the cow's intake should increase to 1.8 percent of her body weight resulting in an intake of 1,400 × 0.018 = 25.2 pound of dry matter per day.

Calculation of the Nutrient Intake with Supplementation:

1. Dry matter intake = 25.2 lbs
2. Protein intake = 25.2 × 0.07 = 1.75 lbs + 0.20 (Supplement) = 1.95 lbs
3. Energy intake TDN basis = 25.2 × (0.46 + (0.46 × 0.20)) = 13.9 lbs
4. Energy intake net energy basis = 25.2 × (0.37 + (0.37 × 0.20)) = 11.2 Mcal
5. Calcium intake = 25.2 × 0.34 = 0.08 lbs
6. Phosphorus intake = 25.2 × 0.0012 = 0.03 lbs

A comparison of the nutrient intake with supplementation to the requirements indicates that the only serious deficiency is probably phosphorus. Also, note that the nutrients provided by the range cubes, other than protein, have not been included in the calculation. The deficiency of phosphorus could be satisfied by providing a free-choice mineral or including phosphorus in the range cube.

Factors Influencing the Difference Between Clipped and Esophageal Samples

1. Plant composition of the pasture – Generally, there is less selective grazing in a pasture consisting of only one species than in a range situation with numerous species of grasses and forbs.
2. The amount of selective grazing is a function of forage availability. In situations where the amount of forage is limited, cattle simply cannot select a much higher quality diet.
3. Intensive grazing systems greatly reduce selectivity meaning that clipped samples will be close to the diet consumed with respect to nutrient content.

Factors to Consider in Assessing the Diet in Grazing Situations

1. As noted above, cattle will selectively graze the forage when adequate forage is available. This applies not only to range and pasture situations but to crop residues as well. For example, when grazing corn stover, cattle eat the leaves, husks, and residual grain before consuming the lower-quality residue.
2. Clipped forage samples give some indication of nutrient content, but positive allowances should be made for selective grazing.
3. The nutrient content of dormant forages, including grasses as well as crop residues, declines over time as a result of weathering.
4. Research data from experiment stations and universities on cattle diets in typical grazing situations may be useful in making educated estimates regarding the actual diet in production situations.
5. Nutrient composition of forages varies from year to year depending on weather and other factors. For example, in dry years, the energy and protein content of forages typically is above average; and conversely, in wetter than average years, the nutrient content typically is below average. Thus, if forage availability is not limited in dry years, cattle performance is usually above average.
6. As noted in the beginning of this chapter, because estimates of the nutrient content of the diet and dry matter intake are frequently inaccurate, monitoring body condition is essential as an indicator of the adequacy of the diet.

Analysis of Fecal Samples as an Aid in Formulating the Supplementation Program

There has been considerable research, especially in Texas, examining the usefulness of fecal samples to determine diet composition. There is a relationship between the composition of the feces and the diet. For example, higher levels of protein in the feces indicate a higher level of protein in the diet. From a practical standpoint, however, analyzing clipped forage samples and adjusting for selective grazing is probably a better approach. Further research may improve the usefulness of fecal sample analysis.

References

Corah, L. and E. Smith. 1978. Feed Supplements for Maximum Use of Native Range. Kansas State University Cooperative Extension Bulletin L517.

Hughes, A.L., M.J. Hersom, J.M.B. Verdramini, T.A. Thrift, and J.V. Yelich. 2010. Comparison of Forage Sampling Methods to Determine Nutritive Value of Bahiagrass Pastures. The Professional Animal Scientist. 26:504-510.

KSU Range Field Day. 1995. 50 Years of Range Research Revisited. Cooperative Extension Service, Kansas State University.

Nutrient Requirements of Beef Cattle. 1984. National Research Council.

Nutrient Requirements of Beef Cattle. 1996. National Research Council.

Chapter 31

Economics of Supplementation

Sometimes the nutrient content of the basal forage will not meet the nutritional requirements of the cattle. During these times, supplemental feeding is often required to avoid reduced reproduction or production. The goal at these times should be to provide the optimal amount of supplement from an economic and nutritional perspective. Accomplishing this goal requires consideration of economic concepts that apply to situations where inputs are added to any production situation.

The Law of Diminishing Returns

This basic law of economics states that past a certain point, as the amount of an input is increased, the incremental response tends to decrease. An example of how this law applies to supplementation is shown in Table 31-1.

The results in Table 31-1 illustrate the law of diminishing returns extremely well. The first input of an additional 50 percent (from 25 percent to 75 percent) of supplement resulted in an increase of 0.3 units of body condition score. Additional inputs of 50 percent of the standard supplement level resulted in 0.2 and 0.1 additional units of body condition score, respectively. Thus, the incremental response declined with additional input. This is a typical response to supplementation of beef cattle. From a practical standpoint, it means that one must recognize that there is an optimal level of supplementation that provides maximum profitability. Some researchers believe that providing 80 to 90 percent of the protein deficiency is probably close to the optimal level with respect to profitability. This percentage will change with the cost of the protein supplement and the expected response in production.

Table 31-1. Results of a Supplementation Study of Beef Cows Grazing Old World Bluestem Pasture.

Supplement Rate, % of Standard	Body Condition Score Change Prepartum	Incremental Body Condition Score Change
25	-1.2	---
75	-0.9	0.3
125	-0.7	0.2
175	-0.6	0.1

Oklahoma State University

Value of Additional Weaning Weight and Weight Gain by Stockers

Another economic concept that must be considered in a supplementation program is the value of additional pounds of weaning weight in a cow/calf operation and gain in a stocker program. It is common for producers to think that if calves are selling for $1.75 per pound, additional weaning weight is worth $1.75 per pound. In fact, additional weaning weight is usually worth much less. Consider an example as shown in Table 31-2.

In this example, each additional pound of weaning weight is worth $0.85, not $1.75 or $1.60. If supplemental feed can increase calf gain for less than $0.85 per pound, it will be profitable to supplement. This concept applies to supplementation for the cow, creep feed, and other feed inputs. Value of gain by stockers is analogous to additional weaning weight since the margin is usually negative (heavier feeders are worth less per pound).

Table 31-2. Calculation of the Value of Additional Weaning Weight.

Weaning Weight, lbs	Selling Price Per lb, $	Total Value, $	Additional Value, $	Value of Additional Weaning Weight, $/lb
500	1.75	875	---	---
600	1.60	960	85	0.85

Why Is the Value of Additional Weaning Weight Typically Less Than the Selling Price?

Stocker operators will typically pay more per pound for lighter calves at weaning because more pounds can be put on lighter calves and they are more efficient than heavier calves (because of lower maintenance requirements). As a general rule, if gain in the stocker or finishing phase of production can be put on cheaper than the purchase price, lighter calves will bring more per pound. High costs of gain tend to narrow the price differences between different weights. The current rapid expansion in ethanol production may increase the price of corn enough to change the historical relationship where selling price per pound declines as weaning weight increases. If this occurs, it will make many supplementation programs, such as creep feeding, more profitable.

Increased Gain in One Phase of Production Tends to Reduce Gain in the Next Phase

Another concept that needs to be considered in evaluating the economic impact of supplementation is the trend for increased weight gain in one phase of production to reduce gain in subsequent phases of production. For example, consider creep feeding. Calves provided unlimited creep feed for several months before weaning would be expected to gain slower in the growing phase compared to calves that did not receive creep feed. Thus, creep feeding may be profitable if the calves are to be sold at weaning but less profitable if they are to be retained through a growing program. Producers must consider the impact of supplementation on the total production system to make sound economic decisions.

Note that the reverse of the point made in the previous paragraph is also generally true, especially with cows. For example, cows in poor condition at calving, when given the opportunity to consume good-quality forage, will consume more and make greater gains than cows that calve in good condition. Very thin stocker cattle also will make compensatory gains.

Biological vs. Economic Response

In many cases supplementation will improve biological measures of production without improving profitability. For example, the relationship between body condition score and reproduction has been noted in previous chapters. However, as also noted, increasing body condition may not result in an increase in reproductive performance. Thus, supplementation may have a positive impact on the biological trait (body condition score) but not on the economically important traits (reproduction and weaning weight).

To illustrate this point, consider the results of a 5-year study of cows grazing corn stalks in Nebraska shown in Table 31-3. The supplemented cows in this study received 2.2 pounds of supplement daily starting 20 days after being placed on corn stalks. They received this supplement until the start of calving on March 1 when the supplemented and control cows were fed together. In this summary, the cows receiving the supplement gained more weight (although the difference was not statistically significant) and body condition as a result of supplementation. However, neither reproductive performance nor calf weaning weight was improved by supplementation. In other words, the biological traits showed a positive response, but the economically important traits were not impacted. It should be noted that there was a slight tendency for the supplemented cows to have a higher pregnancy rate, but the difference was not statistically significant. Additionally, the stocking rate of the cows on the corn

Table 31-3. Effects of Late Gestation Supplementation of Cows Grazing Corn Stalks on Cow and Calf Performance.

Item	Treatment		SEM	P Value
	Supplement	Control		
Oct BW, lbs	1,263	1,265	23.5	0.79
Feb BW, lbs	1,351	1,327	16.5	0.19
May BW, lbs	1,247	1,243	9.7	0.75
Change in BW, Oct-Feb, lbs	89	62	15.0	0.20
Change in BW, Feb-May	-112	-81	12.3	0.14
BCS, Oct	5.4	5.4	0.09	0.89
BCS, Feb	5.6[a]	5.4[b]	0.08	0.02
BCS, May	5.4	5.3	0.07	0.32
Pregnancy rate, %	94	91	0.02	0.18
Calving interval, days	367	366	1.6	0.80
Calf weaning weight, lbs	552	548	11.4	0.35

[ab] *Values with differing superscripts statistically different ($P<.05$). Nebraska Beef Report 2012.*

stalks could have a major impact on the response to supplementation. In the Nebraska study, the level of stocking was conservative.

Unfortunately, too much research has evaluated the impact of supplementation on body weight changes or body condition scores without also measuring the impact on reproduction. Neither body condition score nor cow weight change is of much economic value unless reproductive performance or weaning weights (the economic traits) are improved. The best measure of any supplementation program should be the impact on economically important traits.

References

Paisley, S.I., G.W. Horn, F.T. McCollum III, G.E. Selk, K.S. Swenson, and B.R. Karges. 1996. Supplemental Protein Levels for Spring Calving Cows Grazing Old World Bluestem or Tallgrass Prairie. Oklahoma State University Animal Science Research Report.

Simms, D.D. 2009. Feeding the Beef Cowherd for Maximum Profit. SMS Publishing Inc., Amarillo, TX.

Warner, J.M., J.L. Martin, Z.C. Hall, L.M. Kovarik, K.J. Hanford, R.J. Rasby, and M. Dragastin. 2012. Supplementing Gestating Beef Cows Grazing Cornstalk Residue. Nebraska Beef Report Pg 5.

Chapter 32

Supplementation of Energy and/or Protein

Supplementation of Beef Cows

In production systems across the United States, there are times when the nutrient content of the basal forage or forage availability are inadequate to meet beef cow nutrient requirements. The result is a loss of weight and/or body condition. During these times, it often is profitable to provide some form of supplementation. A profitable supplementation program should have the following goals:

1. Providing Enough Additional Nutrients to Prevent a Significant Reduction in Production or Subsequent Reproduction

An initial thought might be that supplementation should provide nutrients to compensate for all deficiencies in energy, protein, and other nutrients. The most profitable supplementation program, however, may only bring the nutrient levels up to a point where subsequent reproductive performance is not reduced significantly. For example, in many production systems, cows will lose considerable weight during the winter, even with some supplementation. The goal in these production systems might be to prevent the cows from calving in a condition score below a 4.5. These systems take advantage of the fact that the average cow might go into winter as a condition score 6. These cows can lose 100 pounds over the winter and still have a body condition score of 4.5 or better at calving. In these systems, if no supplementation is provided, cows might drop to a body condition score 3 at calving in the spring, resulting in markedly reduced reproductive performance. The most profitable supplementation level is one that allows cows to calve in a body condition that prevents significant reduction in reproduction rate.

2. Increasing the Use of the Basal Forage

A good supplementation program should also increase the use of the basal forage. It should increase the digestibility and/or the intake of the forage. A good example is feeding a protein supplement to cows grazing crop residues, such as corn stover. Feeding a protein supplement typically increases both the intake and digestibility of the stover. The net effect is that the additional energy, protein, etc., that the cow receives is much greater than the level of the nutrients in the supplement alone. Conversely, feeding an energy supplement such as a cereal grain to these same cows would not increase intake and might actually lower the digestibility of the stover. The net effect is that the additional energy and protein available to the cow might be lower than the amount in the supplement.

3. Improving Grazing Distribution

By moving the location where supplements are fed, the grazing distribution in a pasture can be improved. This may result in use of grass that would not be used without supplementation. This may increase the efficiency of the total production system.

4. Reducing Grazing Pressure on Pastures

When the amount of the basal forage is limited, supplements might be fed at fairly high levels to reduce the grazing pressure on the pasture.

For example, feeding a significant amount of hay or a by-product like wheat middlings reduces standing forage intake. This might be a good drought strategy to avoid overgrazing a pasture and reducing future forage production. Feeding strategies for drought conditions will be discussed in Chapter 42.

What Is the Most Common Nutrient Deficiency of Beef Cows?

The most common nutritional deficiency of beef cows is energy — they simply do not have an energy intake sufficient to meet maintenance and production requirements. Consequently, they are forced to use body energy stores, resulting in a loss of condition. As clearly shown in previous chapters, this can result in poor reproductive performance. Given the close relationship between energy and protein, it is often difficult to separate what is cause and effect. Is the lack of energy the result of a deficiency of protein or truly a shortage of energy alone?

How Does a Protein Deficiency Result in an Energy Deficiency?

From a practical standpoint, each feedstuff has a certain amount of energy in its components such as starches, sugars, and fiber. In roughages, much of this energy is in the form of fiber, which can only be digested by the microbes in the rumen. If these microorganisms cannot grow because of a lack of nitrogen, the amount of energy extracted is reduced. The feed composition tables in the back of this book show a NEm of 52 Mcal per hundredweight for mature, fescue hay. But this level of energy assumes that the microbes have adequate nitrogen to digest the fescue hay at the normal level. If protein is deficient, two things happen:

1. Intake is reduced because digestion of the roughage and passage rate through the rumen is reduced; and
2. Total digestibility is decreased, reducing the energy available to the animal. Providing supplemental protein when it is deficient increases intake as well as digestibility, both of which increase total energy available to the animal. This means that in many production situations the most efficient way to supply additional energy is to provide adequate protein for the ruminal microbes so they can extract the available energy from the roughage.

Effect of Supplementation on Forage Intake

Table 27-2 showed a summary of the effect of different types of supplementation on forage intake based on numerous research trials at Oklahoma State University. Protein supplementation increased intake significantly with low-quality forages while energy (grain) supplementation did not influence intake. Even with medium-quality roughages, protein supplementation increased intake. The data in Table 27-2 represent the effect of relatively low levels of supplementation. If high levels of supplementation are fed, it will reduce forage intake.

Protein Supplementation

Research trials show the greatest intake response to protein supplementation occurs when the basal forage is below 7 percent crude protein because both intake and digestibility are substantially increased. When the basal forage is above 7 percent crude protein, protein supplements still increase intake and digestibility but at a much lower level. Thus, the economic response to protein supplementation is greatest when the basal diet is below 7 percent crude protein.

This does not suggest that one should only supplement when the basal forage is below 7 percent crude protein. It only means that the greatest response is below 7 percent. For example, a lactating beef cow needs approximately 11 percent crude protein. If she is being fed an 8 percent crude protein fescue hay, supplemental protein will only slightly increase intake and digestibility. However, the diet should still be balanced to provide approximately 11 percent crude protein to meet the cow's protein requirement to avoid decreased performance.

What Is the Best Type of Protein?

As noted previously, crude protein is based on the amount of nitrogen in a feedstuff. There are considerable differences in the availability and usefulness of different nitrogen sources. Key considerations in evaluating protein sources include:

1. Natural Protein vs. Non-protein Nitrogen

Urea and other nonprotein nitrogen sources are often used to reduce the cost of protein supplements; however, natural protein sources result in better performance, especially when the basal forage is low in quality. As a general rule-of-thumb, a small percentage (20 to 25 percent) of the supplemental crude protein can be in the form of nonprotein nitrogen without reducing performance relative to all natural protein, but higher levels result in lower cow performance, especially with low quality roughages. However, since urea is a relatively low cost protein source, reducing the cost of a supplement by including urea up to the point where it supplies 20 to 25 percent of the total supplemental protein will usually be cost-effective. This is just a rule-of-thumb, which can be adjusted based on price relationships.

2. Degradability in the Rumen

Since a primary goal in supplementation is to meet the nitrogen needs of the ruminal microbes, protein sources for beef cows should be high in degradability.

Importance of Other Nutrients in the Protein Supplement

Research trials show that supplements containing high levels of starch can significantly reduce fiber digestibility of poorer quality roughages because of the competition between fiber digesting and starch digesting microbes. This reduction in digestibility caused by the starch may offset the increase in digestibility normally elicited by the protein supplement. For example, feeding 1 pound of a 40 percent crude protein supplement would provide almost the same amount of protein as 4 pounds of grain, but the 40 percent crude protein supplement would increase forage intake while the 4 pounds of grain would probably have little influence on intake. With low-quality forages, the 40 percent crude protein supplement would give superior cow performance because it increases both forage intake and digestibility.

Research shows that if adequate degradable intake protein is fed, the negative influence of starch on digestibility of forages is greatly reduced. This means that grain can be used successfully as a supplement for roughages if one ensures that the degradable intake protein requirement is met.

Supplementing Both Protein and Energy

In some production situations, providing adequate protein will not solve an energy deficiency. For example, good-quality alfalfa hay can be used to provide supplemental protein for cows consuming low-quality grass hay or grazing crop residues. While the protein requirement might be satisfied, the energy density of the diet could still be inadequate to prevent significant loss of body condition.

Typically producers have used grain to solve this energy deficiency. However, this may not be a good solution given that grain alone decreases forage digestibility if adequate degradable intake protein is not supplied. In the situation described, the alfalfa hay should supply adequate degradable intake protein to make the grain an effective energy supplement.

A good rule of thumb is that 2 to 3 pounds of grain per day can be fed without reducing forage digestibility. Higher levels may result in the additional energy from the grain being offset by less energy availability from the forage. The net result would be a lower performance response than one would anticipate by simply adding the energy values of the forage and grain. If the energy deficiency is severe enough, feeding fairly high levels (8 to 10 pounds) of grain may be warranted, recognizing that the energy from the basal forage will be reduced, but overall the animal's energy requirement will be met. This might be the best solution from an economic standpoint, especially if grain is relatively cheap. Feeding 4 to 5 pounds of grain per day is probably not an effective level unless the degradable intake protein requirement is met. This level of grain alone results in a significant reduction in the energy from the forage, and the grain will not supply enough energy to offset this loss.

Feeding By-Products as Sources of Both Protein and Energy

Several by-products make excellent sources of both protein and energy for beef cows. For example, wheat middlings and corn gluten feed have high energy content from highly digestible fiber, yet their starch levels are fairly low. They also have medium levels of protein. Thus, they make excellent sources of both protein and energy in cow-calf and stocker supplementation programs.

Substitution of Supplement for Forage Intake

One of the goals of effective supplementation is to enhance the intake of the basal forage rather than substitute intake of the supplement for the basal forage. Supplements with a high concentration of protein, fed at a low level of intake, tend to enhance basal forage intake while supplements with less protein often must be fed at levels that result in substitution. For example, when using soybean hulls (12 percent crude protein) as a protein source, the feeding rate required to meet the cow's protein requirement often results in substitution of supplement for forage intake. As a guideline, when low-protein supplements are fed at more than 0.5 percent of body weight (5 pounds for a 1,000-pound cow) substitution occurs at the rate of a 0.5-pound decrease in forage intake for each pound of supplement consumed. The net effect of this substitution may be less total energy and protein intake than desired. As an additional point, when the availability of the basal forage is limited because of drought, high stocking rate, or other reasons, substitution can actually be advantageous.

Frequency of Supplementation

In systems where supplements are hand fed, such as range cubes, the influence of frequency of feeding on cow performance is an important issue. Research has shown that, depending on the type of supplement, a week's allocation of supplement can be fed in equal portions on 2 days with little reduction in performance compared to daily feeding. For example, if the desired supplement intake is 1 pound daily, 3.5 pounds could be fed twice weekly, and performance would probably be similar to daily feeding. In fact, feeding the entire 7 pounds once per week would probably not result in a significant loss in performance compared to 1 pound daily. Feeding less frequently can reduce the cost of fuel, equipment, and labor to provide supplement.

When Should Supplements Be Fed on a Daily or Every Other Day Basis?

1. Supplements Containing High Levels of Starch

Feedstuffs containing high levels of starch, for example, grains, should be fed on a daily basis or at least on an every other day basis. Feeding feedstuffs high in starch frequently rather than in large doses will reduce the effect on fiber digestibility and the possibility of acidosis.

2. Supplements High In Urea

As noted previously, cows can use some urea effectively. However, feeding high levels can result in urea toxicity and possibly death. If supplements contain a significant portion of their protein from urea, the supplements should be fed frequently to avoid potential toxicity. Furthermore, cow performance should be improved by feeding smaller amounts on a daily basis.

References

Bodine, T.N. H.T. Purvis II, M.T. Van Koevering, and E.E. Thomas. 1999. Effects of Supplement Source on Intake, Digestion and Ruminal Kinetics of Steers Fed Prairie Hay. Oklahoma State University 1999 Animal Sciences Research Report.

Bodine, T.N., H.T. Purvis, and D.A. Cox. 2001. Grazing Behavior, Intake, Digestion, and Performance by Steers Grazing Dormant Native Range and Offered Supplemental Energy and(or) Protein. Oklahoma State University 2001 Animal Sciences Research Report.

Bohnert, D.W., C.S. Schauer, and T. DelCurto. 2002. Influence of Rumen Protein Degradability and Supplementation Frequency on Performance and Nitrogen Use in Ruminants Consuming Low-Quality Forage: Cow Performance and Efficiency of Nitrogen Use in Wethers. JAS 80:1629.

Bohnert, D.W., T. DelCurto, A.A. Clark, M.L. Merrill, S.J. Kalck, and D.L. Harmon. 2011. Protein Supplementation of Ruminants Consuming Low Quality Low-quality Cool- or Warm Season Forage: Differences in Intake and Digestibility. JAS 89:3707-3717.

Chase, C.C., Jr. and C.A. Hibberd. 1987. Utilization of Low-Quality Native Grass Hay by Beef Cows Fed Increasing Quantities of Corn Grain. JAS 65:557.

Cochran, R.C. Developing Optimal Supplementation Programs for Cow-Calf Operations. Department of Animal Sciences and Industry, Kansas State University.

DelCurto, T., R.C. Cochran, D.L. Harmon, A.A Beharka, K.A. Jacques, G. Towne, and E.S. Vanzant. 1990. Supplementation of Dormant Tallgrass-Prairie Forage: I. Influence of Varying Supplemental Protein and (or) Energy Levels on Forage Utilization Characteristics of Beef Steers in Confinement. JAS 68:515.

DelCurto, T., R.C. Cochran, L.R. Corah, A.A Beharka, E.S. Vanzant, and D.E. Johnson. 1990. Supplementation of Dormant Tallgrass-Prairie Forage: II. Performance and Forage Utilization Characteristics in Grazing Beef Cattle Receiving Supplements of Different Protein Concentrations. JAS 68:532.

Dhuyvetter, D.V., M.K. Petersen, R.P. Ansotegui, R.A. Bellows, B. Nisley, R. Brownson, and M.W. Tess. 1992. Reproductive Efficiency of Range beef Cows Fed Differing Sources of Protein During Gestation and Differing Quantities of Ruminally Undegradable Protein Prior to Breeding. Proceedings, Western Section American Society of Animal Science.

Huston, J.E., H. Lippke, T.D.A. Forbes, J.W. Holloway, and R.V. Machen. 1999. Effects of Supplemental Feeding Interval on Adult Cows in Western Texas. JAS 77:3057.

Johnson, C.R., D.L. Lalman, J.D. Steele, and A.D. O'Neil. 2002. Energy and Degradable Intake Protein Supplementation for Spring-Calving Beef Cows Grazing Stockpiled Bermudagrass Pasture in Winter. The Professional Animal Scientist 18:52.

Lamm, W.D., J.K. Ward, and G.C. White. 1977. Effects of Nitrogen Supplementation on Performance of Gestating Beef Cows and Heifers Grazing Corn Crop Residues. JAS 45:1231.

Mathis, C.P. 2003. Protein and Energy Supplementation to Beef Cows Grazing New Mexico Rangelands. New Mexico State University Cooperative Extension Circular 564.

Patterson, H.H., D.C. Adams, T.J. Klopfenstein, and G.P. Lardy. 2006. Application of the 1996 NRC to Protein and Energy Nutrition of Range Cattle. The Professional Animal Scientist 22:307.

Chapter 33

Supplementation of Minerals

As with energy and protein, there are many times during the year when basal forages will not provide beef cattle the required minerals. Many producers provide a high quality mineral supplement throughout the year as an insurance policy. This approach typically results in spending much more on minerals than is necessary. Conversely, some producers provide no mineral supplementation, which in many situations results in reduced performance. Mineral deficiencies can directly cause reduced performance, and they can cause poor use of other nutrients in the diet. The goal should be to meet the minimum requirements for all minerals while minimizing the cost of mineral supplementation. Accomplishing this goal requires consideration of both animal requirements and feedstuff mineral content.

Considerations in Developing a Good Mineral Supplementation Program

1. **Mineral Requirements of the Cattle** – Calcium and phosphorus requirements are shown in Appendix Tables 1-9. Other mineral requirements are listed in Appendix Table 10. Mineral requirements change during the year as the cows go through different stages of production.
2. **Typical Mineral Levels in the Feedstuffs** – The level of minerals in forages is typically related to the content of minerals in the soil. However, there are also seasonal trends during the year with higher mineral levels when the plants are rapidly growing and lower levels when they are dormant. Hay made from immature forages will have fairly high mineral levels compared to the same forage harvested at maturity.
3. **Formulation of a Mineral that Will Meet Minimum Requirements at Typical Intake Levels** – Based on the requirements of the animals and the mineral content of the basal diet, formulating a free choice mineral supplement requires an assumption of some average intake level of the mineral. Commercial minerals are typically formulated for a consumption level of 2 or 4 ounces per head per day. There is nothing magical about these levels, but some estimate of intake is required to properly formulate a free-choice mineral.
4. **Monitoring Mineral Intake and Adjusting Intake When Required** – Cattle tend to go on mineral eating binges that often result in overconsumption. If the free-choice mineral has been formulated to meet the animal's needs at a specific intake level, intake should be controlled when the cattle consume more than the target intake. The best way to accomplish this is to add plain salt to the mineral mix. If the intake is below the desired level, reducing the salt level or adding a flavoring agent such as dry molasses should increase intake and prevent caking.

Do Cattle Have the Ability to Balance Their Mineral Needs?

Many producers believe beef cattle have the ability or "nutritional wisdom" to balance their mineral needs and consume a mineral supplement only when needed. Research has consistently shown that this is not true. Beef cattle, like people, eat foods that taste good rather than eating to meet any nutritional requirements. If cattle are severely deficient in certain minerals, they will eat bones or other objects, but they do not have the ability to select materials that contain the specific deficient mineral. Their consumption of minerals is mainly driven by:

1. **Craving for Salt** – Cows crave salt, which is usually the primary factor governing their consumption of minerals.
2. **Flavoring Agents** – Extremely palatable ingredients are often added to mineral supplements to stimulate intake. For example, soybean meal, molasses, or ground grain added to a mineral makes it more palatable to cattle. Feed companies add these and other flavoring agents to minerals to encourage consumption.

Since mineral intake by cattle is not a function of mineral deficiencies and they often overconsume palatable supplements, mineral intake must be managed to avoid spending too much on mineral supplementation programs.

Determining Mineral Adequacy in a Diet

The first issue to address is whether mineral deficiencies exist. The next issue is the severity of the deficiency. Beef cattle have the ability to store most minerals in various organs, which means that marginal deficiencies for short periods may not justify supplementation. Factors to consider in calculating mineral intake include:

1. **Mineral Content of the Forage** – Mineral content can be based on book values or on forage analyses. Since the mineral content of forages is closely related to soil content, forage analyses may be required to establish typical levels. Book values give a general indication of typical mineral content, but there is tremendous variation among soil types and locations. Consequently, having primary feedstuffs analyzed for mineral content is recommended.
2. **Intake of the Forage** – In production situations, forage intake is difficult to determine accurately. The guidelines presented in Table 27-2 provide useful estimates.
3. **Forage Mineral Digestibility** – As with most nutrients, the digestibility or availability of minerals declines as plants mature because a higher percentage of the minerals are bound with the indigestible, fibrous components of the plant. This means that when feed analyses show marginal levels of minerals in low quality forages, the actual levels available to the animals might be inadequate.

Steps in Determining Mineral Intake from the Basal Diet

Step 1 – Estimate the animal's forage intake on a dry matter basis.

Step 2 – Estimate the mineral content using either book values or feed analyses.

Step 3 – Calculate the mineral intake on a daily basis.

Example 1 – Grazing Situation:

1. Cows in late gestation grazing dormant, tall grass pasture in the High Plains during the winter.
2. Table 27-2 shows an estimated dry matter intake of 1.8 percent of body weight on a dry matter basis with protein supplementation.
3. Cows weigh an average of 1,200 pounds.
4. Intake = 1,200 × 0.018 = 21.6 pounds of dry matter.
5. Forage analyses are shown in Table 33-1.

Table 33-1. Forage Analyses.

Nutrient	Concentration	Units
Calcium	0.43	Percent
Phosphorus	0.11	Percent
Potassium	0.8	Percent
Copper	6.0	ppm
Zinc	23.0	ppm

Calculation of Mineral Intake and Comparison to Requirements

1. Calculating mineral intake:

   a. Calcium – 21.6 pounds of dry matter × 0.0043 = 0.093 pounds

   b. Phosphorus – 21.6 × 0.0011 = 0.024 pounds

   c. Potassium – Requirement stated in percent of diet

   d. Copper – Requirement stated in ppm

   e. Zinc – Requirement stated in ppm

2. Based on Appendix Tables 1 and 10, this cow requires the mineral levels shown in Table 33-2.

3. This cow requires phosphorus, copper, and zinc supplementation. Note: These calculations do not include the mineral content of any protein supplement being fed, which in many cases may meet all of the phosphorus needs and even the trace mineral needs. In fact, the best approach would probably be to include the supplemental mineral in the protein supplement.

Table 33-2. Mineral Requirements of the Example Cow.

Nutrient	Amount/ Concentration	Units
Calcium	0.062	Pounds
Phosphorus	0.040	Pounds
Potassium	0.6	Percent
Copper	10.0	ppm
Zinc	30.0	ppm

Example 2 – Hay Diet:

1. The same cows as in Example 1 fed 33 pounds of meadow hay in late gestation with an estimated wastage of 10 percent. Hay is 90 percent dry matter.

   Intake "as fed" considering wastage = 33 × 0.9 = 29.7 pounds

   Dry matter intake = 29.7 × 0.9 = 26.7 pounds

2. Forage analysis is shown in Table 33-3

Table 33-3. Forage Analyses

Nutrient	Concentration	Units
Calcium	0.61	Percent
Phosphorus	0.18	Percent
Potassium	1.58	Percent
Copper	8.0	ppm
Zinc	32.0	ppm

Calculation of Mineral Intake and Comparison to Requirements

1. Calculation of mineral intake from forage:

   a. Calcium – 26.7 × 0.0061 = 0.163 pounds

   b. Phosphorus – 26.7 × 0.0018 = 0.048 lbs

   c. Potassium – 26.7 × 0.0158 = 0.422 lbs

   d. Copper – Requirements are in ppm

   e. Zinc – Requirements are in ppm

2. Based on Appendix Tables 1 and 10, this cow requires the amounts shown in Table 33-2.

3. Mineral intake is marginal for both phosphorus and copper. Also, the zinc level is barely adequate. Thus, a mineral supplement with a low concentration of phosphorus, copper, and zinc should be supplied. Including these minerals in the protein supplement would be a good way of ensuring a more uniform intake than with a free-choice mineral.

Macro Minerals

In most cow/calf production situations, there are only two macro minerals that need to be considered: Salt and phosphorus. The other macro mineral of importance is magnesium, which often needs to be supplemented to beef cows grazing early spring, lush pastures. In addition, typical rations for growing and finishing cattle usually require calcium supplementation.

Salt

Salt should always be provided in a mineral supplement or included in other supplements. As noted above, cattle require some salt, and they have a craving for it. Furthermore, they will exhibit abnormal eating behavior such as eating dirt, bones, or other unusual materials if it is not provided.

Phosphorus

This is the macro mineral that requires the greatest consideration in most production situations because it is often deficient. Moreover, since it is a relatively expensive mineral, meeting requirements without over supplementation is key to a cost effective mineral program.

Typical Deficiency Situations:

1. Marginally deficient in many pasture situations, especially late in the growing season.

2. Deficient in most dormant pastures.

3. Deficient in most crop residues.

Minimizing Expenditures for Phosphorus

1. **Feed a Phosphorus Level that Fits the Situation** – As noted, it is important to determine the severity of the deficiency and feed a mineral that meets the needs of the cattle without feeding an excessive amount. For example, many producers feed a 12 percent phosphorus mineral all year instead of feeding a lower level (4 or 6 percent, for example) during periods where deficiencies should be minimal. Another useful approach is to blend plain salt (40 to 60 percent) with the high-phosphorus mineral during periods where high phosphorus is not required. However, this practice also dilutes all of the other minerals, which should be considered before using this approach. A subsequent section in this chapter provides guidelines for calculating the minimum level of phosphorus necessary in a free-choice mineral.

2. **Consider the Phosphorus Content of By-Products** – The phosphorus content of supplements, especially grain by-products, should be considered when calculating mineral supplementation requirements. For example, wheat middlings, which are often used as an energy/protein supplement, are approximately 1 percent phosphorus. They supply all of the phosphorus needed by beef cows when fed at the typical 2- to 5-pound daily feeding rate.

Guidelines for the Phosphorus Level in a Free-Choice Mineral

Feed companies typically have mineral products formulated with a range of phosphorus levels to fit different production situations. As noted previously, minimizing mineral expenditures means selecting the mineral that fits the situation. Table 33-4 gives general recommendations based on the forage phosphorus

Table 33-4. Minimum Percentage of Phosphorus Required in a Free-Choice Mineral to Meet Requirements in Dry and Lactating Beef Cows Based on the Phosphorus Content of the Basal Forage.

	Dry Cow		Lactating Cow	
Phosphorus Content, % in Basal Forage	After Weaning	Late Gestation	Early Lactation	Late Lactation
0.13	-	2.8	12.1	6.5
0.15	-	0.9	9.9	4.3
0.17	-	-	7.7	2.1
0.19	-	-	5.5	-
0.21	-	-	3.3	-
0.23	-	-	1.1	-
0.25	-	-	-	-

level. These recommendations are based on the following assumptions:

1. Intake of 1.8 percent of body weight for a dry cow.
2. Intake of 2.2 percent of body weight by a lactating cow.
3. Requirements taken from Appendix Table 1.
4. Average consumption of free-choice mineral = 4 oz. per head per day.

Calculating the Required P Level in a Free-Choice Mineral

Step 1 – Calculate the mineral intake from the basal forage.

1,250 lb lactating cow at 2.2 percent of body weight intake = 1,250 × 0.022 = 27.5 lbs.

27.5 lbs × 0.17 percent phosphorus in forage = 27.5 × 0.0017 = 0.047 lbs.

Step 2 – Determine requirement using Appendix Table 1.

Requirement = 0.24% for early lactation

27.5 lbs × 0.0024 = 0.066 lbs of phosphorus

Step 3 – Determine shortage.

0.066 lbs minus 0.047 = deficiency of 0.019 lbs

Step 4 – Determine required phosphorus concentration in mineral.

0.019 lbs ÷ 0.25 lbs
(4 oz consumption) = 0.077 = 7.7% P

Magnesium

During early spring when forages are lush, they are often marginal in magnesium content. Furthermore, the absorption of this mineral is compromised. This can result in grass tetany, which is discussed in detail in Chapter 44. Starting a month or so before going on lush grass, 15 to 25 percent or more of the mineral mix should be magnesium oxide or another magnesium source. A level of 30 percent might be preferable. However, magnesium oxide is unpalatable, which makes it difficult to ensure adequate magnesium intake. Flavoring agents, such as soybean meal or dried molasses should be added to the mix at 6 to 10 percent to stimulate intake. Monitoring intake during these periods is essential to make sure the cattle are consuming the mineral at the desired level.

Trace Mineral Requirements and Maximum Tolerable Concentrations

The requirements for trace minerals listed in Appendix Table 10 should be considered as minimums below which clinical symptoms of deficiency could occur. Many nutritionists believe that even at the recommended minimum levels, there can be a subclinical reduction in performance exhibited as lower reproduction, reduced immune function, and

Table 33-5. Percentage of Forages that Were Adequate, Marginal, or Deficient in Trace Mineral Levels.

	Percentage			Antagonist Levels		
Trace Mineral	Adequate	Marginal	Deficient	High	Marginal	Very High
Cobalt	34.1	17.3	48.6	---		
Copper	36.0	49.7	14.2	---		
Iron	62.8	---	8.4	---	17.0	11.7
Manganese	76.0	19.3	4.7	---		
Molybdenum		---	---	---	48.6	9.2
Selenium	19.7	19.3		44.3		16.7
Zinc	2.5	34.1	63.4	---		

United States National Animal Health Monitoring System (NAHMS) Beef Cow/Calf Productivity Audit (CHAPA), 1996.

Table 33-6. Trace Mineral Content of Forages from Four Pastures on a Southwestern Montana Ranch.

Trace Mineral ppm	Pasture A	Pasture B	Pasture C	Pasture D	1996 NRC Requirements, ppm
Copper	7	10	7	10	10
Iron	457	385	179	136	50
Manganese	57	65	55	41	20-40
Molybdenum	0.62	0.19	4.1	---	---
Zinc	36	36	24	37	30

Montana Livestock Nutrition Conference, 1997

more health problems, especially when the cattle are under stress. There is some evidence to support this belief. Consequently, one should view the NRC requirements in Appendix Table 10 as absolute minimums and provide some margin of safety, perhaps 10 to 20 percent.

Trace minerals represent the most frustrating nutrient with respect to meeting beef cattle needs without spending too much because there are so many variables that influence the need for supplementation. Trace mineral deficiencies are widespread as shown in Table 33-5, which is based on 352 forage samples from 18 states representing the major cow/calf producing areas of the country.

In addition to the variability in mineral levels across regions of the country, there can be tremendous variation among pastures on the same ranch as is shown in Table 33-6 where copper levels were inadequate in two pastures and barely adequate in two others. The levels of zinc also varied from inadequate to adequate. The high level of molybdenum in one pasture illustrates that an antagonistic mineral can be high enough to be a problem in a single pasture but not in other pastures. The high level of molybdenum in pasture C would probably tie up the copper, which is already below the requirement. This level of molybdenum also would tie up supplemental copper if it was not supplied at a high enough level.

Assessing Trace Mineral Status

1. **Analyze Forage Samples** – This is a good way to get a preliminary estimate of trace mineral levels. As with any forage analysis, getting representative samples is crucial. It may be adequate to only analyze forages once because the primary factor dictating trace mineral levels is the soil content, which does not change significantly from year to year.
2. **Blood Levels** – In general, blood levels of trace minerals are of little value in determining status.
3. **Liver Biopsy** – Since the liver is the primary storage organ for copper and zinc, analysis of liver biopsy samples can be useful in assessing the status of these minerals. However, the cost of having a veterinarian take liver biopsy samples and the analyses themselves are expensive, especially considering that a fair percentage of the herd should be sampled. Consequently, liver biopsy analysis should only be used when there is evidence of a severe problem. It may be cheaper to simply supply supplemental trace minerals for a time to see if production improves.

Philosophical Approaches to Trace Mineral Supplementation

There are several basic philosophies of trace mineral supplementation proposed across the industry:

1. **Supply all of the Trace Mineral Required** – This philosophy proposes that a mineral supplement should be provided that meets all of the animal's needs without considering the mineral content of the basal diet. This approach recognizes that because mineral levels in feedstuffs are variable and the availability of minerals in many feedstuffs is also variable, it is difficult to be certain that requirements are being met in all

situations. Proponents of this approach believe that supplying all of the required trace minerals in a supplement or as part of a protein supplement is the only way to assure adequacy. While this approach has some merit, it results in overfeeding minerals, which reduces profitability.

2. **Supply One-half of the Trace Mineral Requirements in a Supplement** – In practice, this approach meets the mineral requirements in most situations, and it certainly reduces the cost compared to the first philosophy. However, it does result in spending too much on mineral supplements in many situations.

3. **Match the Supplement to Specific Deficiencies** – The third major philosophy is to determine the mineral content of the basal diet by feed analysis and provide a supplement to correct any deficiencies. While this approach takes more time and effort, it provides the best chance of meeting the needs of the cattle without wasting money on excess mineral.

Troubleshooting Trace Mineral Problems

Step 1 – Rule Out Energy and Protein Deficiencies – The initial focus should be on the most common causes of poor production and reproduction, which are usually an energy and/or protein deficiency, before evaluating the trace mineral status. If cow body condition is good, but health or reproductive problems exist, then the mineral program should be carefully evaluated. A good mineral supplementation program will not compensate for energy and/or protein deficiencies.

Step 2 – Analyze Forages – Analysis of a representative sample of major forages establishes mineral levels for comparison to requirements, and it indicates the presence of potential antagonists.

Step 3 – Analyze Water – In some environments, the mineral levels in water should also be determined because elevated levels of calcium, sulfates, salt, and iron can be high enough to interfere with the absorption and use of trace minerals.

Step 4 – Liver Biopsy – If the production problem is severe enough and the previous steps have not identified the cause of the problem, have liver biopsy samples analyzed for the minerals stored in the liver such as copper and zinc. As noted, this is an expensive option, but it may be necessary to identify the cause of reduced performance.

Minimum Trace Mineral Levels in a Free-Choice Mineral

Cattle requirements for trace minerals are stated in parts per million (ppm) as discussed previously. Determining the minimum concentration of the trace mineral in a free-choice mineral that results in the total diet meeting NRC requirements necessitates an assumption about intake in addition to knowing the concentration of the mineral in the basal forage.

Assumptions:

1. Intake of 27.5 lbs of hay dry matter with 23 ppm zinc.

2. Requirement from Appendix Table 10 is 30 ppm zinc.

3. Parts per million is the same as mg/kg.

Calculation:

Intake of forage = 27.5 lbs = 27.5 ÷ 2.2046 lbs/kg = 12.5 kg

Intake of zinc = 23 ppm (mg/kg) × 12.5 kg = 288 mg

Required intake of zinc = 30 mg/kg × 12.5 kg = 375 mg

Zinc deficiency = 375 – 288 = 87 mg

4 oz of mineral consumption = 0.25 lbs ÷ 2.2046 lbs/kg = 0.11 kg

Need 87 mg zinc in 0.11 kg

87 mg ÷ 0.11 kg = 791 mg per kg = 791 ppm

Free-choice mineral must be at least 791 ppm zinc

Table 33-7. Relative Bioavailability of Trace Minerals from Different Sources.

	Inorganic Forms				
Mineral	Sulfate	Oxide	Carbonate	Chloride	Organic (Chelated)
Copper	100	0	---	105	130
Manganese	100	58	28	---	176
Zinc	100	---	60	40	159-206

W. Green, et al., 1998.

This level of zinc meets minimum NRC requirements assuming that the cattle consume at least 4 ounces of the mineral. To build in a safety factor, the mineral should probably be at least 1,000 ppm zinc.

Trace Mineral Sources

Trace minerals can be supplied in either the inorganic or organic forms. Inorganic forms are those where the mineral is not attached to a carbon-containing compound such as an amino acid, protein, or carbohydrate. Inorganic minerals are typically bound to a sulfate, oxide, or chloride molecule. In recent years, there has been considerable interest in feeding trace minerals that are attached to carbon containing compounds because they have greater bioavailability, especially in the presence of antagonists. These combinations are referred to as organic trace minerals or chelated trace minerals. There are numerous forms of chelated trace minerals because trace minerals can be bound to a wide variety of organic compounds. Table 33-7 shows the relative bioavailability of various inorganic and organic forms.

As a general rule, the sulfate forms have the highest bioavailability of the inorganic forms. Thus, for each mineral, the sulfate form was assigned a bioavailability of 100 percent. These data show then that zinc chloride is 40 percent as available as zinc sulfate. Most feed companies stopped using the oxide form of most trace minerals years ago, but zinc oxide is still found in some minerals, and it is high in bioavailability. The other exception to the bioavailability of oxides is the use of magnesium oxide, which has been considered fairly bioavailable. However, the sources of MgO have changed in recent years, and some of the current sources are low in bioavailability. Feed companies must list specific sources of minerals on the feed tag, so it is fairly easy to check for sources with low bioavailability.

Common Questions about Sources of Trace Minerals

When Should Organic Trace Minerals Be Used?

For most production situations, the inorganic forms of the trace minerals are adequate, and they are much less expensive. However, in a few situations, it might be worth the extra money to feed organic trace minerals. These situations include:

1. **Antagonists Are Interfering with Trace Mineral Absorption** – The best example of this is where the soil is high in molybdenum, which ties up copper resulting in symptoms of copper deficiency. The higher bioavailability of organic copper might give a better response than inorganic copper.

2. **Reproduction Is Lower Than Expected** – When the reproductive rate is lower than one might expect based on cow body condition, and the mineral supplementation program using inorganic forms appears adequate, it might be worth replacing a portion (40 percent, for example) of the inorganic mineral with chelated minerals.

3. **Health Problems Are Excessive** – Again, if other environmental and nutritional factors have been eliminated as causes of excessive disease, it may be useful to try chelated minerals because some research has shown that their higher bioavailability improves immune function. A blend of inorganic and chelated minerals is often most cost effective.

Table 33-8. Results of a Study at Colorado State University Supplementing Cows with Different Trace Mineral Levels and Sources.

Item	Inorganic[a] Low Level	Inorganic[b] High Level	Organic[c] High Level
No.	99	100	100
Initial Weight, lbs	1,287	1,289	1,278
Final Weight, lbs	1,309	1,274	1,289
Weight Change, lbs	22	15	11
Calf Weaning Weight, lbs	460[d]	447[e]	471[d]
Pregnant to AI, %	61[d]	56[d]	75[e]
Overall, %	88	81	88

[a]Low level = 501 Cu, 2,160 Zn, 1,225 Mn, 11 Co, on a ppm basis. [b,c]High level = 1,086 Cu, 3,113 Zn, 1,764 Mn, 110 Co, on a ppm basis Organic = Same levels as High Inorganic. [d,e]Values in a row with different letters differ (P<.05). Colorado State University

4. **Stress Situations Exist** – Chelated minerals have been shown to be superior during stress because of their higher bioavailability. For example, if sickness has been a problem following weaning, feeding chelated minerals before and following weaning might reduce health problems.
5. **Maximum Reproduction Is Desired** – In purebred herds, it could be beneficial to feed chelated minerals before the breeding season because several research trials have shown that chelated minerals can improve reproduction. However, the results have been variable. Feeding chelates probably is not justified in most commercial herds unless one of the situations discussed above exists.

Is the Relative Level of Various Trace Minerals in a Supplement Important?

In a situation where only copper is deficient, producers may want to feed a mineral high in copper and low in the other trace minerals to reduce cost. There have been several research trials showing a relationship between copper and zinc. Based on this research, it is recommended that the zinc to copper ratio in a mineral should be between 3:1 to 5:1. There are likely other similar relationships that have not been identified.

Can Very High Levels of Trace Mineral Supplementation Be Harmful?

Unfortunately, there is a tendency to think that if a little is good, a lot will be better. With respect to trace minerals, this is not true. There have been trials where copper boluses in calves have reduced weaning weights dramatically. Furthermore, in a study at Colorado State University (Table 33-8), high levels of inorganic minerals reduced weaning weights compared to a lower level. It appears that excessively high levels of some trace minerals can have a deleterious effect on performance. As shown, the organic trace minerals increased the percentage settling to AI supporting their use when artificially inseminating or using embryo transfer. In addition to the effect on weaning weight, there have been cases of beef cows exhibiting symptoms of copper toxicity after consuming large quantities of a highly fortified free-choice mineral. The 2006 Mineral Tolerances of Animals (National Academies Press) publication lists toxic levels for all of the trace minerals. These levels are also shown in Appendix Table 10.

Are All Chelated Minerals the Same?

All chelated minerals are not the same; however, there has not been enough research to clearly define differences among the many forms and brands of chelated minerals. As a recommendation, if the use of a chelate is warranted, then use one from a reputable company that has conducted adequate research to prove their product.

References

Ahola, J.K., J.C. Whittier, S.P. Porter, W.S. Mackay, P.A.G.A. Sampaio, and T.E. Engle. Characterization of Forage Trace Mineral Concentration by Season in Diets of Beef Cows Grazing Native Ranges in Eastern Colorado. Proceedings Western Section American Society of Animal Science 58:335-338.

Ahola, J.K., D.S. Baker, P.D. Burns, R.G. Mortimer, R.M. Enns, J.C. Whittier, T.W. Geary, and T.E. Engle. 2004. Effect of Copper, Zinc, and Manganese Supplementation and Source on Reproduction, Mineral Status, and Performance in Grazing Beef Cattle Over a Two-year Period. Journal of Animal Science 82:2375-2383.

Ahola, J.K., D.S. Baker, P.D. Burns, J.C. Whittier, T.E. Engle. 2005. Effects of Copper, Zinc, and Manganese Source on Mineral Staus, Reproduction, Immunity, and Calf Performance in Young Beef Females over a Two-Year Period. The Professional Animal Scientist 21:297-304.

Corah, L. 1996. Vitamin and Mineral Supplementation. Proceedings Cow/Calf Nutrition Conference Kansas State University.

Gadberry, S. Mineral and Vitamin Supplementation of Beef Cows in Arkansas. 2003. University of Arkansas Cooperative Extension Service Bulletin FSA3035.

Galyean, M.L., G.C. Duff, and A.B. Johnson. 1995. Potential Effects of Zinc and Copper on Performance and Health of Beef Cattle. Proceedings of Southwest Nutrition and Management Conference.

George, M.H., C.F. Nockels, T.L. Stanton, and B. Johnson. Effect of Source and Amount of Zinc, Copper, Manganese, and Cobalt Fed to Stressed Heifers on Feedlot Performance and Immune Function. The Professional Animal Scientist 13:84.

Greene. L.W., A.B. Johnson, J. Paterson, and R. Ansotegui. 1998. The Role of Trace Minerals in the Cow-Calf Life Cycle. NCBA Cattlemen's College Proceedings.

Greene, L.W. 2000. Designing Mineral Supplementation of Forage Programs for Beef Cattle. JAS 77:1-9.

Hale, C. and K.C. Olson. 2001. Mineral Supplements for Beef Cattle. University of Missouri Cooperative Extension Service Bulletin G2081.

Herd, D.B. Mineral Supplementation of Beef Cows in Texas. Texas A&M University Cooperative Extension Bulletin E-526.

Ledoux, D.R. and M. C. Shannon. 2008. Bioavailability and Antagonists of Trace Minerals in Ruminant Metabolism. Proceedings 16th Florida Ruminant Nutrition Symposium.

Mineral Tolerance of Animals. 2005. National Research Council.

Nutrient Requirements of Beef Cattle. 1984. National Research Council.

Nutrient Requirements of Beef Cattle. 1996. National Research Council.

Olson, K.C. Cost-Effective Mineral Supplementation Programs for Beef Cattle. Proceedings 4-Sate Beef Conference.

Paterson, J.A. and T.E. Engle. 2005. Trace Mineral Nutrition in Beef Cattle. Proceedings of 2005 Nutrition Conference at the University of Tennessee.

Spears, J.W. 2008. Trace Mineral Nutrition – What is Important and Where do Organic Trace Minerals Fit in? Proceedings 23rd Southwest Nutrition and Management Conference. Pg 227-236.

Wellington, B., J. Paterson, and R. Ansotegui. The Influence of Supplemental Copper and Zinc on the Performance of Weaned Heifer Calves. Montana State University.

Chapter 34

Supplementation of Vitamins

Vitamin A

As noted in Chapter 20, the only vitamin that usually requires supplementation in beef cattle is Vitamin A. Requirements are expressed in International Units (IU) per day. Appendix Tables 1 through 9 provide minimum levels for various classes of beef cattle. Since vitamin A is relatively inexpensive, it should be provided in some form whenever a potential deficiency exists.

Common Situations Where Vitamin A Could Be Deficient

1. In drought conditions.
2. When forages are bleached or weathered.
3. When feedstuffs have had excess exposure to sunlight.
4. When feedstuffs have been stored for long periods.
5. When supplements (mineral or protein supplements as the source of supplemental vitamin A) have been stored for long periods, especially at high temperatures.

Storage of Vitamin A in the Liver

When cattle are consuming lush pasture and other green, leafy forages, they store significant quantities of vitamin A in the liver. In addition, during these periods the cattle will store carotene in their fat. This causes the yellow color in the fat of cattle consuming diets high in carotene. However, there is some question about how much of the carotene stored in fat is actually converted to vitamin A. Table 34-1 shows general guidelines for how long it takes for the liver to become depleted when the diet is inadequate in vitamin A. These guidelines assume that the liver stores are at their maximum when the cattle are placed on a deficient diet.

Table 34-1. Time it Takes for the Liver to Become Depleted of Vitamin A for Various Classes of Beef Cattle.

Class of Cattle	Days
Light feeder steers	40 – 80
Heavy feeder steers	80-140
Yearling steers	100-150
Cows	120-180

University of Missouri Bulletin G2058.

Typical Methods of Supplementation of Vitamins

1. **In a free-choice mineral** – Vitamin A is typically included in free-choice minerals at a level that meets beef cow and stocker requirements if the mineral is consumed at the recommended level.
2. **Include in a protein supplement** – If cattle are receiving a protein supplement, including vitamin A in the supplement is an effective way to meet the requirement.
3. **Intramuscular injection** – Injection of vitamin A effectively supplements cattle. Injected vitamin A is more efficiently used than vitamin A in the diet. Generally, an injection of vitamin A will ensure adequacy for about 2 months.
4. **Feeding leafy, green forages** – Feeding forages such as alfalfa and high-quality grass hays can be an effective way of supplementing vitamin A, but providing preformed vitamin A is more effective.

Special Considerations

1. Since vitamin A is destroyed by exposure to sunlight, air, or heat, the biological activity of vitamin A in supplements decreases over time. In addition, the biologically active vitamin A precursor (carotene) declines over time in stored forages. These losses of activity during storage must be considered and supplementation provided to avoid deficiencies.
2. The instability of vitamin A in forages means that the level is often the lowest in late winter and early spring (before green-up) when beef cow requirements are at their highest due to calving and lactation in spring calving herds.

Vitamin D

Because vitamin D is synthesized by cattle exposed to sunlight, it is rarely deficient. However, vitamin D is usually included in free-choice mineral and protein supplements as a precaution. It is typically included at 10 percent of the level of vitamin A.

Vitamin E

While vitamin E is stored in many tissues in the body, body stores are actually minimal. Consequently, the diet must contain adequate vitamin E to prevent signs of deficiency.

Beef Cow Herds

In most cases, roughages contain enough vitamin E to prevent deficiency symptoms. However, cattle consuming dormant forages for long periods could benefit from vitamin E supplementation through a free-choice mineral, protein supplement, or by intramuscular injection.

Stocker Cattle

The importance of vitamin E supplementation of stocker cattle is largely a function of the prior diet and level of stress. For example, the 1996 Beef Cattle NRC recommends 400 to 500 IU per day of vitamin E for highly stressed calves during the starter period. Recommendations for growing cattle are generally 50 to 100 IU per day.

Finishing Cattle

The recommendation for finishing cattle is also 50 to 100 IU per day. Feeding high levels (300 IU per day) to finishing steers increases the shelf life of the meat by slowing discoloration.

Vitamin K

Because rumen bacteria make adequate quantities of vitamin K, there is no need for supplementation under most conditions. When vitamin K is deficient as a result of feeding moldy sweet clover hay, the solution is to change feedstuffs rather than supplement with vitamin K.

B Vitamins

As with vitamin K, the rumen microorganisms produce adequate quantities of the B vitamins under most conditions. However, a deficiency of cobalt results in a deficiency of B_{12} because cobalt is a major component of B_{12}. High energy, finishing rations interfere with the production of thiamine by rumen microorganisms resulting in polioencephalomalacia, a central nervous system disorder. Most feedlot nutritionists add supplemental thiamine with inconsistent results.

References

Nutrient Requirements of Beef Cattle. 1984. National Research Council.

Nutrient Requirements of Beef Cattle. 1996. National Research Council.

Parish, J. and J. Rhinehart. 2008. Mineral and Vitamin Nutrition for Beef Cattle. Mississippi State University Extension Service.

Sewell, H.B. Vitamins for Beef Cattle. University of Missouri Extension Bulletin G2058.

Chapter 35

Minimizing Supplementation Costs

Supplementation to meet deficiencies in the basal forage is often a significant portion of total feed cost. In many situations, changing management practices significantly reduces supplementation requirements and total feed costs. Take a holistic approach to evaluating management changes since there are often multiple effects from what appears to be a simple change. For example, changing the time of calving may seem like a relatively simple modification, but it may influence labor requirements and the marketing program because of changes in calf weights at weaning. Consequently, careful consideration must be given to any major changes in overall management. The following sections discuss a few changes in management, which might reduce the need for supplementation.

Matching the Calving Season to the Forage Resource

Forages used for grazing in cow/calf operations tend to have periods when they have high nutritional value followed by periods of relatively low nutritional quality. The timing of these periods of high value varies with the type of forage and the environment. Matching the highest nutrient requirements of the cow to the highest nutritive value of the forage offers a way of reducing supplementation costs and increasing profitability. To illustrate, consider Figure 35-1, which shows cow requirements compared to grass nutritive values for February, March, and April calving, respectively, in a typical High Plains situation. In most of the High Plains, native grass starts growing vigorously in early May; quality peaks in late May or early June; and quality declines rapidly starting in July.

With all three calving dates, there are times when the cow's requirements exceed the nutritive value of the forage. As a general statement, some supplementation is required to avoid negative impacts on performance. As the calving date is moved closer, however, to the time of peak grass quality, the magnitude of the deficiency is reduced, which means that less supplementation is required. It is important to note that as the calving date is moved closer to peak forage quality, the number of days the cows will be on high quality forage before breeding increases. This means that getting the cows rebred will be easier.

Changing calving seasons influences other aspects of the business besides supplementation costs. Calving later with the same weaning date will result in smaller calves at weaning, which could reduce revenue if the calves are sold at that time. This may necessitate a post-weaning growing program to allow the sale of the same number of pounds. Another potential problem for producers who also farm is that late-spring calving can cause conflicts between labor requirements for farming and the cow/calf enterprise. Another consideration is that calving too late in the spring results in breeding occurring in the heat of the summer, which may result in reduced fertility.

Each producer must weigh all of the pros and cons for their operation and decide which time of calving offers the greatest potential for profit. Conceptually, matching calving time to the grass resource offers the potential for reducing the cost of supplementation and improving profitability.

The example used to illustrate this concept is representative of warm-season grass pastures in the High Plains. In other sections of the country, the dominant forages are cool-season grasses. Examples include fescue and bromegrass. These grasses show bimodal peaks in nutritive value with one peak in the spring and another in the fall. The concept of matching the

Figure 35-1. Comparison of Energy Requirements by Beef Cows and Grass Energy Content.

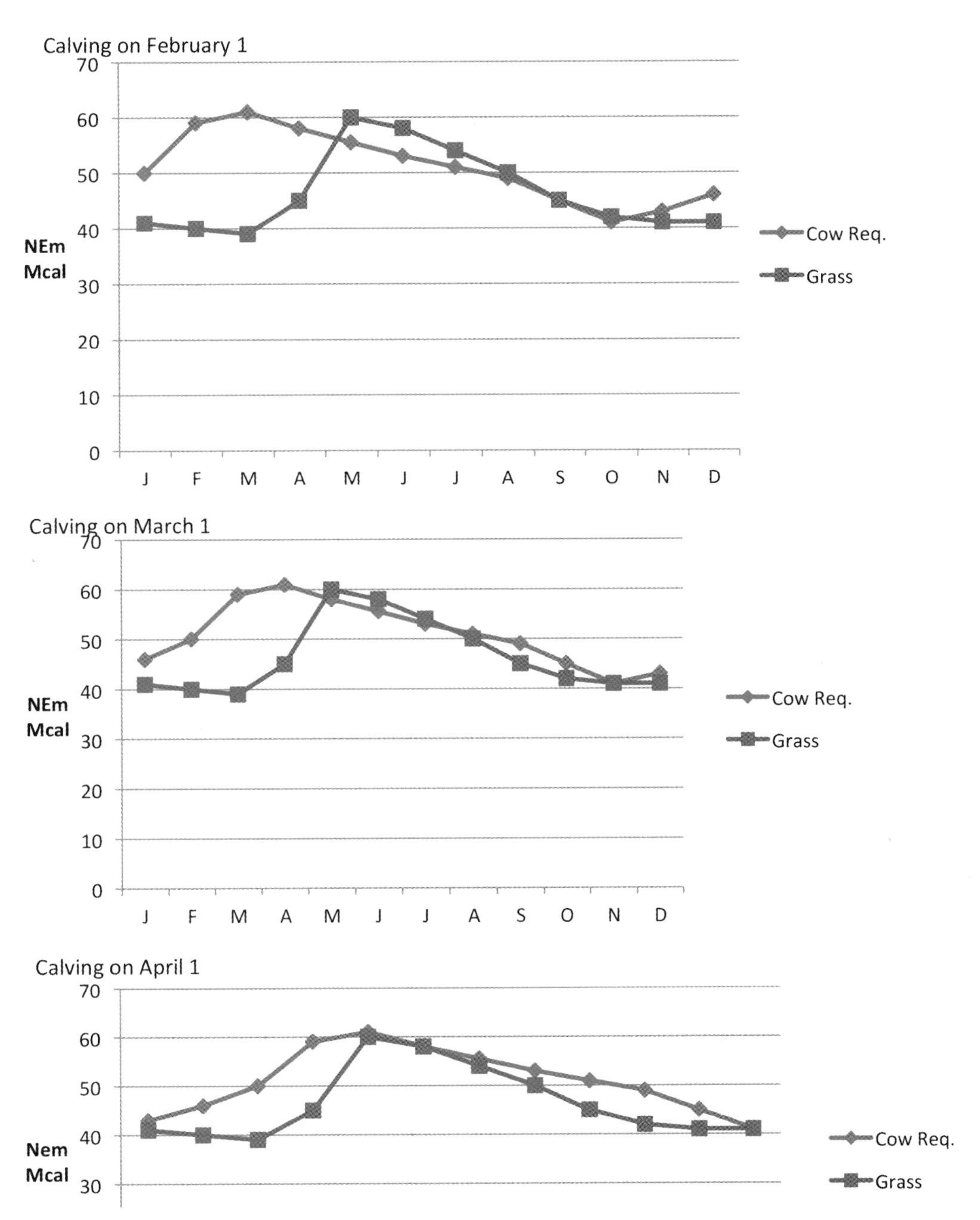

calving season to one of these peaks still applies, but the timing may be different. For example, since fescue breaks dormancy earlier in the year than a typical warm-season grass, the optimal time to calve may be earlier in the spring to take advantage of the peak nutritive value and rebreed before the nutritive value of the fescue declines in the summer. Lush, green grass probably has the greatest ability to stimulate breeding of any feedstuff — the key is to maximize its benefits on reproduction.

Weaning Early to Reduce Nutrient Requirements

As noted previously, lactation greatly increases the nutrient requirements of a beef cow. In a three-state study summarized in Table 35-1, time of weaning significantly influenced body condition of the cows at weaning. For example, weaning in August resulted in an increase of almost one body condition score compared to weaning in November. Even though spring-calving cows are not giving much milk in the

Table 35-1. Summary of the Effect of Weaning Time on Cow Weight Change and Body Condition Scores in a Three State (ND, SD, and WY) Study.

	NDSU Dickinson REC		SDSU Antelope Station		UW Beef Unit	
Item	Early	Normal	Early	Normal	Early	Normal
August Weight, lbs	1,299	1,336	1,343	1,329	1,239	1,250
November Weight, lbs	1,314	1,199	1,376	1,283	1,356	1,276
Cow Weight Change, lbs	15	-137	33	46	117	26
August BCS	5.2	5.3	5.6	5.6	5.5	5.6
November BCS	6.1	4.7	6.0	5.6	5.7	5.1
BCS Change	0.9	-0.6	0.4	0	0.2	-0.5

Proceedings of Colorado Nutrition Roundtable, 2006.

fall, the nutrient requirements for lactation still put these cows in a negative energy balance resulting in a loss in condition. On average, studies show that delayed weaning results in a loss of 0.5 pounds of cow body weight for every pound increase in weaning weight. Higher cow body condition scores going into winter mean that it will take less supplementation during the winter to arrive at the same cow body condition at calving the following spring. Weaning earlier results in lighter weaning weights, which may necessitate a growing program to avoid a reduction in revenue.

Maximize the Quality of Harvested Forages

In many cases, harvesting forages at higher quality can significantly reduce the need for supplementation. For example, sorghum-sudan hay cut early will often have 8 to 10 percent crude protein, while the same hay cut later in maturity will have only 5 percent CP. The early-cut hay will have adequate protein for a dry, pregnant cow whereas the late-cut hay will be inadequate. Thus, while the late-cut hay may yield a little more per acre, considerable supplementation is required to achieve the same animal performance compared to the early-cut hay. As a general statement, all forages show a significant decline in protein and energy as they approach maturity.

Purchasing Supplements Using Bids

Sending a set of nutritional specifications to several feed companies for bids can reduce the actual price per ton for a supplement. The following items must be considered to ensure that the bidding process results in the correct supplement at the lowest price.

1. **Nutrient Specifications Must be Clearly Defined** – Defining the nutrient specifications for a supplement requires an understanding of animal requirements, nutrient content of the available forages, and expected intake. It may be advisable to have someone with training in nutrition develop the specifications. The specifications should list both minimums and maximums where necessary.

2. **Define the Acceptable Ingredients** – The ingredients that are acceptable should be listed with a minimum and maximum inclusion level. The ingredients that cannot be included should be clearly noted. For example, bids for a creep feed should indicate that corn gluten feed is unacceptable because of its poor palatability.

3. **Demand a Listing of the Ingredients in the Formula as Well as the Nutrient Specifications** – Unless the bid clearly states that the ingredient composition cannot change during the bid contract period, feed companies use the cheapest combination of ingredients that meet the nutrient specifications. This may result in rapid changes in the ingredients from load to load with possible palatability changes. To

avoid these changes, the bid can be written to require that the ingredients are "fixed" for the bid period. Demanding a list of ingredients and their inclusion rates should prevent the inclusion of filler ingredients with little nutritive value.

References

Nutrient Requirements of Beef Cattle. 1984. National Research Council.

Nutrient Requirements of Beef Cattle. 1996. National Research Council.

Simms, D.D. 2009. Feeding the Beef Cowherd for Maximum Profit. SMS Publishing.

Chapter 36

Supplementation – Delivery Systems

There are essentially two approaches to delivering supplement: 1) Hand-feeding systems where supplement intake is controlled by the amount delivered, and 2) Limited-intake systems where intake is controlled by a limiting agent or physical form.

Comparison of Hand-Feeding vs. Self-Feeding (Limited-Intake Systems)

Advantages of Hand-Feeding Delivery Systems

1. Cost of feed per ton is usually lower.
2. Variability in consumption across the herd is usually lower. Research in Montana indicates that the variability in intake is one-half as much with hand feeding compared to limit-feeding systems.
3. Hand feeding provides a better opportunity to control the location and movement of cattle to improve grazing distribution.

Advantages of Limited-Intake Delivery Systems

1. Lower cost of delivery of the supplement to the cattle including labor and fuel.

Hand-Feeding Systems of Delivery

There are a number of hand-feeding systems for delivering supplements to beef cattle. Common systems include range cubes and bunk feeding. While hand-feeding results in less variation in consumption, there is still considerable animal-to-animal variability. Typically, boss cows get more than their share while younger, meeker cows get less than their share. The cows that need the supplement the most often get the least, which leads to loss of body condition in younger or less aggressive cows. While it is impossible to ensure that all cows consume the desired amount of supplement in a hand-feeding situation, several management practices minimize the variability in intake of supplement.

Minimizing Variability in Consumption in Hand-Feeding Situations

1. Sorting cows into groups based on age and body condition is a good way of reducing variability in intake. While there may be some differences within each group, the variability across the entire herd will be reduced significantly.
2. When using feeders, the optimal number of feeders exists when every animal can eat at the same time without excess feeder space. Having too much feeder space results in boss cows forcing meeker cows away from the feeder, and obviously, too little space results in the younger, less aggressive cows not having a chance to eat.
3. Feeding larger amounts at each feeding will reduce variability. For example, feeding 6 pounds of range cubes every third day will result in less variation in intake over time than feeding 2 pounds every day.

Limited-Intake Systems

Several limited-intake systems help control intake and reducing the cost of delivery. These systems are based on adding chemical agents that reduce the palatability of the supplement or by making the supplement difficult to consume based on its physical characteristics. As a general statement, a system that limits intake to an exact and uniform level over time has not been developed. Consumption will change based on the quality and quantity of the basal forage and on the size and nutritional demands of the cattle.

Consider cows grazing corn stover in the fall. When they are first placed on the stalks, they typically consume little supplement. As the quality of the available forage declines, they increase consumption of the supplement; and if they are kept on the corn stalks too long, they consume the supplement in excessive quantities. This intake pattern is not bad since consumption parallels the need for supplemental nutrients to some extent. However, the consumption of supplement may be more than desired from an economic standpoint. The key point is that there is not a limited-intake system that consistently keeps intake at the desired level — all systems require monitoring and periodic adjustments to prevent over consumption.

It would seem that supplement intake in self-fed systems would be less variable than that observed with hand-fed systems. However, the research conducted to date shows just the opposite. Some animals never consume the supplement, while others consume at much higher than desired levels. It appears that cattle exhibit a learning curve with respect to consuming supplement. For example, a lower percentage of 2-year-old cows consume supplement than 3-year-old cows, and a higher percentage of mature cows consume supplement compared to younger cows.

Common Intake Control Systems

Limiting Feed Intake with Salt

In many areas, "salt mixes" are used to provide a free-choice supplement while limiting intake. As with other limiters, salt is not a perfect regulator of intake, but by adjusting the ratio of salt to supplement periodically, it is possible to effectively limit the intake to a desired level. As a rule of thumb, cattle eat about 0.1 pound of salt per 100 pounds of body weight per day. Research has not shown any adverse effects of high salt levels on production or health. Cattle have a high tolerance for salt as long as they have adequate, clean water. They will consume 50 to 75 percent more water than normal or approximately 5 gallons of water for each pound of salt. Salt mixes should not be used when the total dissolved solids in the water are over 5,000 ppm.

Key Considerations in Using Salt to Limit Intake

Factors Influencing Intake

1. Cattle typically adjust to salt and intake increases over time.
2. Cattle eat more of the salt-limiting supplement when forage quality or quantity declines.
3. If the salt content of the available water is significant, supplement intake will be below the expected level.

Starting Cattle on Salt – If cattle are accustomed to eating supplements, but not accustomed to self-feeding, they may overeat at first. One method of avoiding this problem is to start with a high salt level (50 percent) and reduce it until the desired intake is achieved. On the other hand, if cattle are not accustomed to eating supplement, it may be necessary to hand-feed the supplement without salt until the cattle are readily consuming the supplement before adding salt to the mix.

Preventing Separation – The particle size of the salt should be similar to that of the other ingredients in the supplement to prevent separation. Otherwise, the animals may sort the supplement and over consume the actual protein/energy portion. The supplement with salt can be pelleted to avoid sorting. If the particle size of the other ingredients is similar to the salt in a loose mix, pelleting is not necessary. It is much harder to manage intake when using pelleted supplements compared to loose mixes.

Use Plain Salt – Trace mineralized salt should not be used in salt mixes because of potential toxicity or mineral imbalances. Plain white salt or even rock salt should be used.

Importance of Having Adequate Clean Water – Having an adequate supply of clean water is essential when using salt as a limiter. In addition, if there is potential for a significant level of dissolved solids, it should be tested.

Determining the Amount of Salt in the Mixture

Table 36-1 presents general guidelines for the amount of salt that cattle will consume. As shown, there is a considerable range in intake because of differences in forage quality and adaptation to salt. Table 36-2 shows the percentage of salt that would need to be in a mix to give a desired level of intake of the actual supplement based on the average expected intake of salt as shown in Table 36-1.

Using Tables 36-1 and 36-2, it is possible to determine the combination of salt and supplement necessary to obtain intake close to the desired level. For example, if the average cow weighs 1,100 pounds, the average salt consumption should be about 1.1 pounds per day (Table 36-1). If the target intake is 3 pounds of the actual protein supplement, Table 36-2 indicates that at 25 percent salt, the intake of the protein supplement should be about 3.3 pounds. At 30 percent salt, the protein intake should be about 2.6 pounds. Thus, the mix should contain about 27.5 percent salt. This cow would consume a total of about 4.0 pounds of the salt mix per day.

Fishmeal/Fish Oil Limiters

Several companies have fishmeal or fish oil products to mix with a supplement to limit intake. These products are generally effective, but as is the case with all limiters, the inclusion rate of these products has to be adjusted to keep intake at the desired level. Fishmeal is fairly expensive, which can make this approach an expensive way to limit intake.

Blocks and Tubs

This general category includes the 250-pound blocks and tubs as well as the smaller 33.33-pound blocks. Intake of these products is typically controlled by hardness. They have become popular in recent years because they are effective in controlling intake. Moreover, by moving the location of placement, they can be used to improve grazing distribution. However, they can be an expensive way to provide supplemental nutrients. Furthermore, some brands limit intake to such low levels that small quantities of supplemental protein and other nutrients are actually provided. With these low intakes, the cost per head is low, but in many cases they do not provide sufficient nutrients to prevent a reduction in performance. The use of blocks and tubs needs to be evaluated on a cost versus convenience basis. As the cost of fuel to deliver hand-fed supplements has increased, blocks and tubs have become more viable as supplement delivery systems.

Table 36-1. Estimated Salt Intake of Cattle Fed Salt-Limiting Supplements

	Salt Consumption lbs/day		
Body Wt., lbs	Low	Average	High
300	0.3	0.5	0.6
500	0.5	0.6	0.7
700	0.6	0.7	0.9
900	0.7	0.9	1.1
1,100	0.8	1.1	1.3
1,300	0.9	1.3	1.5
1,500	1.0	1.5	1.6

Great Plains Beef Cattle Handbook, GPE 1950.

Liquid Supplements

Another convenient way to supplement beef cows is liquid supplements. Intake of the supplement is typically limited by reducing the palatability of the liquid. A problem with liquid supplements is that, in most cases, the protein is in the form of non-protein nitrogen because it is difficult to suspend sources of natural protein. This limits the effectiveness of liquid supplements, especially when the basal forage is of low quality. There are some liquid supplements available that contain natural protein, but most contain primarily NPN. Some manufacturers of liquid supplements make claims that they are using natural protein when the protein source is actually non-protein nitrogen—it just is not urea. Consequently, it is important to evaluate the type of protein in a liquid supplement.

Limiting Intake by Limiting Access Time

Another approach to limiting the intake of supplemental feed is by limiting access time. This approach is typically implemented by limiting access to high-quality forage used as a supplement to lower-quality forage. For example, cows grazing dry winter grass might be allowed access to small grain (wheat or rye, for example) pastures for a few hours every few days to provide supplemental protein. Another example is allowing limited access to high-quality hay

Table 36-2 Estimated Level of Salt to Add to a Mixture to Give a Desired Intake of Supplement.

Salt Intake, lbs		Percentage of Salt in the Supplement												
		6	8	10	12	14	16	18	20	25	30	35	40	50
0.3	Total	5.0	3.7	3.0	2.5	2.1	1.9	1.7	1.5	1.2	1.0	0.9	0.7	0.6
	Nonsalt	4.7	3.4	2.7	2.2	1.8	1.6	1.4	1.2	0.9	0.7	0.6	0.4	0.3
0.4	Total	6.7	5.0	4.0	3.3	2.9	2.5	2.2	2.0	1.6	1.3	1.1	1.0	0.8
	Nonsalt	6.3	4.6	3.6	2.9	2.5	2.1	1.8	1.6	1.2	0.9	0.7	0.6	0.4
0.5	Total	8.3	6.2	5.0	4.2	3.6	3.1	2.8	2.5	2.0	1.7	1.4	1.2	1.0
	Nonsalt	7.8	5.7	4.5	3.7	3.1	2.6	2.3	2.0	1.5	1.2	0.9	0.7	0.5
0.6	Total	10.0	7.5	6.0	5.0	4.3	3.8	3.3	3.0	2.4	2.0	1.7	1.5	1.2
	Nonsalt	9.4	6.9	5.4	4.4	3.7	3.2	2.7	2.4	1.8	1.4	1.1	0.9	0.6
0.7	Total	11.7	8.7	7.0	5.8	5.0	4.4	3.9	3.5	2.8	2.3	2.0	1.7	1.4
	Nonsalt	11.0	8.0	6.3	5.1	4.3	3.7	3.2	2.8	2.1	1.6	1.3	1.1	0.7
0.8	Total	13.3	10.0	8.0	6.7	5.7	5.0	4.4	4.0	3.2	2.7	2.3	2.0	1.6
	Nonsalt	12.5	9.2	7.2	5.9	9.4	4.2	3.6	3.2	2.4	1.9	1.5	1.2	0.8
0.9	Total	15.0	11.2	9.0	7.5	6.4	5.6	5.0	4.5	3.6	3.0	2.6	2.2	1.8
	Nonsalt	14.1	10.3	8.1	6.6	5.5	4.7	4.1	3.6	2.7	2.1	1.7	1.3	0.9
1.0	Total	16.7	12.5	10.0	8.3	7.1	6.2	5.5	5.0	4.0	3.3	2.9	2.5	2.0
	Nonsalt	15.7	11.5	9.0	7.3	6.1	5.2	4.5	4.0	3.0	2.3	1.9	1.5	1.0
1.1	Total	8.3	13.7	11.0	9.2	7.9	6.9	6.1	5.5	4.4	3.7	3.1	2.7	2.2
	Nonsalt	17.2	12.6	9.9	8.1	6.8	5.8	5.0	4.4	3.3	2.6	2.0	1.6	1.1
1.2	Total	20.0	15.0	12.0	10.0	8.6	7.1	6.7	6.0	4.8	4.0	3.4	3.0	2.4
	Nonsalt	18.8	13.8	10.8	8.8	7.4	6.3	5.5	4.8	3.6	2.8	2.2	1.8	1.2
1.3	Total	21.7	16.2	13.0	10.8	9.3	8.1	7.2	6.5	4.2	4.3	3.7	3.2	2.6
	Nonsalt	20.4	14.9	11.7	9.5	8.0	6.8	5.9	5.2	3.9	3.0	2.4	1.9	1.3
1.4	Total	23.3	17.5	14.0	11.7	10.0	8.7	7.8	7.0	5.6	4.6	4.0	3.5	2.8
	Nonsalt	21.9	16.1	12.6	10.3	8.6	7.3	6.4	5.6	4.2	3.2	2.6	2.1	1.4
1.5	Total	25.0	18.7	15.0	12.5	10.7	9.4	8.3	7.5	6.0	5.0	4.3	3.7	3.0
	Nonsalt	23.5	17.2	13.5	11.0	9.2	7.9	6.8	6.0	4.5	3.5	2.8	2.2	1.5

Great Plains Beef Cattle Handbook, GPE-1950.

such as alfalfa for cows consuming low-quality hay. The frequency of access depends on the quality of both the basal forage and the high-quality forage. The practicality of this type of program depends on the location of pastures relative to one another. This can be a labor intensive but cost effective way to deliver supplementation.

References

Bowman, J.G. and B.F. Sowell. 1997. Delivery Method and Supplement Consumption by Grazing Ruminants: A Review. JAS 75:543.

Rich, T.D. S. Armbruster, and D.R. Gill. Limiting Feed Intake with Salt. Great Plains Beef Cattle Handbook. GPE 1950.

Sawyer, J.E. and C.P Mathis. Supplement Delivery Systems. New Mexico State University Cooperative Extension Circular 571.

Chapter 37

Creep Feeding

Creep feeding is providing supplemental feed for calves in an enclosure that cows cannot enter. The economic principles discussed in Chapter 31 should be considered when evaluating creep feeding because both the production response and the value of the additional gain influence profitability. Additional considerations when considering creep feeding include the:

1. Effect of creep feeding on forage intake.
2. Effect on cow weight and condition.
3. Post-weaning performance of the calves.

Effect of Creep Feeding on Milk Intake and Cow Condition

Many producers believe creep feeding benefits the cows by reducing the nutritional demands for milk production. Research has shown that calves consuming creep feed reduce their intake of forage, but they will not reduce milk consumption. Consequently, creep feeding typically does not benefit cow condition or reduce the cow's nutritional requirements.

As a general statement, creep feeding can be divided into two types:

1. Free-choice.
2. Limit-fed.

Free-Choice Creep Feeding

Free-choice creep feeding programs provide unlimited access to a creep feed that does not contain a limiting agent. A summary of 31 university trials evaluating free-choice creep feeding is shown in Table 37-1. The average conversion of 9.0 pounds of creep feed for each pound of gain makes free-choice creep feeding of questionable profitability in most situations.

Table 37-1. Summary of 31 Trials Evaluating Free-choice Creep Feeding.

Item	Creep	No Creep
Total gain, lbs	279	221
Daily Gain, lbs	1.83	1.45
Total creep per calf, lbs	524	---
Lbs creep per lb of added gain	9.0	---

Free-choice creep-fed calves often are too fleshy, resulting in a reduction in sale price even greater than what might be expected because of the additional weight. When the quality of the basal forage is low or the quantity is inadequate, free-choice creep feeding might be warranted. For example, free-choice creep feeding might be profitable with fall-born calves grazing dormant winter pasture or crop residues with their dams. It should be noted that the research summary in Table 37-1 is fairly old, and the increased growth rate of today's cattle may make full creep feeding more economical.

Limit-Fed Creep Feeding

Because free-choice creep feeding has been shown to result in poor conversions, many researchers have focused on limiting daily intake to improve conversions and the profitability of creep feeding. Studies evaluating limit-fed creep show conversions from approximately 2.0 pounds of feed per pound of extra gain to 5.5 pounds per pound of additional gain. As a general rule, the greater the restriction of intake, the better the conversion. The economics of limit-fed creep feeding are much better than those of free-choice creep based on the improved conversions.

Table 37-2. Comparison of a High Protein vs. High Energy Creep Feeding Program.

Item	Control	Protein	Energy
No. Calves	15	14	15
Initial Weight, lbs	201	205	200
Calf Gain, lbs (6/4 – 10/15)	230[a]	260[b]	309[c]
Creep Feed per Calf, lbs	---	99	614
Lbs Creep per lb of Added Gain	---	3.3	7.8
Cow Weight Change, lbs (6/4-10/15)	101	88	89

Oklahoma State University. [a,b,c]Means with different superscripts differ significantly (P<.05)

Protein vs. Energy Creep Feed

A key question is whether creep feed should be high in protein or energy. The answer should be based on which nutrient gives the greatest response with consideration of cost differences. Table 37-2 shows the results of a study at Oklahoma State University comparing cottonseed meal limited at less than 1 pound per day to grain with an intake of 4.6 pounds per day. While total gain favored the grain creep, conversion was much better with the high-protein creep feed.

These results indicate that the protein creep was probably increasing calf forage intake and digestibility while the energy creep was being substituted for forage intake. The lack of any positive effect on cow weight change is in line with other research. In general, research has shown that creep feed should be at least 16 percent crude protein.

Response to Limit-Fed Creep

Table 37-3 shows the results of four trials where high-protein, limit-fed creep was supplied to calves in Florida. In these trials, straight cottonseed meal was fed until the calves started eating and then 8 percent salt was mixed with 92 percent cottonseed meal. The researcher noted that the calves receiving the limit-fed creep had more eye appeal and slicker hair coats. The average feed conversion is similar to that observed in the Oklahoma study (Table 37-2).

Effect of Limit-Fed Creep Feeding on Post-Weaning Performance

There have been a number of studies where limit-fed creep has improved performance during the first few weeks following weaning. This improved performance has been attributed to the calves being accustomed to eating from a feeder and adaptation of the digestive system. Thus, weaning stress should be reduced, along with post-weaning health problems.

The influence on performance through the entire growing period has been more variable with some trials showing an improvement and other trials showing a decrease in performance. Since the additional gain in a limit-fed creep feeding program is usually not enough to significantly change condition on calves, one would

Table 37-3. Summary of Four Field Trials Evaluating Limit Feeding of a High Protein Supplement to Nursing Calves.

Forage	Creep Consumption, lbs/ Day	Average Daily Gain, lbs			
		Control	Protein	Increase	Feed/Gain
Bahia-Native	0.40	1.95	2.14	0.19	2.0
Bahiagrass	0.84	1.60	1.74	0.14	6.0
Bahiagrass	0.95	1.20	1.65	0.45	2.1
Pangola-clover	0.75	1.59	1.95	0.36	2.1
Average	0.74	1.58	1.87	0.29	3.0

The University of Florida.

Table 37-4. Effect of Limited-Creep Feeding on Pre- and Post-Weaning Calf Performance.

Item	No Creep	Limit-fed Energy Creep	Limit-fed Protein Creep
Pre-weaning Calf Performance in 61 Days			
Daily Gain, lbs	1.16	1.44	1.50
Daily Creep Intake, lbs	0	1.91	1.82
Creep Per Added Gain, lbs	0	6.8	5.4
Post-weaning Performance in 43 Days			
Daily Gain, lbs	2.1	2.6	2.3
Daily Dry Matter Intake, lbs	12.8	14.2	13.4
Feed Conversion, lbs	6.1	5.6	5.8

Kansas State University

not expect a significant decrease in performance in the growing period. The tendency for limit-fed, creep-fed calves to eat more and gain more early in the post-weaning period might also be expected.

Table 37-4 shows the results of a Kansas study that supports these trends. In this study, the creep-fed calves gained more during the post-weaning period because they consumed more feed. This study also shows that the calves receiving the high-protein creep gained slightly more and converted slightly better pre-weaning than calves receiving an energy creep. The added response might not justify the higher cost of the high protein creep.

Use of Limit-Fed Creep as Part of a Preconditioning Program

In addition to the improved performance during the post-weaning period observed in the Kansas study, there are several other potential benefits of limit-fed creep feeding. For example, limit-fed creep before weaning provides an opportunity to easily ensure that trace minerals and vitamin A are adequate before the stress of weaning. Furthermore, supplemental vitamin E might be beneficial during the stress of weaning. This could be especially beneficial when the quality of forage is low in late summer or early fall.

Creep Feeding Heifer Calves

Several research trials have shown that free-choice creep feeding, which often results in increased deposition of body fat and udder fat can have a adverse effect on the milk production of beef heifers when they are kept as replacements. This is evidenced by reduced weaning weights of calves from these heifers. There seems to be little reason to increase the weaning weight of a heifer calf destined to become a replacement. Consequently, free-choice creep feeding of heifers, at least those that might become replacements, should be avoided. On the other hand, research on the effects of limit-fed creep feeding on future production by heifers is lacking. Since the total number of additional pounds is not large and body condition is not changed significantly, limit-fed creep feeding probably will not affect subsequent milk production.

Starting Calves on Limit-Fed Creep Feed

Typically, it takes some time for calves to start eating creep feed, and if the initial creep feed contains salt or another limiter, it will extend the time before calves consume the creep feed. Consequently, the initial creep feed shouldn't contain a limiter. Then, as soon as the calves start consuming the creep at 1 pound per day, the limiter can be added. In the case of salt, 5 to 10 percent should be added. This level of salt will hold the calves to approximately 1 pound per day until close to weaning, when they will increase consumption. The actual consumption will be influenced significantly by the quality and quantity of the basal forage. Creep feeders should be placed close to water and areas where the cattle typically congregate.

Feed Additives in Creep Feeds

There are several additives that increase the response to creep feed including Aureomycin (CTC), Rumensin®, and Bovatec®. As a general recommendation, one of these additives should be included in a creep feed. Almost all feed companies offer creep feeds with these additives.

Pasture Creep

Another option for creep feeding is to provide access to high-quality forage through a creep-fence. The high-quality forage could be a grass-legume mix, wheat or rye pasture, or simply a cool-season grass adjacent to a pasture of lower quality forage.

Creep Feeding Effects on Genetic Evaluations

Creep feeding might mask differences in milk production reducing the accuracy of genetic evaluations in seedstock herds. It might be advisable for seedstock producers to not creep feed.

References

Bagley, C.P., R.R. Evans, and R.L. Ivy. 1993. Creep Feeds for Spring-Born, Fall-Weaned Beef Calves. Mississippi Prairie Research Unit Report 1993.

Buskirk, D.D., D.B. Faulkner, and F.A. Ireland. 1996. Subsequent Productivity of Beef Heifers that Received Creep Feed for 0, 28, 56, or 84 Days Before Weaning.

Hummel, D.F., D.B. Faulkner, D.F. Parrett, and D.D. Buskirk. 1990. The Effects of Corn vs. Soyhulls Limited vs. Ad Libitum Intake as a Creep Supplement on Calf Performance, Subsequent Performance During Growing and Finishing and Carcass Characteristics. University of Illinois Animal Science Report 1990.

Kunkle, B., Pat Hogue, E, Jennings, and S. Sumner. 1990. Creep Feeding is Still a Good Practice for Some. The Florida Cattlemen, June 1990.

Kunkle, B., J. Fletcher, and D. Mayo. 2002. Florida Cow-Calf Management, 2nd Edition – Feeding the Cow Herd. Florida Cooperative Extension Service Publication AN117.

Lusby, K.S. and D.R. Gill. Creep Feeding. Beef Cattle Handbook BCH-5476.

Lusby, K.S. 1989. Limit Fed Creep Feeds for Nursing Calves. Oklahoma State University.

Mayo, S.L., D.L. Lalman, C.R. Krehbiel, G.E. Selk, R.P. Wetteman and D.R. Gill. 2002. Effect of Fall Calving Cow Nutrition and Calf Creep Feeding on Subsequent Feedlot Performance and Carcass Traits. Oklahoma State University Animal Science Report.

Nelson. L.A. and K.S Hendrix. 1985. Creep Feeding of Beef Calves. Purdue University Cooperative Extension Service Publication AS-415.

Sewell, H.B. Creep-feeding Beef Calves. University of Missouri Extension Guide G2060.

Sexten, W.J., D.B. Faulkner, and F.A. Ireland. 2004. The Professional Animal Scientist 20:211.

Chapter 38

Managing Rangelands and Introduced Pastures

Managing Rangelands

Rangelands are typically unsuitable for cultivation because of soils, topography, or climate. They produce native vegetation that is grazed by domestic animals and wildlife. Rangelands make up about 50 percent of the land in the United States. Introduced pastures are land that has been seeded with non-native species. They are intensively managed using cultivation, fertilization, and/or irrigation.

The very essence of beef production is the conversion of forages into beef. Managing rangelands and introduced pastures for sustainable production is one of the most important roles of the manager. The decisions regarding stocking rate and grazing system are complicated by changes in rainfall from year to year and by periods of drought. A major goal should be to protect the forage resource because even one year of excessive grazing can reduce forage production for many years.

Range Plant Physiology

An understanding of a few basic concepts about range plant physiology is essential for managing pastures and ranges. These concepts include:

1. Range plants do not get their food from the soil. They obtain some basic nutrients like nitrogen from the soil, but they depend on photosynthesis to produce sugars, starches, proteins, and fiber.
2. Photosynthesis occurs primarily in the leaves. When the leaves are removed, photosynthesis is greatly reduced. And cattle prefer the leaves to the stems and other fibrous parts of the plant.

Range plants differ in the amount of leaf material that can be removed and still remain vigorous. As a general statement, 50 to 70 percent of the leaf and stem production should be left after grazing to serve as a metabolic resource with the remaining 30 to 50 percent available for livestock and other herbivores. Plants that are overgrazed cannot maintain their root system, which gradually shrinks. If overgrazing occurs over long periods, rangeland plants will die. The time that the leaf material is removed also influences plant vigor:

1. Removal of the leaves during the period from floral initiation to seed development has the greatest influence on plant vigor because the demand for photosynthetic products is high and the opportunities for regrowth are often low.
2. Removal of the leaves after the plant has gone dormant has the least effect on plant vigor. However, complete removal of all of the plant material during the dormant stage adversely affects plant vigor, especially for some plant species.
3. Removal of the leaves during the period after the plant initiates growth has an intermediate effect on plant vigor because the plant still has an opportunity for regrowth if adequate moisture is available and the temperature is favorable.

Overgrazing results in:

1. Decreased photosynthesis.
2. Reduced root growth.
3. Reduced carbohydrate storage.
4. Reduced seed production.
5. Reduced ability to compete with ungrazed plants, which results in a proliferation of undesirable species and a reduction in species preferred by cattle.

Table 38-1. Summary of 25 studies on Effects of Grazing Intensity on Native Vegetation and Livestock Production in North America.

Item	Grazing Intensity		
	Heavy	Moderate	Light
Average use of forage, %	57	43	32
Average forage production (lbs/acre)	1,175	1,473	1,597
Forage production drought years (lbs/acre)	820	986	1,219
Range trend in ecological condition	Down(92%)[1]	Up (52%)	Up (78%)
Average calf crop, %	77	84	82
Average lamb crop, %	78	82	87
Calf weaning weight, lbs	422	454	431
Gain per steer, lbs	158	203	227
Steer/calf gain per day, lb	1.83	2.15	2.30
Steer gain per acre, lbs	40.0	33.8	22.4
Net returns per animal, $	29.00	39.71	58.89
Net returns per acre, $	1.29	2.61	2.37

Holechek et al. (1999a). [1]Percentage of studies showing this trend.

The Importance of Stocking Rate

The most important decision relative to range and pasture management is stocking rate, which is the amount of land available for each animal for a given time. It is typically expressed as the number of acres required for a cow or stocker animal during the year or grazing season. Inherent in this concept is the optimal stocking rate, which is the maximum stocking rate possible year after year without harming the vegetation. In some areas, the stocking rate is expressed as the carrying capacity of a specific ranch. Stocking rates are also expressed as animal units per section of land in some areas with extensive production. In parts of the western United States, stocking rates are expressed in terms of animal units. An animal unit is one mature cow (1,000 lb) either dry or with a calf up to 6 months of age. This cow would be expected to consume 20 pounds of forage per day. An animal-unit-month is the amount of forage (20 × 30 days) required by one animal unit for one month.

While there has been considerable interest in grazing systems and their impact on carrying capacity in recent years, research has consistently shown that stocking rate is the most important factor in productivity and profitability from rangeland grazing. Research on stocking rates has typically compared low, moderate, and heavy stocking rates. With low stocking rates, palatable species are allowed to maximize their herbage producing ability. Moderate stocking rates allow the palatable species to maintain themselves but not increase in herbage producing ability. Heavy stocking rates do not allow the most desirable forage species to maintain themselves. As shown in Table 38-1, a summary of research trials shows an average forage use of 32, 43, and 57 percent for low, moderate, and heavy stocking rates, respectively.

The summary in Table 38-1 also shows the effect of stocking rate on forage production and animal performance. Moderate stocking rate resulted in the highest returns per acre.

Optimal Stocking Rate Based on Percentage of Forage Harvested by Livestock

For many years, the old adage, "take half leave half," has been used as a rule-of-thumb for setting the optimal stocking rate. This rule may fit in areas of the country with high rainfall like the South and Midwest, but in arid regions, moderate grazing should remove less of the forage produced. This concept is supported by the research shown in Table 38-2 where a conservative stocking rate with 33 percent of the forage used was the most profitable level in an arid environment.

Table 38-2. Influence of Grazing Intensity on Cow/Calf Production at the Chihuahuan Desert Rangeland Research Center in South-Central New Mexico.

Item	Grazing Intensity	
	Conservative	Moderate
Duration of study, years	7	7
Average annual precipitation, in.	8.6	8.6
Utilization of forage, %	33	45
Average forage production in drought yrs, lbs/acre	122	64
Average forage production, lbs/acre	156	125
Calf crop, %	85	78
Calf weaning weight, lbs	485	476
Net income, $/acre	0.52	0.31

Holechek et al. (1999b).

Evaluating Grazing Intensity

Grazing intensity can be visually evaluated using the following guidelines:

1. **Low intensity** – The range appears undisturbed. There is no use of poorer quality plants.
2. **Moderate intensity** – Most accessible range shows some evidence of grazing. There is little evidence of trailing. Highest quality plants show some use and fair quality plants show some use.
3. **Heavy intensity** – The range has a "mowed" appearance. There is evidence of livestock trailing. Even the least desirable plants show evidence of grazing.

Table 38-3 shows general guidelines for converting stubble heights into percentage of use for different types of grasses.

Grazing Systems

Over the last 30 to 40 years, there has been considerable interest in the effect of grazing systems on animal performance and the carrying capacity of cattle operations. In many cases, there have been impressive claims about the benefits from rotational grazing systems compared to the more traditional continuous grazing system. A review of the applicable research indicates that there are both advantages and disadvantages of all grazing systems, and the "best" system for a particular operation depends on the goals and objectives of the manager.

Most grazing systems fit into one of four categories:

1. Season-long Continuous Grazing
2. Rest-Rotation Grazing
3. Deferred-Rotation Grazing
4. Intensively Managed Grazing (Short Duration Grazing or Cell Grazing)

Table 38-3. General Grazing Intensity Guide for Converting Stubble Height of Shortgrasses, Midgrasses, and Tallgrasses into Percentage Use.

Grazing Intensity	Stubble Height Guide			% Use of Forage
	Shortgrasses	Midgrasses	Tallgrasses	
	---------- Inches ----------			%
Light use or no use	2.5+	9+	16+	0-30
Conservative	2.0-2.5	8-9	14-16	31-40
Moderate	1.5-2.0	6-8	12-14	41-50
Heavy	1.0-1.5	4-6	10-12	51-60
Severe	<1.0	<4	<10	>60

Based on Holechek and Galt (2000).

Season-long Continuous Grazing

As the name implies, with this system the cattle graze the same pasture for the entire grazing season.

Advantages:

1. Maximizes average daily gain of the cattle.
2. Easy to manage because of few decisions regarding cattle movement.
3. Minimal expenses for fencing and water.

Disadvantages:

1. Grazing distribution is often a problem, especially in large pastures.
2. Minimal potential for improving range condition.
3. May require considerable time to "check" cattle in large pastures.
4. Cattle may apply too much grazing pressure on preferred species reducing their presence in future years.

Rest-Rotation Grazing

With this grazing system, one pasture is rested for the entire year to promote improved range condition. For example, in a four-pasture rest-rotation system, one of the four pastures would be rested each year and the cattle rotated through the other three pastures. The pasture being rested would be rotated from year-to-year. Effectively, this system increases the grazing pressure in the pastures grazed in any single year if the number of cattle grazed is not reduced.

Advantages:

1. Improved grazing distribution compared to continuous grazing.
2. Animal management is easier than with continuous grazing.
3. Improves range condition compared to continuous grazing.

Disadvantages:

1. Poorer gains than with continuous grazing.
2. May require a reduction in cattle numbers because only three-fourths of the land is being grazed.

Deferred-Rotation Grazing

With this system, the rangeland is divided into four or more pastures and the cattle are rotated through the pastures. Each pasture is only grazed once during the grazing season, and the length of time in each pasture may need to be adjusted based on forage availability. The order of grazing should be rotated from year-to-year so that the same forage species in a pasture is not grazed at the same time each year.

Advantages:

1. Improves range condition compared to continuous grazing.
2. Increases the vigor of preferred plant species.
3. Facilitates cattle management.
4. Improves grazing distribution compared to continuous grazing.

Disadvantages:

1. Reduced cattle gains compared to continuous grazing because cattle are forced to graze mature plants when the last pasture is grazed.
2. Increased cost of fencing and water development compared to continuous grazing – lower cost than with intensively managed grazing.
3. More risk of making mistakes in pasture moves than with continuous grazing – less risk than with intensively managed grazing.

Intensively Managed Grazing (Short Duration Grazing or Cell Grazing)

With this system, the rangeland is divided into a large number (at least eight) of pastures or paddocks. The cattle are rotated through the paddocks rapidly. The time spent grazing in each paddock should be flexible depending on forage availability.

Advantages:

1. Maximum grazing distribution.
2. Facilitates livestock management.
3. Flexible with respect to accomplishing individual pasture management objectives.

Disadvantages:

1. Requires high level of management.
2. Significant fencing and water development expenses.
3. High risk for mistakes with respect to making pasture moves.
4. Does not improve range condition as well as the deferred-rotation system.
5. Does not increase the vigor of preferred species as well as the deferred-rotation system.

Selecting a Grazing System

Reviewing the advantages and disadvantages of the common grazing systems makes it clear that there is not a "best" system for all situations. The objectives in rangeland use dictate the "best" system. For example, if maximum livestock performance is the goal, continuous grazing is probably the "best" system. If improving range condition is a priority; one of the rotation systems might be preferable. Another factor to consider when deciding on a grazing system is grazing distribution. If it is a problem, the rotational grazing systems have the added benefit of improving grazing distribution compared to continuous grazing. It may be possible, however, to solve a grazing distribution problem in a continuous grazing system as discussed later in this chapter. Other factors to consider include the cost of developing water and fencing required for the rotational grazing systems. Finally, consider the time required to manage rotational grazing.

Table 38-4 shows a summary of trials comparing rotational grazing systems to continuous grazing on rangelands. While there did appear to be slightly more forage produced with rotational grazing, animal performance appeared to be poorer. This reduction in animal performance under rotational grazing occurs because with continuous grazing, animals select a higher-quality diet throughout the grazing season. To illustrate, consider a four-pasture deferred-rotation system compared to continuous grazing. With the deferred-rotation system, when the cattle are rotated to the last pasture, the plants are already mature and of low quality. Meanwhile, the cattle in the continuous grazing system can still select less mature, higher-quality forage from regrowth.

Short Duration Grazing (Cell Grazing)

Because proponents of this grazing system have made such impressive claims about its benefits, this system deserves additional discussion. Cell grazing was first proposed by Allan Savory and evaluated in Zimbabwe starting in the 1960s. The idea is that the cattle will always be eating immature, high-quality forage during the growing season. A research review indicates that this appears to be true when cattle are grazing seeded pastures with a long growing season, but grasses in native pastures and ranges have a relatively short growing season, which reduces the potential for keeping the plants in an immature, high-quality state.

Proponents of short duration grazing claim that stocking rate can be significantly increased using this system. Research in the United States and Africa does not support these claims. Table 38-5 shows a summary of research trials comparing short duration grazing to continuous grazing. While the results are somewhat variable, there certainly does not seem to be an economic advantage for short-duration grazing.

Table 38-4. Summary of 15 Studies on Effects of Rotational Grazing Systems on Native Rangeland Vegetation and Livestock Production in North America.

Item	Season-Long or Continuous Grazing	Rotation Grazing
Average use of forage, %	41.8	42.4
Increase in forage production, %		+7%
Range Trend	Up=61%, Stable=31%, Down=8%	Up=69%, Stable=8%, Down = 23%
Average calf crop, %	89.4	85.9
Calf weaning weight, lbs	504.6	494.1
Net returns, $/acre	6.60	6.37

Holechek et al. (1999c).

Table 38-5. Financial Returns from Studies Comparing Short-Duration Grazing and Continuous Grazing Systems.

Location	Type of Livestock	Net Returns Short-Duration Grazing	Continuous Grazing
		$/Acre	
North Central Texas	Cow/Calf	6.36	5.25
South Central Texas	Cattle-Sheep	7.39	7.20
Southeastern Wyoming	Yearling Cattle	12.07	15.20
North Central Oklahoma	Yearling Cattle	2.83	8.50
Average		7.16	9.04

Rangelands 22(1):18-22.

Short Duration Grazing Compared to Deferred-Rotation Grazing

If a producer wants to take advantage of the benefits of using a rotational grazing system, the question becomes which of the rotational grazing systems offers the greatest benefit. Starting in 1999, the University of Nebraska conducted a 10-year study comparing deferred-rotation to a cell grazing system. In summary, the system used had little effect on the botanical composition changes during the study. In addition, the crude protein content and digestibility of the diets did not differ, and animal performance was similar. Based on these results, the researchers concluded that the deferred-rotation system was the most practical because of the lower cost of developing the system and the reduced management requirements.

Grazing Distribution

Grazing distribution is essentially a measure of the evenness of grazing in a pasture. The major causes of poor grazing distribution include:

1. Long distance to water
2. Uneven topography
3. Vegetation type

Distance to Water

The distance to water is the major cause of poor grazing distribution. The relationship between the percentage of use and the distance from water is illustrated in Table 38-6.

Table 38-6. Average Percent Use of Forage by Cattle of Flat Good Condition, Chihuahuan Desert Rangeland with Varying Distances from Water Under Moderate Continuous Grazing.

Distance from Water (Miles)	% Utilization
0 – 0.5	50
0.5 – 1.0	38
1.0 – 1.5	26
1.5 – 2.0	17
2.0 – 2.5	12

Journal of Forestry 10:749-754.

The recommended distance to water is also a function of the type of terrain. Bell (1973) in his range management book recommended maximum travel distance to water as follows:

Rough country ....0.5 miles

Rolling, hilly country....................1.0 mile

Flat country.................................2.0 miles

Smooth, sandy country.................1.5 miles

Undulating, sandy country ....1.0 mile

In addition to causing uneven grazing, long distances to water influence performance because of the energy expended traveling to and from water.

Topography

The second most important cause of poor grazing distribution is topography. Cattle simply will not graze steep slopes as much as gentle slopes or flat ground as shown in Table 38-7. A preference index greater than one indicates more use by cattle than expected by random use. Values less than one indicate a particular slope was not preferred.

Vegetation Type

Another cause of poor grazing distribution is the different plant species found in different parts of a pasture. For instance, the plants found in draws or bottomland are often different from those found on the slopes or tops of hills. There also may be a seasonal effect with cool-season grasses in the bottoms in early spring and warm-season grasses later in the summer. Cattle graze areas where the plants are most palatable more than other areas in the pasture.

Improving Grazing Distribution

The development of additional water sources, placement of supplements, and fencing can improve grazing distribution. However, other approaches are used in some situations. For example, spot burning an underused area often increases use because cattle like the succulent plants that grow in a recently burned area. Another approach used in some of the large rangeland areas is driving cattle to areas that are being underused. In seeded pastures, applying fertilizer to underused areas also tends to increase use.

Water

Since the distance from water is the major cause of poor grazing distribution, developing new sources of water is the best way to improve grazing distribution. Adding additional water sources often increases the carrying capacity of a ranch significantly.

Placement of Supplements

Many producers improve grazing distribution by feeding supplements in different areas in a pasture to encourage cattle to graze in these areas. Mineral feeders also can be moved to new locations periodically. Placement of self-fed supplements like tubs also can be used to attract cattle to underused areas.

Fencing

In many situations, cutting a large pasture into smaller pastures greatly improves grazing distribution. These smaller pastures can then be continuously grazed or rotationally grazed depending on the number of cows and bulls.

Season of Use

Another factor influencing livestock and vegetation response to grazing is season of use. Using native rangeland or forages when they are of highest quality enhances livestock performance. Sequence grazing of different types of forages when they are productive and high quality lengthens the grazing season and can have a major influence on livestock performance and profitability. For example, grazing cool-season forages in the spring followed by grazing native warm-season dominated rangeland during the summer and

Table 38-7. Preference Indices for Slope Gradient Classes During Three Grazing Periods as Determined by Direct Cattle Observations on Mountain Rangelands in Northeastern Oregon.

Slope Gradient %	Season-Long Grazing	Early Summer Only	Late Summer Only
5 or less	3.0	3.7	4.0
6-10	5.3	2.9	4.8
11-15	0.8	1.6	1.6
16-20	0.8	1.1	0.6
21-30	0.4	0.5	0.5
31-45	0.3	0.4	0.4
Greater than 45	0.1	0.1	0.1

Journal of Range Management 37:549-553.

then returning to the cool-season forages in the fall can greatly extend the grazing season. This type of sequence grazing allows a rest period for both the rangeland and pastures that maintains plant vigor.

Key Points Related to Rangeland Management

1. Adequate leaf material must be left after grazing to allow the plants to build reserves.
2. Stocking rate is the most important factor in managing the forage resource.
3. Grazing systems probably will not allow major increases in stocking rate unless they solve a grazing distribution problem.
4. Continuous grazing can be an effective system if the stocking rate is at an acceptable level and steps are taken to ensure that grazing distribution is not a problem.

Managing Introduced Pastures

In many areas of the United States, the basal forage consists of introduced grasses rather than native plants. For example, fescue is widely used for cattle grazing across the South. Numerous other grass species have been planted on land that was farmed in the early 1900s. In many cases, poor farming practices resulted in erosion and abandonment. Subsequently, much of this land was seeded to grasses and other forage species including various legumes. In other areas, forests have been cleared and planted to forage species. In many areas of the West, cattle are grazed on irrigated pastures.

Many of the management concepts that were discussed in the rangelands section apply to these introduced pastures. For example, stocking rate is still the major factor influencing long-term productivity of the pasture. Overgrazing results in the proliferation of weeds and loss of plant vigor similar to that observed on rangelands. Prolonged overgrazing greatly reduces the long-term productivity of the pasture. However, there are some basic differences between managing rangelands and managing introduced pastures. For example, research has shown that fertilization usually is not profitable on native range, but applying fertilizer is often profitable on introduced pasture. Another difference is that introduced pastures respond differently to grazing systems.

Fertilization of Introduced Pastures

Most introduced pastures are fertilized during the year to increase forage production. The amount and timing of the fertilization depends on the species and the expected cost/return scenario. The response in forage production to fertilization follows the law of diminishing returns with the greatest response to the first few units of fertilizer. Consequently, a key management decision is the amount of fertilizer to apply in any given year. Factors to consider include the cost of fertilizer, expected forage production response, expected cattle production response, and the value of the additional gain or increase in carrying capacity of the pasture. The Cooperative Extension Service in most states has conducted research trials that provide information on the forage and cattle production responses on introduced pastures in the state. All of the variables involved in establishing the optimum level of fertilizer change from year to year making it difficult to make an accurate decision. For example, applying fertilizer before a drought means the typical production response probably will not be achieved. All producers can do is use "rules-of-thumb" values and hope rainfall is close to normal and production responses and price estimates are fairly accurate.

Rotational Grazing Introduced Pastures

While research on rotational grazing of rangelands has not shown a consistent benefit compared to continuous grazing, the response on introduced pastures has been more positive, although the results have been variable. A review of the research comparing rotational grazing to continuous grazing indicates that rotational grazing on introduced pastures has the following advantages:

1. Increased forage production, which may result in more grazing days or allow higher stocking rates.
2. Fewer weeds and greater incidence of more desirable plant species.
3. Less forage waste.

The influence of rotational grazing on average daily gain and gain per acre has been variable. A review of research trials indicates rotational grazing does not improve average daily gain significantly, but some studies have shown a significant improvement in gain per acre especially with high stocking rates because of the increase in forage production provided by rotational grazing.

Rotational grazing certainly is not a panacea. It requires considerable investment in fences and water development and time to manage the system. However, it does appear to offer the potential to increase the gain per acre if managed properly.

Rotational grazing systems can be simple or complex. They may consist of only three or four pastures (paddocks) or many more paddocks. The key to making rotational grazing work is to move the cattle to a paddock containing high-quality forage and provide a rest period for the paddock recently grazed. Because of year-to-year variation in forage production, flexibility is required to effectively manage a rotational grazing system. This may require adjusting the number of cattle in the system to the level of forage production. Moreover, a successful program may require developing ways of using excess forage such as haying or stockpiling the forage produced in one or more paddocks.

The influence of a rotational grazing system on profitability depends on many factors including the plant species and the ability to take advantage of the additional forage produced with rotational grazing compared to continuous grazing. Research comparing continuous grazing to rotational grazing shows that there is significant year-to-year variation. For example, in some years there will be little difference in gain per acre while in other years there will be considerable difference.

It is beyond the scope of this text to provide specific recommendations for all introduced pasture situations given the large number of plant species used across the United States and the diversity of environments. Check with the Cooperative Extension Service in states with significant acreages of introduced pastures for information on the management of these pastures to optimize production.

References

Bell, H.M. 1973. Rangeland Management for Livestock Production. University of Oklahoma Press, Norman, OK.

Bidwell, T.G. Management Strategies for Rangeland and Introduced Pastures. Oklahoma State University Cooperative Extension Bulletin NREM-2869.

Gerrish, J. Impact of Stocking Rate and Grazing Management System on Profit and Pasture Condition. Forage Systems Research Center – University of Missouri.

Gillen, R.F., W.C. Krueger, and R.F. Miller. 1984. Cattle Distribution on Mountain Rangeland in Northeastern Oregon. Journal of Range Management 37:549-553.

Gillen, R.L. and P.L. Sims. 2002. Stocking Rate and Cow/Calf Production on Sand Sagebrush Rangeland. Journal of Range Management 55(6):542-550.

Heitschmidt, R. and J. Walker. 1983. Short Duration Grazing and the Savory Grazing Method in Perspective. Rangelands 5(4):147-150.

Holechek, J.L. 1988. An Approach for Setting the Stocking Rate. 1988. Rangelands 10(1):10-14.

Holechek, J.L., H. Gomez, F. Molinar, and D. Galt. 1998. Grazing Intensity: Critique and Approach. Rangelands 20(5):15-18.

Holechek, J.L., H. Gomez, F. Molinar, and D. Galt. 1999a. Grazing Studies: What We've Learned. Rangelands 21(2):12-16.

Holechek, J.L., M. Thomas, F. Molinar, and D. Galt. 1999b. Stocking Desert Rangelands: What We've Learned. Rangelands 21(6):8-12.

Holechek, J.L., H. Gomes, F. Molinar, D. Galt, and R. Valdez. 1999c. Short-Duration Grazing: The Facts in 1999. Rangelands 22(1):18-22.

Holechek, J.L. and D. Galt. 2000. Grazing Intensity Guidelines. Rangelands 22(3):11-14.

Holechek, J.L., R.D. Pieper, and C.H. Herbel. 2004. Range Management Principles and Practices. Pearson Education, Inc.

Joseph, J., F. Molinar. D. Galt, R. Valdez, and J. Holecheck. 2002. Short Duration Grazing Research in Africa. Rangelands 24(4):9-12.

Lalman, D. and D. Dye. Eds. 2008. Beef Cattle Manual. Oklahoma State University Extension Service, Stillwater, OK.

Lauriault, J.E. Sawyer, and R.D. Baker. 2003. Grazing Systems and Management for Irrigated Pastures in New Mexico. New Mexico Cooperative Extension Service Circular 586.

McKown, C.D., J.W. Walker, J.W. Stuth, and R.K. Heitschmidt. 1991. Nutrient Intake of Cattle on Rotational and Continuous Grazing Treatments. Journal of Range Management 44(6):596-601.

Olson, K.C. 2005. Range Management for Efficient Reproduction. Journal of Animal Science 83:E107-E116.

Reece, P.E., J.D. Volesky, and W.H. Schacht. 2008. Integrating Management Objectives and Grazing Strategies on Semi-arid Rangeland. University of Nebraska Cooperative Extension Bulletin EC158.

Rouquette, F.M. Management Strategies for Cool-Season Annual Pastures and Stocker Cattle. Texas Agri-life Research and Extension Center at Overton.

Schacht, W.H., J.D. Volesky, D.E. Bauer, and M.B. Stephenson. 2011. Grazing Systems for Nebraska Sandhills Rangeland. University of Nebraska Cooperative Extension Bulletin EC127.

Valentine, K.A. 1947. Distance to Water as a Factor in Grazing Capacity of Rangeland. Journal of Forestry 45:749-754.

Walker, J. W., R.K. Heitschmidt, E. A. De Moraes, M.M. Kothmann, and S.L. Dowhower. 1989. Quality and Botanical Composition of Cattle Diets Under Rotational and Continuous Grazing Treatments. Journal of Range Management 42(3):239-263.

Chapter 39

Feeding Replacement Heifers

Developing replacement females is an expensive part of any cow/calf operation. Economic analyses of factors affecting profitability in cow herds have identified replacement rate as a key factor. In many herds, the percentage of heifers kept as replacements that wean a calf and rebreed for a second calf is too low. The net result is excessive cost per replacement.

Consequently, feeding replacement heifers to maximize the chance that they will rebreed a second year should be an important part of a cow herd nutrition program. The result of poor heifer development is shown in Table 39-1.

Inadequate development has long-term effects beyond just getting heifers to cycle and breed for their first calf. As shown in Table 39-1, poorly developed heifers rebred at a lower rate for their second calf. Even the highest level of nutrition in this study is probably too low because only 72.7 percent (0.835 × 0.871) of the heifer calves kept for breeding had a chance to calve as a 3-year-old. A good target is for more than 80 percent of the heifer calves kept as replacements to eventually calve as 3-year-olds.

Weaning to Breeding

Nutrition provided to replacement heifers during their first year has longer term effects than illustrated in Table 39-1. Research has shown heifers calving early with their first calf tend to calve early the rest of their lives and wean heavier calves. Conversely, heifers that calve late with their first calf tend to calve late the rest of their lives and have a much greater chance of failing to conceive in subsequent breeding seasons. This makes sense considering a late-calving heifer or cow has fewer opportunities to conceive in a breeding season of limited length. With year-round breeding, the effect of poorer reproduction and lower weaning weights are masked to some extent, but the overall effect is the same — fewer calves weaned in a lifetime and fewer pounds sold per cow per year.

Table 39-1. The Effect of First Winter Nutrition on Subsequent Performance of Replacement Heifers.

Item	Lbs Grain/Day with Hay 0.0	3.0	6.0
Number of Heifers	112	113	112
Fall Weight, lbs	496	502	493
Winter ADG, lbs	0.07	0.50	0.80
Weight at Breeding, lbs	506	577	613
Percent Conceiving as Yearlings	69.2	73.9	83.5
Rebreeding after First Calf, %	67.3	75.4	87.1
Weaning Weight First Calf, lbs	405	433	443

Purdue University.

The Goal – Puberty by the Target Date

The first step in developing a good nutrition program for the replacement heifer following weaning is to establish the desired calving date. Based on this date, the desired breeding date, or target date, can be established. For example, assume the heifers are to be bred to start calving on February 1. Using an average gestation length of 282 days, breeding must be initiated on April 20, which becomes the target date. The goal, then, is to get a high percentage of heifers cycling by the target date.

Factors Influencing the Initiation of Cycling Activity (Puberty) In Heifers

Puberty is reached when a heifer becomes sexually mature and starts cycling (releasing an egg every 21 days). The most visible outward sign of puberty is standing when mounted by other females or a bull. There are a number of factors that influence when a heifer reaches puberty including:

1. **Age** – Heifers typically have to reach a certain age before they will cycle, even if they grow rapidly following weaning. They must be physiologically mature enough to start cycling, regardless of body weight.
2. **Weight** – Heifers must reach a certain percentage of their mature weight before they will start cycling. A good rule-of-thumb is that they should weigh 65 percent of their mature weight at the start of breeding, as will be discussed in more detail later.
3. **Breed** – Breed differences for age at puberty were discussed in Chapter 8. There are significant differences among breeds in the average age at puberty in a similar environment. This difference must be accounted for in any plan to have heifers reach a target weight by a specific date. These data also indicate that with a good nutrition program, it is possible to get all breeds to reach puberty early enough to calve by 23 months of age, if desired.

Target Weights at Breeding

For many years, a rule-of-thumb was that yearling heifers needed to weigh 65 percent of their mature weight at breeding. Recent research shows this percentage may be slightly higher than necessary, but it has worked successfully for many years and is still a good target. Table 39-2 illustrates a comparison of developing heifers for 55 vs. 65 percent of their mature body weight. As shown, inadequate development reduced the percentage of heifers becoming pregnant, increased calving difficulty, increased calf death loss, and reduced rebreeding success for the second calf. All of these results have a negative influence on profitability.

Establishing the Weaning to Breeding Average Daily Gain

Using the rule-of-thumb of 65 percent of mature weight, a heifer that will weigh 1,250 pounds as a mature cow in body condition 5, needs to weigh 1,250 × 0.65 = 813 pounds at breeding. Using this rule-of-thumb and the target date, a rate of gain from weaning to breeding can be established for this heifer as follows:

1. Age at Weaning = 215 days
2. Age at Breeding = 13.7 months = 416 days
3. Weight at Weaning = 500 pounds
4. Days to Gain Desired Weight = 416-215 = 201 days
5. Weight Gain Required = 813-500 = 313 pounds
6. Average Daily Gain Required = 313 ÷ by 201 = 1.56 pounds per day

Table 39-2. Effect of Pre-breeding Development on Subsequent Heifer Performance.

Percent of Mature Weight at Breeding	No.	Pre-Breeding Weight, lbs	Calving Weight, lbs	Calf Birth Weight, lbs	Calving Difficulty, %	Calf Death Loss, %	Fall Preg., %
55	60	600	834	70.9	52.3	6.2	85.0
65	61	683	897	73.3	28.8	4.5	93.4

Kansas State University Extension Bulletin C-841.

Table 39-3. Effect of Group Feeding on Reproduction in Replacement Heifers.

Item	Fed Together		Fed Separately	
	Heavy	Light	Heavy	Light
Initial Weight, lbs	476	377	463	375
Ave. Weaning Age, Days	204	187	200	185
Projected Ending Winter Weight, lbs	650	650	650	650
Actual Ending Winter Weight, lbs	719	588	670	675
Puberty Age, Days	404	423	389	405
October Pregnancy, %	80	60	90	79

Fort Keogh Range Livestock Research Center, Miles City, Montana.

Based on Appendix Table 6, a 500-pound heifer requires a diet with 0.36 NEg and 12.9 percent crude protein to gain 1.5 pounds per day. The procedures for balancing this diet are illustrated in Chapter 46.

Separating Heifers into Weight Groups from Weaning to Breeding

Competition among a group of heifers can result in fewer heifers reaching puberty by the desired date. This is illustrated in Table 39-3 where heifers above the weaning average and below the weaning average were fed together or separated. When fed together, the heavier heifers at weaning continued to gain faster and the difference between the heavy and light groups was even greater after the winter feeding period.

When the heavy and light groups were fed separately, the lighter heifers at weaning caught up to the heavier group by the end of the winter feeding period. The net effect was a significant reduction in the average days of age at puberty and an increase in the percentage pregnant in the fall. It is not always practical to separate heifers for feeding because of a lack of pen space, but if practical, it should result in superior development of herd replacements.

Effect of the Time of Gain on Replacement Heifer Reproductive Performance

As noted, the goal is to get yearling heifers to a body weight that is adequate for puberty by the target date. However, an important question is whether the timing of this gain affects subsequent performance. In other words, does it matter when the growth occurs relative to the target breeding date? Research has shown that it does not seem to matter when the gain occurs as long as the desired body weight is attained. For example, a University of Nebraska study evaluating timing of gain is shown in Table 39-4.

The differences in pregnancy rates were not significant.

Table 39-4. Effect of Time of Gain on Replacement Heifer Reproductive Performance.

	Treatment		
Gain 1st Half of Development Period, lbs/Day	0	1.0	2.0
Gain 2nd Half of Development Period, lbs/Day	2.0	1.0	0
Number of Heifers	60	60	60
Initial Weight, lbs	408	406	408
Final Weight, lbs	619	613	628
Puberty Age, Days	403	394	392
% Exhibiting Estrus by Start of Breeding	85	90	90
Pregnancy Following Breeding, %	75	82	73
Calf Weaning Weight, lbs	300	304	300

University of Nebraska.

Feeding Ionophores to Heifers between Weaning and Breeding

Ionophores are polyether antibiotics approved as supplements for beef cattle. The two most commonly fed ionophores are Rumensin® and Bovatec®, both of which have been shown to be beneficial when fed to heifers from weaning to breeding. These products increase gain and improve feed conversion, reducing the cost of feed to get the heifer to target weight. Furthermore, the benefits of these products extend beyond what one might expect from the additional gain and improved feed conversions. It appears some hormonal effect of these ionophores stimulates cycling activity. It is recommended that 150 mg per day of either of these products be fed from weaning to breeding.

Breeding to Calving

As a general rule-of-thumb, pregnant heifers should be fed to weigh approximately 85 percent of their mature weight at calving and be in a body condition score of 5 to 6, preferably 6. Table 8 in the Appendix shows the nutrient levels required to achieve the 85 percent target. Numerous studies have shown that low body condition at calving will delay return to estrus, and heifers typically take at least 20 days more than mature cows to start cycling following calving. This is the basis for the recommendation to breed yearling heifers 20 to 30 days ahead of the cow herd.

Feeding Replacement Heifers Separate from the Cow Herd

Yearling replacement heifers should be fed separately from the mature cows because of their higher nutrient requirements on a concentration basis. Furthermore, they are not able to compete with mature cows in a hand-feeding situation or when feeder space is limited. Even if adequate hay is supplied for all animals, for example, the mature cows may eat the higher-quality hay leaving the lower-quality hay for the heifers. If it is not practical to feed the heifers separately, it may be necessary to feed the entire herd a higher plane of nutrition to obtain satisfactory reproductive performance from the heifers.

Impact of Pre-calving Nutrition on Calving Difficulty

Many producers restrict the growth and deposition of fat by bred yearling heifers in an attempt to reduce calving problems. This is usually unsuccessful — it only results in poorer reproductive performance. This is illustrated in the summary of 3 years of research at Miles City, Montana, shown in Table 39-5. The low feed level was 7.1 to 7.9 pounds of TDN per day while the high level was 13.7 to 15.0 pounds. These diets were fed for the last 90 days of gestation. Pregnancy rate was based on a 45-day AI period.

This research clearly shows that trying to "starve" calving difficulty out of heifers does not work. This and other research have shown that unless the heifers are severely starved, birth weight is only reduced slightly and calving difficulty is usually increased because the heifers are weaker. Furthermore, inadequate nutrition before calving in yearling heifers consistently results in poorer reproductive performance, higher calf mortality, and reduced weaning weights.

Effect of High Protein Levels during Gestation on Calf Birth Weights and Calving Difficulty

There have been a couple of studies where high protein levels (well above requirements) have resulted in a significant increase in birth weight and calving

Table 39-5. Effect of Pre-calving Nutrition on Calf Birth Weights, Calving Difficulty, and Subsequent Pregnancy Rate.

Precalving Gestation Feed Level	Calving Body Weight, lbs	Calf Birth Weight, lbs	Calving Difficulty, %	Pregnancy Rate, %
Low	694	68.3	61	65
High	794	70.5	56	83
Difference	+99	-2.2a	-5	+18[a]

Fort Keogh Range Livestock Research Center, Miles City, Montana. [a]Difference significant at the P<.05 level.

difficulty. However, this effect has not been consistent in research trials. But, there have been numerous reports where heifers consuming forages high in protein before calving have exhibited high birth weights and severe calving problems. Consequently, it is wise to avoid high protein levels late in gestation.

Calving to Breeding

The period between calving and breeding is crucial to the development of replacements, and certainly a period where nutritional requirements should be met. This group of females has requirements for maintenance, growth, lactation, and reproduction with reproduction only occurring if the other production functions are satisfied. Inadequacies during this period can have a significant influence on reproduction as shown earlier in Table 22-3. Appendix Table 4 shows the nutrient requirements of lactating first-calf-heifers.

References

Bellows, R.A. Managing Heifers for Optimum Reproduction. Fort Keogh Livestock and Range Research Laboratory.

Beverly, J.R. and J.C. Spitzer. 1980. Management of Replacement Heifers for a High Reproductive and Calving Rate. Texas A&M University Cooperative Extension Service Bulletin B-1213.

Bolze, R. and L.R. Corah. 1993. Selection and Development of Replacement Heifers. Kansas State University Cooperative Extension Service Bulletin C-841.

Fox, D.G. 1986. Feeding Herd Replacement Heifers. Cornell Beef Production Reference Manual Fact Sheet 1300a.

Kunkle, W.E. and R.S. Sand. 1993. Nutrition and Management for the Replacement Heifer. Proceedings Northwest Florida Beef Production Conference.

Kunkle, W.E., R.S. Sand, and P. Garces-Yepez. 1995. Applying New Strategies for Raising Beef Heifers. 1995. 6th Annual Florida Ruminant Nutrition Symposium.

Larson, D.M., J.L. Martin, and R.N. Funston. 2008. Effect of Wintering System and Nutrition Around Breeding on Gain and Reproduction in Heifers. Nebraska Beef Cattle Report MP 91.

Patterson, D.J., K.D. Bullock, W.R. Burris, and J.T. Johns. 1995. Management Considerations in Beef Heifer Development: Breed Type, Weight & Height, Reproductive Tract Score. Cooperative Extension Service University of Kentucky Bulletin ASC-144.

Summers, A.F., S.P. Weber, T.L. Meyer, and R.N. Funston. 2012. Late Gestation Supplementation Impacts Primiparous Beef Heifers and Progeny. Nebraska Beef Cattle Report.

Wiltbank, J.N. 1985. Beef Cattle Reproduction and Management. Red Poll News Summer 1985.

Chapter 40

Feeding and Managing Herd Bulls

Breeding bulls are overlooked in many cow herd feeding programs. Unfortunately, in some cases this results in lower than desirable fertility and libido. Over the last 40 years, there has been a shift from purchasing 2-year-old bulls to purchasing yearling bulls. This has increased the importance of feeding programs for adequate bull development on most cow/calf operations.

Weaning to Breeding Age

The goal during this period is to grow bulls at a rate of gain that allows them to reach puberty and adequate size for use in breeding at 13 to 15 months of age. The rate of gain needed to accomplish this goal is a function of weaning weight and breed. Excessively low rates of gain delay puberty. Conversely, there is some evidence that rates of gain that result in excessive fat deposition can lower semen production and quality. Unfortunately, purebred breeders recognize that "fat sells," which results in most yearling bulls being overly fat before their first breeding season. From a developmental standpoint, bull calves should probably be grown at moderate rates of gain (2.5 to 3.25 pounds per day) following weaning rather than gains approaching or exceeding 4.0 pounds per day. Nutrient requirements for various rates of gain are shown in Appendix Table 7.

Yearling Bulls during the Breeding Season

Typically, yearling bulls lose a great deal of weight during their first breeding season. Supplementing these bulls during this period is advisable, but is rarely practical. If they are not kept in the breeding pastures for too long (more than 60 days), they can be fed adequately after removal from the breeding pasture to regain some condition and maintain structural growth. If they are left in the pasture for prolonged periods, there is a tendency for some reduction in mature size. In some cases, if the forage base is of poor quality or of limited supply, "stunting" may occur. Unfortunately, there has been little research studying the effect of stunting on future reproductive performance. The eye appeal of the bull is reduced, but the influence on performance as a breeding bull is unclear.

Yearling Bulls between the First and Second Breeding Seasons

The goals during this period should be to provide adequate nutrition for skeletal growth and overall development. A good target is to have the bull in at least a body condition score of 5 and preferably a 6 by the start of the next breeding season. Appendix Table 9 provides some general guidelines for nutritional requirements sufficient to meet these goals.

Mature Bulls

Body condition score has been shown to have an influence on the percentage of bulls that pass a breeding soundness exam (BSE) as shown in Table 40-1, which represents more than 3,000 exams conducted

Table 40-1. Relationship of Body Condition Score to the Percentage of Bulls Passing a Breeding Soundness Exam.

Body Condition Score	BSE Pass Rate, %
<4	55
4	70
5	72
6	67
>6	62

Texas Cooperative Extension Service.

Table 40-2. Relationship between Body Condition Score, Loss in Scrotal Circumference, and Breeding Soundness Examination (BSE) Pass Rate in November on a Texas Ranch.

Number of Bulls	Body Condition Score	Average Scrotal Circumference Change, cm	BSE Pass Rate %
6	4	-3.2	33
4	5	-1.9	25
4	6	-0.8	75

Food Animal – The Compendium 1997.

on more than 400 bulls in the Texas Department of Corrections beef herds. As shown, the highest rates of passage were when the bulls were in body condition score 4 to 6. Extremes in either direction tended to reduce semen quality and BSE scores.

A good goal when feeding mature bulls is to have them in a body condition score of 5 or 6 at the start of the breeding season. This score optimizes the number passing a breeding soundness exam and provides the bulls with energy reserves to be used during the breeding season. Bulls starting the season below a BCS of 5 usually become thin before the end of the breeding season, which has been shown to reduce semen production and quality. This is shown in Table 40-2, which presents the results of breeding soundness exams conducted at the start of a fall breeding season on a Texas ranch. The bulls had all passed a BSE in March when they had an average condition score of 6. The high failure rate of the bulls that were a condition score 5 is surprising and may not represent a typical situation.

Bulls in a body condition score over 6 often exhibit reduced physical ability to breed and lower libido. Furthermore, deposition of fat in the scrotum and in other areas of the reproductive tract has been shown to reduce semen quality. Appendix Table 9 shows the nutrient requirements for mature bulls at maintenance and at a low rate of gain. These requirements should be adjusted to allow for higher rates of gain if necessary to ensure that bulls are in good condition at the start of breeding.

Bull Management

Breeding Soundness Examinations

Bulls should be given a breeding soundness examination at least 2 months before the start of the breeding season. This examination consists of:

1. **Physical Exam** – The feet, legs, testicles, and other reproductive organs are examined for defects. This includes a rectal examination of the internal organs.
2. **Scrotal Circumference** – A yearling bull should have a minimum circumference of 32 centimeters (12.5 inches). A 2-year-old should have a scrotal circumference of at least 32 centimeters, and a mature bull should have a circumference of at least 34 centimeters. Body condition has some effect on scrotal circumference since there is some deposition of fat in the scrotum.
3. **Semen Evaluation** – The primary requirement for passing this part of the BSE is at least 70 percent normal sperm. Yearling bulls tested at 13 to 14 months of age often fail the initial BSE. They should be retested 30 to 45 days later; many of them pass their second BSE.

Bull:Cow Ratio

A number of factors influence the optimal bull:cow ratio such as:

1. **Terrain, Water Availability, and Stocking Rate** – In rugged country and areas with low stocking rates, the number of bulls per 100 cows should probably be slightly higher than in smaller pastures. In pastures with numerous water sources, the ratio should be higher because in pastures with only a few water

Table 40-3. Recommended Bull:Cow Ratios for Different Ages of Bulls and Management Scenarios.

Age of Bull and Management Scenario	Recommended Ratio
Yearling bulls	1:25
Yearling bulls, nature service with estrous synchronization	1:15
Clean-up following artificial insemination	1:60
Rugged terrain or low stocking rate	1:35
Mature bulls	1:40

Food Animal – The Compendium 1997.

sources, bulls often congregate around the water sources enabling them to breed cows without excessive travel. If numerous water sources are available, bulls must travel more to find cows in heat requiring more bulls per 100 cows.

2. **Age of Bulls** – Mature bulls breed more cows than yearling bulls. Table 40-3 shows some guidelines for bull:cow ratios with yearling and mature bulls in different management scenarios. These recommendations are based on using bulls that are in good condition at the start of the breeding season, have passed a BSE, and have demonstrated good libido and ability to mate.
3. **Management Decisions** – Several management decisions influence the optimal bull:cow ratio. For example, long breeding seasons mean that the bull:cow ratio should be lower because bulls lose considerable condition during the breeding season, and there is evidence that their fertility is reduced. With long breeding seasons some type of bull rotation program is warranted. The level of observation possible during the breeding season is also a factor in the optimal bull:cow ratio. Bulls often get injured during the breeding season effectively increasing the bull:cow ratio in multi-sire pastures. Furthermore, bulls differ greatly in their level of libido. If frequent observation is possible, the bull:cow ratio could be lower. Conversely, when frequent observation is not possible, the bull:cow ratio should probably be higher.

Managing Social Dominance in Bulls

Bulls, especially mature bulls, will exhibit social dominance where the dominant bull breeds most of the cows preventing the lower ranking bulls from breeding many cows. Table 40-4 shows the results from a study in Australia looking at the effect of social dominance on the number of cows bred by bulls. Clearly, the oldest bull bred most of the cows until the bull became too old to dominate the next oldest bull. These data indicate that in multi-sire breeding situations a yearling or two-year-old bull will sire few calves when a mature bull is present. Considering the current cost of quality yearling bulls, it is crucial that they are used in a way that allows them to contribute genetics to the herd.

Social dominance has no relationship to libido, mating ability, or quality. In some situations, the dominant bull will spend more time and energy keeping the

Table 40-4. Reproductive Performance of Bulls Over a 5-Year Period.

		Percentage of Calves Sired				
Bull	Age Year 1	Year 1	Year 2	Year 3	Year 4	Year 5
A	10	70	76	12	0[a]	0[a]
B	4	17	18	63	73	25
C	3	7	6	12	13	63
D	2	6	0[a]	12	15	12

[a]Bull absent from the herd. Oklahoma State University Extension Bulletin ANSI-3254.

other bulls away from the cows than actually breeding. Yearling and two-years-old bulls tend to exhibit less social dominance with respect to breeding activity than mature bulls. This allows the use of these younger bulls in multiple sire pasture without the level of social dominance exhibited by mature bulls.

Bull Rotation

One approach that has proven valuable in some situations is to rotate bulls during the breeding season rather than leaving all of the bulls with the cows for the entire breeding season. Bull rotation can assist in avoiding social dominance issues and gives all of the bulls a chance to breed cows. The bull:cow ratio is typically high (1:60, for example) with each bull left in the breeding pasture for a maximum of 3 weeks at any one time. A mature bull might be used for the first 21 days followed by a yearling or 2-year-old for the next 21 days. The mature bull could then be rotated back into the breeding pasture for the next 21 days. During the "rest" periods the bull should be fed to regain as much of the weight lost while in the breeding pasture as possible.

References

Barling, K., S. Wikse, D. Magee, J. Thompson, and. R. Field. 1997. Management of Beef Bulls for High Fertility – Food Animal The Compendium, July 1997.

Coulter, G.H. and G.C. Kozub. 1984. Testicular Development and Epididymal Sperm Reserves and Seminal Quality in Two-Year-Old Hereford and Angus Bulls: Effects of Two Levels of Dietary Energy. JAS 59:432.

Coulter, G.H., T.D. Carruthers, R.P. Amann, and G.C. Kozub. 1987. Testicular Development, Daily Sperm Production and Epididymal Sperm Reserves in 15-Mo-Old Angus and Hereford Bulls: Effect of Bull Strain Plus Dietary Energy. JAS 64:254.

Kirkpatrick, F.D. Management of the Beef Bull. Agricultural Extension Service University of Tennessee.

Pruitt, R.J. and L.R. Corah. 1985. Effect of Energy Intake after Weaning on the Sexual Development of Beef Bulls. I. Semen Characteristics and Serving Capacity. JAS 61:1186.

Selk, G. Management of Beef Bulls. Oklahoma Cooperative Extension Bulletin ANSI-3254.

Wohlgemuth, K and L.W. Insley. 1984. Breeding Soundness Evaluation of Beef Bulls. North Dakota State University Cooperative Extension Service Bulletin V-833.

Chapter 41

Feeding High-fat Diets to Improve Reproduction

Many research studies have examined the influence of high-fat diets pre- or post-calving on subsequent reproduction. While the results have been variable, most of the studies show high-fat diets improve reproductive performance. The variability in response should be expected considering reproduction is influenced by many variables such as different feedstuffs, environments, and biological types of cows. These studies indicate the effect of added fat is not due to energy alone but to hormonal effects because in most of these studies the diets were formulated to have the same energy level. As with any input that increases cost, the question is, does added fat increase net profit?

The Effect of Fat on Reproductive Performance

As mentioned above, the results of studies with high-fat diets have been quite variable, and there is a great deal unknown about how to use this nutritional tool. There are still numerous questions about how much fat is necessary, when to feed the fat, and what type of fat works best. It is clear, however, that feeding fat can improve reproductive performance.

It appears that high fat levels improve reproductive performance by reducing early embryonic loss. The importance of this is clear considering a bull typically settles only 60 to 65 percent of the females he breeds. In many cases, an embryo is formed, but it fails to properly attach to the uterus or it dies when the corpus luteum (CL) on the ovary regresses. It appears that high-fat diets reduce the number of CL that regress, resulting in increased embryonic survival. One of the most consistent results in studies evaluating fat supplementation is a higher conception rate. Improving the conception rate can influence profitability in two ways:

1. **Increased Pregnancy Rate** – In systems where the breeding season is relatively short (60 days or fewer), improved conception rates could increase the overall pregnancy rate. In many cases, this increase may be only a couple of percentage points. With short breeding seasons, such as 45 days, the influence on overall pregnancy could be significant. With long breeding seasons (over 60 days), the influence on pregnancy rates will probably be minimal.
2. **Improved Calving Sequence** – The most common effect of feeding high-fat diets has been to increase the percentage of females conceiving early in the breeding season. This has a couple of positive economic benefits. First, it results in heavier weaning weights in the following year because of the increase in the average age of the calves. Secondly, a cow that calves early in the breeding season has a greater chance of getting pregnant and of getting pregnant earlier in the coming breeding season. Thus, improving conception rates and calving distribution can have long-term benefits. This also means that high-fat diets may be especially valuable in herds with a poor calving sequence.

Feeding High-fat Diets to Yearling Heifers

The response to feeding fat to yearling heifers has been inconsistent. There have been several trials where feeding fat improved the pregnancy rate and the calving distribution. As a general recommendation, if the fat source is relatively inexpensive, it should be fed to yearling heifers.

Feeding High-fat Diets to First-Calf Heifers

Research indicates that fat has the greatest potential to improve reproductive performance in first-calf-heifers. Since this group of females often exhibits the poorest reproductive performance, any improvement has the potential to significantly influence profitability. As pointed out previously, it takes longer for a first-calf heifer to cycle after calving than a mature cow. This means that they often have fewer chances to settle than a mature cow. Thus, any management practice that increases conception rates can be of great value. Feeding fat to first-calf heifers is highly recommended.

Feeding High-fat Diets to Cows

High-fat diets also have improved reproductive performance in mature cows but with less consistency than shown with first-calf-heifers. Again, the cost of the supplemental fat is the key to its cost effectiveness.

Common Questions Regarding the Feeding of High-fat Diets

When Should High-fat Diets Be Fed?

There have been numerous studies evaluating high-fat diets fed either pre- or post-calving or during both periods, and while the results are variable, it appears that feeding high-fat diets before calving gives a more consistent reproductive response.

How Much Supplemental Fat Should Be Fed?

Unfortunately, there have not been sufficient studies conducted to determine the optimum level of fat. However, a good rule-of-thumb is that a minimum of 0.4 pound of added fat should be fed daily. It is likely that future research will change this recommendation, but at present it appears to be a good minimum level. It is important to note that fat can have a harmful effect on fiber digestion. As a general rule, fiber digestion is reduced when the overall fat level in the diet exceeds 7 percent. Consequently, it is recommended that the total added fat should not exceed 0.6 pound. A good recommendation is to feed between 0.4 and 0.6 pound of added fat.

What Are the Best Sources of Fat?

While there has been some research evaluating sources of fat, there are questions that remain unanswered. The consensus at this time is that vegetable sources of fat give a better response than animal sources. There also has been considerable research trying to determine whether a specific fatty acid gives superior response to other fatty acids. While it is likely that future research will determine that specific fatty acids give a superior response, at this time it appears that the fatty acids in common oil seeds are effective. For example, safflower, soybean, sunflower, and cottonseeds have all been used as sources of fat. Furthermore, by-products from processing these seeds have been used with success. The most notable example is the use of soapstock, a by-product of refining soybeans for soy oil. Fish oil and fishmeal have also been used with some success. These fish by-products are unpalatable, however, and usually more expensive than oil seeds as sources of fat.

How Long Should High-fat Diets be Fed?

While adequate studies have not been conducted to show the optimal length of the feeding period, it appears that feeding high-fat diets for 50 to 60 days before calving is sufficient. Feeding for longer periods simply has not resulted in any significant additional benefit.

Economics of Fat Supplementation

In general, fat is an expensive ingredient when evaluated simply as an energy source. Since it appears to have some beneficial hormonal and reproductive effects, the economics improve. As with any supplementation program, the question is, will the added benefit more than offset the added cost? Fat supplementation might be advisable in the following situations:

1. If the current levels of reproduction (measured by either pregnancy rates or calving sequence) are deemed poor, fat supplementation might be warranted.
2. If an inexpensive source of fat is available, it might be advisable to routinely feed high-fat diets for 50 to 60 days pre-calving to the entire herd. In areas where one of the oil seeds is grown, using these seeds as a source of fat could

be economical. These seeds tend to have fairly high levels of protein, which adds significantly to their value. For example, soybean seeds are 18.8 percent fat and 40 percent CP (DM basis). Feeding a little more than 2 pounds of these seeds per head per day would supply the minimum level of fat and probably supply adequate supplemental protein. Currently, it appears that the growth of the biodiesel industry will make the price of oil seeds relatively high, which greatly reduces the potential for feeding them as a source of fat.

3. If the source of fat is relatively expensive, it might be advisable to target the group of females that offers the greatest potential for improvement — the first-calf heifers.

The inclusion of fat into self-limiting products, such as tubs and blocks, often reduces the intake-limiting ability of these products. In other words, adding fat to these products often allows cattle to consume more supplement than desired. Obviously, this may negatively influence the economics of feeding the fat. To avoid this problem, some feed companies market tubs and blocks under the label of "High Fat" that probably do not have enough fat to actually influence reproductive performance. To evaluate a product promoted as "High Fat," calculate the expected fat intake by multiplying the level of fat by the expected intake. Again, the target should be approximately 0.4 pounds of added fat per day. As an example, consider a 4 percent fat tub with a 2-pound-per-day expected consumption. This results in 0.08 pounds (0.04 × 2) of supplemental fat daily. This level is probably inadequate to significantly improve reproduction. Unfortunately, these products are often sold at significantly higher prices than traditional "low-fat" products.

References

Alexander, B.M., B.W. Hess, D.L. Hixon, B.L. Garrett, D.C. Rule, M. McFarland, J.D. Bottger, D.D. Simms, and G.E. Moss. 2002. Influence of Prepartum Fat Supplementation on Subsequent Beef Cow Reproduction and Calf Performance. Prof. Anim. Sci 18:351.

Bader, J.F., E.D. Felton, M.S. Kerley, D.D. Simms, and D.J. Patterson 2000. Effects of Postpartum Fat Supplementation on Reproduction in Primiparous 2-yr-old and Mature Cows. JAS 83(Suppl 1):224 Abstract.

Bellows, R.A., E.E. Grings, D.D. Simms, T.W. Geary, and J.W. Bergman. 2001. Effects of Feeding Supplemental Fat During Gestation to First-Calf Beef Heifers. Prof. Animal Sci. 17:81.

Burns, P.D., E.R. Downing, T.E. Engle, J.C. Whittier, and R.M. Enns. 2002. Effect of Fishmeal Supplementation on Fertility in Young Beef Cows Grazing Pastures. Proc. Western Sec. Amer. Soc. Animal Sci. 53:392.

Grings, E.E., R.E. Short, M. Blummel, M.D. McNeil, and R.A. Bellows. 2001. Prepartum Supplementation with Protein or Fat and Protein for Grazing Cows in Three Seasons of Calving. Proc. West. Sec. Amer. Soc. Anim. Sci. 52:501.

Hess, B.W., D.C. Rule, and G.E. Moss. 2002. High Fat Supplements for Reproducing Beef Cows: Have We Discovered the Magic Bullet? Proceeding Pacific Northwest Nutrition Conference.

Long, N.M., G.M. Hill, J.F. Baker, W.M. Graves, and M.A. Froetschel. 2007. Reproductive Performance of Beef Heifers Supplemented with Corn Gluten Feed and Rumen-Protected Fat Before Breeding. The Professional Animal Scientist 23:316.

Willaims, G.L. and R.L. Stanko. 2000. Dietary Fats as Reproductive Nutraceuticals in Beef Cattle. Proc. Amer. Society of Anim. Sci.

Chapter 42

Feeding During Drought

Unfortunately, drought is a common occurrence across the United States. It often seems to be the norm rather than a rare occurrence in some areas of the country. In most cases, the goals during drought should be avoiding damaging the forage resource, maintaining the nucleus of the cow herd, avoiding spending too much on hay or other feedstuffs, and maintaining good performance in the cows and heifers that are retained.

Drought Management Strategies

Early Wean Calves

One of the first steps in a drought is to wean the calves as early as possible. This will greatly reduce the nutritional requirements of the cows and the demands on the forage resource. Calves can be weaned at 90 to 120 days easily if adequate nutrition is provided.

Cull Early

Cull more of the old cows and reduce the number of heifers retained.

Supplement the Cows on Pasture

Supplementing the cows on pasture may be necessary to prevent long-term damage to the forage resource. It is important to provide enough supplementation to adequately reduce the grazing pressure on the pastures. Otherwise, the long-term productivity of the pastures can be reduced significantly. It can take many years for a pasture to recover from overgrazing. What is the best supplement in this situation?

1. **Grain** – Grain is usually the cheapest feedstuff on a cost per unit of energy basis, and has been widely used as a supplement when forage supply is limited. Because grain contains a high level of starch, and starch reduces fiber digestion, grain probably will not be the best supplement unless care is taken to make sure that DIP is adequate. As a general rule, when feeding grain to cows grazing low-quality forages, the grain should be fed at either relatively low levels (2 to 3 pounds) or at high levels (greater than 7 pounds). With levels in the 4- to 7-pound range, much of the additional energy from the grain is offset by the loss of energy from the forage due to reduced digestibility. Again, if the DIP supplied is adequate, the loss in digestibility will probably be minimal. The low levels of grain mentioned above (2 to 3 pounds) will not reduce the forage intake enough to provide much relief on the grazing resource. If high levels of grain are to be fed, the grain should be gradually introduced to the cattle.
2. **Harvested Feed** – If hays or silages are available at reasonable prices, they are good feedstuffs for drought conditions. Unless the drought is confined to a small area, these feeds usually are not available at affordable prices.
3. **By-Product Feeds** – Many of the by-products from grain processing make excellent supplements for beef cows. In most cases, much of the starch has been removed, but they still have relatively high energy content. For example, corn gluten feed has little or no starch, yet approximately 80 percent the energy value of corn. Wheat middlings and soybean hulls also make excellent supplements. All of these by-products could be useful as supplements for cows in drought situations. Because of their low starch content, they are preferable to grain as supplements for cows grazing low-quality forages.

Long-Term Drought Management

In many areas where drought is common, producers develop long-term strategies to avoid some of the consequences of drought. For example, producers often maintain enough forage to feed the herd for a year as a backup. Other producers bale large quantities of wheat straw and other residue as a backup. These residues can then be chemically treated (Chapter 58) to improve their quality if needed as feedstuffs. Other producers have reduced the size of their cow herds to below the average carrying capacity of the pasture resource. In years where moisture is average or above average, they graze stockers on the forage not needed for the cow herd. In years where moisture is below average, they do not graze any stockers. In areas where droughts are common, a long-term plan should be implemented.

Limit-Feeding Cows in Drylot

In severe drought situations, weaning the calves, and putting the cows in drylot can be an effective program. This minimizes the cow's nutrient requirements because both the grazing activity requirement and the lactation requirement have been removed. Feeding in drylot allows the feeding of high-grain (or by-product), low-roughage diets. This is probably the cheapest way to maintain cows through a drought. Cows in this program should be placed in drylot and not left on pasture. If the cows are left on pasture, they will continue to graze resulting in possible damage to the resource.

Because drought is so common, there have been many university studies evaluating high-grain, low-roughage, limit-fed rations for beef cows. While several levels of roughage have been evaluated, the most cost effective rations minimize roughage inclusion. Table 42-1 shows an example of this type of ration compared to the nutritional requirements of dry, gestating beef cows.

With the ration shown in Table 42-1, the cows should gain some weight unless the crude protein level is inadequate to support gain. In fact, the level of crude protein is probably low relative to the energy level. Thus, a little more crude protein might improve performance slightly. The 32 percent crude protein supplement could be formulated to contain the required minerals and vitamins along with an ionophore such as Rumensin®or Bovatec®.

The performance in some trials with all concentrate, limit-fed diets has been satisfactory. Diets with some roughage are easier to manage and safer. Several nutritionists have suggested a minimum roughage level of 0.5 percent of body weight. The level (3 pounds of roughage daily) shown in Table 42-1 has been fed in several studies with good results. Rations with even more roughage are easier to manage but have a higher cost per cow per day in typical drought situations.

Managing High-Grain, Low-Roughage, Limit-fed Diets

Consider the following items when feeding high-grain, low-roughage, limit-fed diets:

1. Limit-fed programs require accurate information on the nutritive content of feedstuffs and intakes to be successful. Furthermore, feed wastage, if significant at all, must be considered. In summary, limit-feeding requires a high level of consistent management.
2. Cattle should be adjusted to the diets gradually by increasing the grain portion and reducing the roughage portion over 10 to 14 days. Cows should be fed at the same time every day.

Table 42-1. Evaluation of the Nutrient Composition of a High Grain, Low Roughage, Limit-fed Ration Compared to a Beef Cow's Requirements.

Item	Energy, Mcal NEm	Crude Protein, lbs
Requirements of a 1,300 lb Cow, Mid-Gestation	9.51	1.54
11 lbs corn	9.84	0.95
3 lbs Brome Hay	1.41	0.26
1 lb 32% CP Supplement	0.61	0.32
Total	11.68	1.53

3. Adequate bunk space must be provided since hunger will cause aggressiveness, especially during the first week or so. It has been recommended that at least 30 inches of feeder space be allowed for each animal. Experience has shown that 2 feet of bunk per cow is probably a minimum. Cows tend to adjust to the low level of intake and settle down after a couple of weeks.
4. Fences must be in good condition because the cattle will challenge them, especially during the first week.
5. A complete mixed ration works better than feeding the concentrate and roughage portions separately.

Mineral and Vitamin Considerations in High Grain, Low Roughage Diets

With typical high roughage diets fed to beef cows, calcium is of little concern. With limit-fed, high concentrate diets, calcium will often be deficient. For example, the diet shown in Table 42-1 will be deficient in calcium because of the low level of calcium in corn. Thus, either calcium should be added to the 32 percent supplement or a free-choice supplement high in calcium should be provided. A good supplement for this situation might be one-third limestone, one-third dicalcium phosphate, and one-third trace-mineralized salt. Alternatively, a commercial "feedlot" mineral supplement could be used. Vitamin A should always be provided at 15,000 to 20,000 IU per day to cows and heifers on limit-fed rations, since grains and by-products are low in this nutrient.

Use of Urea in High Grain, Low Roughage Diets

As discussed in previous chapters, natural protein typically gives better performance in cow diets when low-quality roughages are the basal diet. However, in high-grain, low-roughage diets, more of the protein can come from urea, and since urea is a cheaper source of protein, substituting some urea for natural protein should reduce the cost of the feeding program. Supplements containing 30 to 50 percent of the supplemental protein from urea have been used with satisfactory results in limit-fed rations.

Using By-Products as Energy Sources Instead of Grain

Most studies evaluating limit-feeding beef cows have used grain as the energy source. There has been some research using by-products, specifically wheat middlings. It appears that by-products can replace grain in rations similar to those shown in Table 42-1. Other by-products could be substituted, but consideration should be given to nutrient levels in the by-product. Remember by-products from grains have concentrated levels of protein, minerals, fat, and other nutrients. Consider the potential for harmful effects from these concentrations of nutrients. For example, the high level of sulfur in corn gluten feed may be a problem. Use caution when attempting to limit-feed a by-product in the manner described above. In many situations substituting by-products for a portion—say 40 to 60 percent—of the grain is the safest option.

References

Brethour, J.R., John Jaeger, and K.C. Olson. 1990. A Strategy for Feeding Cows During Drought. Fort Hays Experiment Station Roundup 1990.

Brownson, R. Emergency Rations for Wintering Beef Cows. Beef Cattle Handbook. BCH-5460.

Corah, L., D.D. Simms, and G.L. Kuhl. 1989. Alternative Feeding Strategies for Cows and Calves Due to Drought Related Forage Shortages. Kansas State University Publication.

Cunningham, D.B. Faulkner, A.J. Miller and J.M. Dahlquist. 2005. Restricting Intake of Forages: An Alternative Feeding Strategy for Wintering Beef Cows. The Professional Animal Scientist 21:182.

Faulkner, D.B. and L.L. Berger. Limit Feeding Beef Cattle. Beef Cattle Handbook. BCH-5175.

Feeding and Managing Livestock During Feed Shortage. Agriculture Canada.

Ritchie, H. and S. Rust. 1993. Limit-Feeding of Grain to Beef Cows. Michigan State University Extension Animal Science Newsletter.

Tjardes, K.E., D.B. Faulkner, J.W. Castree, and D.D. Buskirk. 1995. Limit Feeding Whole or Cracked Corn-Hay Diets Compared to an Ad Libitum Hay Diet for Beef Cows. Illinois Beef Research Report.

Chapter 43

Managing Feed Costs

As discussed in Chapter 2, feed costs are the largest single expense on most cow/calf operations. Consequently, managing feed costs should be a major priority. Note the use of the word "manage" rather than minimize. This choice reflects the fact that the goal is not to minimize feed expenses but to maximize profitability. As the old adage goes, "Sometimes you have to spend some to make some."

The focus in managing feed costs should be on using lower-cost feedstuffs whenever possible and on ensuring that expenditures on supplements result in returns greater than the investment. The manager's responsibility is to know when to supplement, how much to supplement, and what type of supplement. The information presented in previous chapters should be helpful in answering these questions.

This chapter summarizes typical opportunities to reduce feed costs while maintaining high levels of productivity. The order of presentation does not reflect their possible effect on profitability, which will vary among operations.

Common Methods of Reducing Feed Costs

Minimize the Feeding of Harvested Forages

Roughages that are harvested as hay or silage are usually more expensive than grazed roughages, and increases in the prices of fuel and equipment make this truer today than it was a few years ago. Consequently, every effort should be made to reduce the use of harvested feed. For example, it is common in the southeastern states to "stockpile" fescue by applying fertilizer in the early fall to produce forage to be grazed during the winter. In the mountain states, some producers windrow hay and allow the cows to graze these windrows in the winter. In other areas, crop residues are grazed during the winter to reduce the need for hay. Other producers raise a forage sorghum crop and graze it rather than harvesting it as hay or silage. There are numerous other examples where producers have developed innovative systems to reduce the need for harvested forages. Maximizing profitability requires constantly evaluating ways to reduce feed costs. Minimizing the use of harvested feed usually increases profitability.

Group Animals by Requirements

In a typical cow herd, there are usually distinct groups of females with different nutritional requirements. For example, in a spring calving herd, during the winter, there are heifer calves, pregnant yearling heifers, 2-year-old cows, 3-year-old cows, and mature cows. In addition, there is normally a range of body conditions within each age group. The nutritional requirements of each of these groups are quite different as shown in Appendix Tables 1 through 9. Too often, these females are fed together resulting in lower productivity by the younger or thinner females or excessive feed costs for the mature cows in good body condition.

Grouping cows according to their nutritional requirements provides the opportunity to supply adequate nutrition to all groups without overfeeding any group. While it would be nice to feed all of the groups listed above separately, it often is not practical because of the number of pens available, especially in small herds. Because of their high nutritional requirements compared to the other groups, heifer calves should be fed separately. The other group that should be fed separately, if possible, is the pregnant yearling heifers. The next priority for separation from the main herd should be the 2-year-old cows. And, finally, if possible, the mature cows should be separated into groups based on body condition. In most situations, the number of pens available will not allow this many groups. Consequently, it is often necessary to combine some groups based on nutritional requirements. For example, the thin, mature cows might be combined with the 2-year-old cows. The key point is that grouping according to nutritional requirements provides the opportunity to optimize the feeding program.

Match the Genetic Potential of the Herd to the Feed Resource

Too many producers try to adjust the feeding program to the genetic potential of their cattle rather than matching the genetic potential of the cattle to their resources. It is more profitable to fit the cattle to the resource base than visa versa.

The two most important factors influencing cattle nutritional requirements are milk production and cow body weight as shown in Appendix Table 1. The fact is that some feed resources will not support large and/or heavy milking cows without expensive supplementation.

Consider the results of a Kansas State University study where cows of two breeds were wintered on milo stover, a low-quality feed. One breed was large with high milk production potential while the other was moderate in both size and milk production potential. On the same basal feed, the large breed required considerable protein and energy supplementation to avoid major weight and body condition losses. The other breed required minimal supplementation to maintain body condition. While the large, high milk production breed produced heavier calves; the extra weaning weight was not enough to pay for the additional supplement. The key point is that matching the size and milk production potential of the herd to the available feed resources is usually the most profitable approach.

When matching the genetic potential of the cattle to the resource, producers tend to focus on cow size because it is something that can be visually assessed. The potential for milk production is a more important factor than cow size.

How Do We Determine the Genetic Potential for Milk Production?

Because milk production is not measured in beef cattle, weaning weight must be relied on as an indicator of milk production because it is highly correlated to milk production. Table 17-7 should be useful in assessing milk production potential. The expected progeny differences for milk (Milk EPD) of the bulls used in the herd gives some indication of the milk production potential.

How to Assess the Match between the Genetic Potential in the Herd and Feed Resources?

Reproductive performance is the best indicator of the match between the genetic potential and the feed resources. There are many examples of producers selecting for increased growth rate and milk production to the point where reproductive performance declined markedly.

If the cattle are reproducing at a high rate, there is a good match. Conversely, poor reproductive performance can indicate that the genetics of the herd and the feed resources are not matched. The other variable that needs to be considered is supplementation cost. High levels of protein and energy supplementation overcome a poor match between genetics and the feed resources, but the cost of high levels of supplementation usually results in lower profitability.

Minimize the Expenditures for Protein and Mineral Supplements

Since these supplements represent the largest out-of-pocket feed expenses for most producers, considerable management focus should be placed on a cost/return analysis for these expenditures. The importance of providing adequate protein has been stressed throughout this book, but there is still a cost/return relationship for protein supplementation. As covered in Chapter 24, cow body condition is a powerful tool for assessing the protein/energy supplementation program. Because there are usually significant differences in the cost per unit of protein from different sources, comparing protein sources on a price per unit of protein is advisable. For example, several common by-products, such as wheat middlings and corn gluten feed, are excellent sources of protein and usually much cheaper per unit of protein than commercial supplements. However, commercial supplements will typically have other nutrients like minerals and vitamins that are not contained in by-products feeds. These additional nutrients must be considered in making a comparison of supplements.

Some mineral supplementation is required in all cow herds, but it is easy to spend too much for mineral supplements. The high cost of mineral supplements makes it essential that careful consideration is given to meeting the needs of the cattle without feeding more mineral than necessary. The guidelines presented in Chapter 33 should be helpful in evaluating the mineral supplementation program.

Determine the Optimum Body Condition for the Operation

The reproductive performance exhibited by cattle with different body condition scores is not the same in all environments. As discussed in Chapter 24, in some environments, cows must calve in a body condition score 5 to reproduce at a high rate while in others a 4 is adequate. There is typically a marked difference in the feed cost to either reach or maintain a body condition score 5 compared to a 4. Knowing the minimal body condition score that produces good reproductive performance in a specific environment can be a powerful tool for maximizing profitability.

Pregnancy Test and Cull Open Females

Feeding a female that is not going to produce a calf is a waste of feed. Consequently, culling open cows is highly recommended. With respect to profitability, however, it may be best to only pregnancy check the yearling heifers and 2-year old cows, especially if the percentage of mature cows that fail to get pregnant is traditionally low. This minimizes the expenses for pregnancy checking while identifying most of the open females for culling. Whether an open 2-year-old should be culled is debatable because considerable expense has been incurred in her development with only a single calf produced to offset these expenses. The need for replacements, genetic merit of the 2-year-olds, and sale value of the 2-year-olds are key considerations in making the decision to sell the 2-year-olds or not.

Maximize the Use of By-Products

As noted, feeding by-products from the grain processing industries can be an effective way to reduce feed costs. A number of by-product roughages can greatly reduce feed costs. For example, low-quality roughages like cotton burrs and wheat straw can be a major part of cow diets, especially when the cow's nutritional requirements are low. These and other low-quality roughages can be blended with high-quality feeds such as corn gluten feed or distiller's grains to make good rations. Making maximum use of a low-quality feedstuff like wheat straw may require taking care to maximize its quality. For example, cattle will consume much more wheat straw if it is baled right after harvest while the straw is still bright, and the nutrient content will be higher if it is baled before it is exposed to weather.

Improve Grazing Distribution in Pastures

In areas of the country where pastures are large, grazing distribution is often a problem. It is typified by grazing pressure being greatest near water sources and lowest away from water sources. In many cases, there will be areas that will not be grazed at all.

Poor grazing distribution means more pasture is required per cow, increasing feed costs. Developing water sources throughout the pasture is usually the cheapest way of improving grazing distribution. Splitting the pasture by cross fencing also is effective. Another low-cost approach to improving grazing distribution is using the location of mineral feeders and placement of supplement to increase grazing in underused areas.

Stock Pastures at Moderate Rates

Over time, stocking density has a tremendous influence on the amount of forage produced in a pasture. Plants must replenish the nutrient stores in their roots periodically or their root system gradually becomes smaller and shallower. This reduces their ability to extract nutrients and water from the soil. If plants are repeatedly cropped close to the ground, there will not be enough leaf area to produce the nutrients required for root growth and development. For many years, the old adage "Take half, leave half" has been promoted as a guideline, and it probably is correct in some environments. In other environments, however, the use rate should only be 25 percent to allow for long-term productivity of the pasture. The plant composition in the pasture is a good indicator of the influence of grazing because overgrazing results in an increase in less-desirable species and weeds. More information on stocking rates is available in Chapter 38.

Another benefit of stocking at a moderate rate is that it allows cows to select a higher-quality diet resulting in better performance. This should result in cows in higher body condition at the end of the grazing season, which reduces the need for supplemental feed during the winter.

Fencing a small area of the pasture to prevent grazing is a useful tool for assessing the total amount of forage produced, which allows the estimation of the percentage being harvested by the cattle. The fenced area should be changed each year, and it should be fairly representative of the entire pasture.

Adjust the Stocking Rate for Cow Production Level

The potential for growth and milking ability of beef cattle has increased markedly in the last 30 years, and in many cases this has resulted in increased grazing pressure on pastures. To produce heavier calves, cows must eat more to produce more milk. If stocking rate is not adjusted for the increase in forage intake, the net effect can be reduced forage production in the pastures and/or reduced production, especially reproduction. For example, the pasture resource might have been able to provide adequate forage for 100 head of cows of moderate size and milking ability, but can only provide forage for 85 head of higher-producing cows.

Control Parasites

Parasites can influence the amount of feed required in several ways. They can directly compete for nutrients, as is the case with internal parasites. They can increase maintenance requirements by reducing insulation as is the case with lice. They can consume blood, removing previously digested nutrients as is the case with lice and flies. They also can modify grazing behavior resulting in lower intakes, as is the case with flies. Minimizing the influence of parasites can reduce feed costs.

Match the Quality of Harvested Feed to Nutritional Requirements

There are often marked differences in quality between forages harvested at different times or from different fields. Having each lot of feed analyzed separately permits matching feed quality to animal requirements, which can eliminate the need for supplementation or at least reduce the amount of supplementation required.

Harvest Feeds to Maximize Their Usefulness

One way of reducing the need for supplement, especially protein supplement, is to harvest feeds when their protein content meets a beef cow's nutrient requirements. Consider prairie hay in Kansas. Cut in mid-July, this hay could be expected to exceed 7 percent crude protein, which is close to adequate for a nonlactating, beef cow.

Conversely, this same hay cut in August would probably only have 5 percent crude protein, well below the dry cow's requirement. Harvesting forages at higher nutrient levels reduces supplementation costs, but note that the amount of hay harvested would be slightly lower in July than from the same field cut in August. As with many things in the cattle business, there is a trade-off between quality and quantity, and optimizing this combination maximizes profitability.

Optimize Fertilizer Expenses

In some regions, fertilizer for pastures is a major expense, and increases in fertilizer prices make this expense even more important. Since there is a direct relationship between the amount of fertilizer applied and the amount of forage produced, fertilizer expense should be considered a part of the total feed cost. Furthermore, the law of diminishing returns (Chapter 31) applies to fertilizer application. In other words, there will be a greater response to the first few units of fertilizer than to subsequent units. The rising prices for fertilizer make determining the optimal fertilization rate a key management decision. Furthermore, the stocking rate should be reduced in proportion to the reduction in fertilizer to avoid overgrazing.

Minimize the Cost of Delivering Supplements

Rising fuel and equipment costs make the cost of delivering supplements an important part of the total cost of supplementation. Consequently, the frequency of supplementation (Chapter 32) is an important consideration, and in some situations self-limiting supplements should receive serious evaluation.

Calve at the Right Time

The potential effect of calving time on supplementation requirements was discussed in Chapter 35. It is mentioned here because of its potential to markedly reduce feed costs. The key point is that for each operation there is a calving time that produces maximum profitability; establishing that time is an important management decision.

Wean at the Right Time

As with the previous topic, the effect of weaning time on feed costs was discussed in Chapter 35. It is listed here because it can be a powerful tool for managing feed costs.

Use Only Proven Products

There are many feed products offered to cow/calf producers that do not fit into the traditional nutrient categories. For example, there are microbial additives of many types. Use of these products should be based on sound research. A good source of information on the efficacy of these products is the state beef cattle extension specialist. Typically, these individuals are knowledgeable regarding research on all types of feed additives. Unfortunately, it is easy to spend a lot of money on feed additives that provide little benefit.

Minimize Feed Wastage

Not only is harvested feed usually the most expensive feed in a cow/calf operation, but wastage of harvested feed is often a key factor in total feed expense. Losses during storage and feeding can result in the loss of more than 30 percent of the dry matter harvested. For example, dry matter loss observed in a Missouri study was 24 percent for big round bales stored on dirt without cover.

Placing the bales on a rock base and covering them with plastic reduced the loss to less than 3 percent. The stacking system used for big round bales also influences the amount of loss. South Dakota State University researchers compared the loss in bales stacked in a pyramid, stored individually, or end-to-end in rows with space between the rows. Dry matter losses were 10.3, 4.0, and 0.8 percent, respectively, after one year of storage. It should be noted that this study was conducted in a moderate rainfall area, and losses would be expected to be higher in an area with more rainfall.

Wastage during feeding also can be significant. Table 43-1 shows estimates of losses for different types of bales and feeding systems. These estimates show that wastage can greatly influence total feed costs.

References

Kallenbach, R. Reducing Losses When Feeding Hay to Beef Cattle. University of Missouri Extension Publication G4570.

Kamstra, L.D., P.K. Turnquist, C.R. Krueger, C.E. Johnson, E.A. Dowling, and T.S. Chisholm. Changes in Field Stored Large Packages. South Dakota State University Extension Bulletin TB53.

Simms, D.D. 2009. Feeding the Beef Cowherd for Maximum Profit. SMS Publishing. Amarillo, TX.

Table 43-1. Estimated Losses (Percentage of Hay Offered) with Different Hay-Feeding Systems.

Bale Type	With Rack 1-Day Supply, %	With Rack 7-Day Supply, %	Without Rack 1-Day Supply, %	Without Rack 7-Day Supply, %
Small Square Bales	3.9	4.1	6.7[a]	---
Large Round or Square Bales	4.9	5.4	12.3[a]	43.0[a]
Formed Hay Stacks	8.8	15.0	22.6	41.0

University of Missouri Extension Publication G4570. [a]Bales spread across pasture

Chapter 44

Beef Cow Herd Health

Common Diseases in U.S. Cow Herds

A number of diseases affect cow herds across the United States. The following discussion covers only the most common.

Anaplasmosis

Anaplasmosis is caused by a parasite that destroys red blood cells. It is similar to malaria in humans. Blood-sucking insects like mosquitoes, flies, and ticks typically spread it. It also can be passed during processing where blood can be passed from animal to animal by contaminated instruments. For example, blood can be passed during dehorning and vaccination if the same equipment or needles are used for numerous animals without sterilization. Anaplasmosis typically occurs in the late summer or fall in older cattle. It rarely occurs in younger cattle. The affected cattle become anemic and lethargic. One early sign of anaplasmosis is animals that fall behind the herd during movement. In areas where anaplasmosis is prevalent, feed companies have free-choice minerals that contain oxytetracycline or chlortetracycline to control anaplasmosis. There is not a vaccination for anaplasmosis.

Blackleg

Blackleg is one of the clostridial complexes of diseases caused by bacteria that live in the ground. These organisms enter the animal through ingestion or wounds, and they produce strong toxins that result in death. Affected animals are often found dead without exhibiting any symptoms. Younger cattle, 6 months to 2 years of age are mostly commonly affected. The 7-way clostridial vaccine is effective in preventing blackleg, and it is inexpensive.

Bovine Respiratory Syncytial Virus

This viral disease is usually observed in calves, but it can affect older animals. It is one of the organisms often found in cases of "shipping fever" or pneumonia; consequently, it is included in the viral vaccines.

Bovine Viral Diarrhea (BVD)

While BVD is also one of the viruses associated with shipping fever, it also causes numerous other problems in beef cow herds including abortions and weak calves at birth. In recent years, the existence of persistently infected calves has been identified. These calves occur when their dam is infected with BVD from day 40 to day 125 of pregnancy. The persistently infected calf is born with immunity to BVD and it sheds the organism throughout its life. Persistently infected calves may or may not show symptoms, but a high percentage die before one year of age. Persistently infected animals can be identified, and herds with BVD problems should identify and remove persistently infected cattle. These calves become the most common source of transmission in a herd. Most of the viral vaccines for respiratory disease contain BVD. These vaccines can be either killed or modified-live viral products. The modified-live version results in stronger and longer immunity, but it should not be given to a pregnant animal. Consequently, if the modified-live version is used, it is typically given between calving and breeding.

Brucellosis

Brucellosis (Bang's Disease) causes late-term abortion in beef cows. While a major effort has been conducted to identify and slaughter carriers, a few outbreaks occur each year. Cows that abort will become immune, but they shed the organism, a bacteria, infecting other members of the herd. The most common symptom is a significant number of late term abortions. One method of prevention is to vaccinate heifers when they are between 4 and 10 months of age.

Coccidia

Coccidiosis is caused by a protozoan parasite that enters the cells of the digestive tract. This organism is common wherever cattle are found. Many times, it will exist in a subclinical form in a group of growing animals. In this situation, it reduces feed conversion and performance. When these growing animals are stressed, clinical symptoms (watery feces with or without blood) often occur. If unchecked, the stool will eventually contain a significant amount of blood. Unfortunately, some of the damage to the intestine can be permanent resulting in poor performance even after the infection is terminated. Infected animals shed large numbers of oocysts that contaminate the ground. These oocysts are picked up by other animals, which become infected. As noted, stress like weaning and shipping often sets off an outbreak, and muddy pens regularly result in an outbreak.

Monensin (Rumensin®), Lasalocid (Bovatec®), or Decoquinate (Deccox®) are commonly fed to prevent an outbreak of coccidiosis. However, these drugs are ineffective as a treatment in case of an outbreak. When an outbreak occurs, Amprolium (Corid®), sulfaquinoxaline, or sulfamethazine should be fed. Amprolium also may be given through the drinking water if feasible.

Infectious Bovine Rhinotracheitis (IBR)

This virus is another one commonly associated with shipping fever. In addition, it can cause abortions and eye inflammation similar to pinkeye. The vaccines for IBR are available in both a killed and modified-live form. The modified-live version should not be given to pregnant animals because it can cause abortion. The vaccine is also a routine part of a preconditioning program for calves.

Leptospirosis

This bacterial disease causes abortions, stillbirths, and weak calves. The organism is shed in the urine, contaminating ponds. All breeding-age female cattle should be vaccinated at least annually. Moreover, in some areas, cattle should receive a booster vaccination approximately one month after the initial vaccination.

Parainfluenza Type 3

This is another virus associated with shipping fever. Consequently, it is included in the typical viral vaccine.

Pinkeye

The common symptom of pinkeye is a discharge from the eye(s). In advanced cases, there may be damage to the cornea and even blindness. Pinkeye causes a significant reduction in performance. It typically occurs in dry, dusty conditions when the eyes become irritated. The primary causative organism is a bacterium, but IBR virus also may be involved. Several systemic antibiotics are effective in treating pinkeye, and patches are often placed over an infected eye to serve as protection from further irritation. Controlling face flies reduces pinkeye problems, and feeding a free-choice mineral with oxytetracycline or chlortetracycline reduces the incidence of pinkeye.

Scours/Diarrhea

Scours is one of the most common disease problems in cow herds. A number of organisms, including viruses, bacteria, and protozoa, can be involved. To some extent, scours can be thought of as an environmental disease because the environment where the calves are born and raised is a key factor in the incidence and severity of scours. For example, calves born in drylot pens routinely exhibit more scours problems than calves born on pasture. Pens or areas that are used for calving year-after-year become heavily infected with the organisms that cause scours. These organisms are then ingested when the calf nurses the udder of its dam. Conversely, the udders of cows on pasture tend to have much less contamination. Another factor in the prevalence and severity of scours is the age of the dam. Calves born to older cows typically exhibit less scours than calves born to first-calf-heifers. This occurs to some extent because older cows pass more immunity to their calves through the colostrum. Another factor contributing to more scours in calves from first-calf-heifers is that producers tend to concentrate first-calf-heifers in small areas so they can assist with calving problems. This concentration in a relatively small area results in much higher levels of scours causing organisms on the heifer's udders.

Prevention of scours is best accomplished by providing a clean calving area and vaccinating the heifers and cows for the causative organisms which include *E. coli,* rota and corona virus, BVD, and clostridium perfringens (included in 7-way clostridial vaccine). These vaccines should be given to the heifers and cows late in pregnancy so they can pass immunity to the calf through the colostrum.

Trichomoniasis

Trichomoniasis is a venereal disease of cattle caused by a protozoan, *Tritrichomonas foetus.* This organism lives in the reproductive tract of both bulls and cows. Bulls usually remain lifetime carriers, while cows usually clear the disease within 3 months after infection. However, some cows will remain infected indefinitely, becoming a source of re-infection for the herd. *Trichomoniasis foetus* causes embryonic death resulting in delayed breeding and/or open females. Herd bulls, especially non-virgin "experienced" bulls should be tested for "Trich" before use. In addition, ensuring that both cows and bulls in the herd do not commingle with neighboring cattle reduces problems with "Trich." There is not an effective treatment, so testing and culling infected cattle is the only way to eliminate Trich.

Vibriosis

Vibrio is a sexually transmitted disease that results in loss of the fetus early in pregnancy. Infected animals will become pregnant and then exhibit heat approximately 40 days later. Cows that have previously been exposed may develop immunity and become asymptomatic carriers, but virgin heifers and cows that have not been previously exposed will abort fetuses. The net effects of vibriosis are delayed breeding resulting in long calving seasons or a significant number of open cattle. Precautions should be taken to avoid bringing an infected animal into the herd. Additionally, a vaccine is available that should be given 30 to 60 days before breeding.

Beef Cow Herd Vaccination Schedule

The vaccination schedule outlined below is provided as a guideline. It should be modified in consultation with a local veterinarian because of variation in disease problems across the country.

Vaccinations for Cows and Bulls

Before breeding:

IBR, BVD, PI_3, BRSV – Preferably modified live – cows must be open

Leptospirosis/Vibriosis

Before calving in herds with a history of scours:

1. 7- or 8-way clostridial

2. Rota and Corona viruses

3. BRSV

Vaccinations for Calves

At processing (2 to 3 months):

7-way clostridial

Preweaning:

7-way clostridial

IBR, BVD, PI_3, BRSV

At weaning:

IBR, BVD, PI_3, BRSV

Vaccinations for Replacement Heifers

From 4 to 10 months:

Brucellosis

Prebreeding:

IBR, BVD, PI_3, BRSV

Leptospirosis/Vibriosis

7-way Clostridial

Nutritional Diseases in Beef Cow Herds

A number of cattle diseases have a nutritional basis. In some cases, a nutritional deficiency or excess of some compound is the direct cause of the disease, while in other cases the deficiency is a predisposing factor for the disease. For example, high nitrates in feeds can be the direct cause of death, while a trace-mineral deficiency can reduce immune function allowing disease-causing organisms to proliferate. Because of the relationship between nutrition and immune function, there are numerous diseases where nutrition plays a role. It is beyond the scope of this book to cover all of these diseases. There are, however, a few nutritional diseases that occur often enough and across a broad enough area that they warrant coverage.

Bloat

Bloat in cows, calves, and stocker cattle normally results from the consumption of alfalfa or wheat pasture, especially when these pastures are lush. This type of bloat, termed frothy bloat, results when a stable foam is formed in the rumen. This foam does not allow the gas produced in the rumen to be "belched" off. The pressure from the gas distends the area between the last rib and the hip on the left side. When the pressure becomes great enough, the animal is not able to breathe, resulting in death. The pressure can usually be relieved by passing a hose or tube into the rumen, but it may be necessary to drench with vegetable oil or dishwashing detergent to reduce the foaming. In an emergency, an incision may be made in the distended area to relieve the pressure. When the bloat has subsided, this incision should be sutured and an antibiotic administered. The incidence of bloat can be reduced by feeding Bloatguard® when cattle are grazing forages that cause frothy bloat. Making sure that cattle are full before going on lush legume or wheat pastures also reduces the problems with bloat.

Fescue Toxicosis

Tall fescue is often contaminated with an endophyte fungus, *Acremonium coenophialum,* which reduces intake and performance in both cows and calves. Affected cattle will exhibit rough hair coats and a tendency to stay in the shade or in ponds. Hot, humid weather increases the effects. Because the fungus is most concentrated in the seed head, clipping the pasture to remove seed heads reduces the problem. Inter-seeding with a legume also increases performance by diluting the amount of toxin ingested. Replacing the infected tall fescue with low-endophyte varieties also has been successful. However, low-endophyte varieties tend to produce less forage than infected varieties.

Nitrate Poisoning

High nitrate levels are a common problem in many forages. Nitrate, by itself, is not toxic, but when bacteria break it down to ammonia in the rumen, an intermediate product, nitrite, is formed. Nitrite attaches to hemoglobin in the blood to form methemoglobin, which can't carry oxygen. If enough nitrite enters the blood, the animal dies of asphyxiation. Since methemoglobin is brown, a brownish colored blood is good evidence of nitrate poisoning.

Causes of High Nitrate in Forages

Nitrate is the primary source of nitrogen for plants, and thus nitrate is normally found at some level in all plants. Under some conditions, however, plants pull nitrate from the soil, but they cannot convert the nitrate into protein. The nitrate then accumulates often reaching toxic levels. When plant growth is slowed dramatically, nitrate can accumulate. The following conditions are typical of situations that result in high nitrate levels.

Drought – Forages grown in drought conditions often exhibit high nitrate levels because the plants extract nitrate from the soil but cannot convert it to protein. Drought-stressed plants usually have high nitrate levels for a couple of weeks following a significant rain.

Cloudy Weather – Rapidly growing plants that experience an extended period of cloudy weather often have high nitrate levels until the resumption of sunshine, which is required for photosynthetic activity and the conversion of nitrate to protein.

Low Temperatures – Cold temperatures reduce the rate of photosynthesis and can cause an accumulation of nitrate.

Nitrogen Fertilization – As might be expected, the amount of nitrogen applied to the soil has some bearing on the level of nitrate in the plants. If plant growth is not reduced by drought or other factors, the nitrate absorbed by the plant is rapidly converted to protein. Thus, high levels of fertilization do not typically cause high nitrate levels unless something else reduces plant growth.

Species of Plant – Some plants are much more prone to develop high nitrate levels than others. For example, oats, sudans, sorghums, sorghum hybrids, and millet all have a tendency to accumulate high levels of nitrate. Some weeds like pigweed also are prone to accumulate nitrate. Conversely, legumes and most native grasses have little tendency to accumulate nitrates.

Part of the Plant – The level of nitrate tends to be highest in the stalk and lowest in the leaves. Consequently, the height of cutting at harvest can influence the overall level of nitrate in the forage. This concentration of nitrate in the lower portion of the plant explains why cattle forced to graze the stalks of forages high in nitrate may exhibit nitrate toxicity even though they have been grazing the forage for an extended period.

Stage of Growth – Immature plants tend to have higher levels of nitrate than mature plants. And immature plants that suddenly have their growth rate reduced are prime candidates for high nitrate levels.

Ensiling Forages Reduces Nitrate Levels

During fermentation, the nitrate content of forage is reduced. As a rule, 30 to 50 percent of the nitrate is converted to other nitrogen compounds by bacteria during the ensiling process. This normally reduces the levels of nitrate below toxic levels. In extreme stress conditions, the level of nitrate in forage may be high enough that even with a 50 percent reduction, a toxic level may exist. In other words, if silage was made from a severely stressed crop, it should be tested for nitrates.

Feeding High Nitrate Feedstuffs

Feedstuffs high in nitrate can still be fed, but the overall level of nitrate in the diet must be managed. The most common approach to reducing the overall level of nitrate is to mix the high nitrate feed with a low-nitrate feed. Typically, the high and low nitrate feeds are ground and mixed in a ratio that results in a sufficient reduction in overall nitrate level. This results in the added cost of grinding, but it allows the use of the high nitrate feedstuff. Alternate day feeding of high- and low-nitrate roughages in a self-feeding situation is not advised because nitrate toxicity can occur from a single high dose of nitrate. Limiting intake of the forage by restricting the access time or the amount delivered can be successful, but care should be taken to prevent "boss" cows from overconsuming the forage. Grain feeding both dilutes the level of nitrate and provides energy for the ruminal microbes to rapidly convert nitrate to protein. Thus, more of a high nitrate feedstuff can be fed safely when grain or other high-energy feedstuffs are also fed. The nitrate content of livestock water should also be considered when using high-nitrate feedstuffs.

Methods of Reporting Nitrate Levels

Unfortunately, there is inconsistency in how nitrate levels are reported and in what levels are considered toxic. Table 44-1 shows the three common ways of reporting nitrate and the levels that are generally considered lethal. It should be noted that some veterinarians and nutritionists believe that toxic levels are actually lower than those shown. The levels shown in Table 44-1 are good rules-of-thumb.

The different methods of reporting nitrate levels often cause confusion, especially when producers are communicating with nutritionists or veterinarians. For example, a level greater than 9,000 ppm is toxic if reported as nitrate, but not toxic if reported as potassium nitrate. When communicating with others regarding nitrate levels, the chemical form should be clearly stated. Most laboratories include the levels that they consider toxic based on the chemical form shown on the

Table 44-1. Methods of Reporting Nitrate Levels Commonly Used by Feed Analysis Laboratories.

	Potentially Lethal Level	
Chemical Form	**%**	**ppm**
Nitrate (NO_3)	Over 0.9	9,000
Nitrate Nitrogen (NO_3N)	Over 0.21	2,100
Potassium Nitrate (KNO_3)	Over 1.5	15,000

Table 44-2. Conversion Coefficients to Convert between Chemical Forms of Nitrate.

From	To	Formula
Potassium Nitrate (KNO_3)	Nitrate	$KNO_3 \times 0.6$
Potassium Nitrate (KNO_3)	Nitrate Nitrogen	$KNO_3 \times 0.14$
Nitrate Nitrogen (NO_3N)	Nitrate	$NO_3N \times 4.4$
Nitrate Nitrogen (NO_3N)	Potassium Nitrate	$NO_3N \times 7.0$
Nitrate (NO_3)	Potassium Nitrate	$NO_3 \times 1.6$
Nitrate (NO_3)	Nitrate Nitrogen	$NO_3 \times 0.23$

analysis. Table 44-2 shows conversion coefficients useful in converting from one method of reporting to another.

Grass Tetany

Grass tetany, also called grass or blind staggers, wheat pasture poisoning, winter tetany, and hypomagnesemia, occurs in many areas across the United States. As one of its names implies, grass tetany is caused by low blood magnesium.

Symptoms:

In the early stages, affected animals are excitable and aggressive. As the disease progresses, they stagger and stumble. In the latter stages, they exhibit violent convulsions and go into a coma. Since the progression of the disease is relatively rapid, often the only sign of a problem is a dead animal. In most cases, older cows are more affected than younger cows. Lactating cows are more susceptible because of their higher magnesium requirement. Diagnosis is usually based on the time of year and on the diet being conducive to grass tetany. However, low blood or urine magnesium levels in the herd are also good diagnostic tools. Measuring the magnesium level in the blood of a dead animal is of little value because magnesium is released into the blood from damaged tissues before death.

Situations Conducive to Grass Tetany

There is considerable variability in the prevalence of grass tetany even in situations that appear to be quite similar. For example, adjacent ranches with similar soil types often experience different levels of grass tetany. Although grass tetany occurs in many situations, the following conditions are the most common:

1. Most cases occur in the spring in lactating cows grazing pastures that have recently come out of dormancy. The high level of nitrogen, potassium, and calcium in this early grass appears to tie up the magnesium making it less available to the cow. Furthermore, lactating cows have a higher requirement for magnesium.
2. Cows consuming cool-season grass hay in the winter also are susceptible. Cold weather also appears to reduce magnesium absorption. Again, lactating cows have a higher requirement.
3. Cows grazing wheat pasture also are susceptible to grass tetany. In this case, the problem may be a calcium deficiency rather than a magnesium deficiency.
4. Applying high levels of nitrogen and potassium fertilizer before the growing season increases

the incidence of grass tetany. It may be wise to apply these fertilizers after the peak period for tetany has passed.

Preventing Grass Tetany

Grass tetany is more prevalent in some pastures than others because of low soil magnesium levels or plant species. If possible, pastures where tetany problems have occurred should not be grazed during peak tetany periods. However, in many cases this simply is not practical. In these situations, the best approach is to feed a mineral high in magnesium that prevents grass tetany.

Mineral Supplementation to Prevent Grass Tetany

Providing adequate magnesium in a mineral supplement is more difficult than expected because the most common source of magnesium (magnesium oxide, MgO) is unpalatable and the requirements for supplemental magnesium are relatively high. Furthermore, there is not a readily available storage organ for magnesium, meaning that cows must consume the mineral almost on a daily basis. Consequently, during situations where grass tetany is likely to occur, supply a palatable mineral with a high magnesium level. Also, multiple mineral feeders should be placed in locations where cows congregate.

How Much Magnesium Should Be Supplied?

A lactating beef cow has a requirement for 0.2 percent magnesium in the diet. Assuming 25 pounds of dry matter intake, she requires $25 \times 0.002 = 0.05$ pounds of magnesium × 16 ounces per pound = 0.8 ounce per head per day. Magnesium oxide contains 54.0 percent magnesium. Meeting the cow's entire magnesium requirement would require $0.8 \div 0.54 = 1.48$ ounces of MgO per day. This calculation does not take into account the amount of magnesium from the basal forage, which is difficult to determine.

MgO Levels in High Magnesium Minerals

Because MgO is not palatable, it is usually included at a maximum level of 25 percent. Thus, total mineral consumption must be $1.48 \div 0.25 = 5.9$ ounces per head per day to meet the total magnesium requirement. To get this level of consumption, the remaining ingredients in the free-choice mineral must be palatable. For example, soybean meal and dried molasses are often used to improve the palatability of the mineral mix.

Evaluating Commercial High Magnesium Mineral Supplements

Most feed companies have mineral supplements that have been formulated for use when the potential for grass tetany is high. Generally, the magnesium content in the mineral should be at least 10 percent. In addition, the intake of the mineral should be monitored closely. At 10 percent magnesium concentration, it will take a consumption of $0.8 \div 0.10 = 8$ ounces to meet a lactating cow's total requirement.

References

Coccidiosis of Cattle. The Merck Vet Manual.

Lalman, D. and D. Doye, Eds. 2008. Beef Cattle Manual, 6th Ed. Oklahoma State University Extension Service.

Pence, M. Coccidiosis in Cattle. University of Georgia Publication.

Powell, J., S. Jones, S. Gadberry, and T. Troxel. Beef Cattle Herd Health Vaccination Schedule. University of Arkansas Extension Bulletin FSA3009.

Sharko, P. and R. Burris. University of Kentucky Extension Website.

Striegel, N., R. Ellis, and J. Deering. 2009. Trichomoniasis Prevention: The Cost Per Cow to Prevent. Colorado State University Extension No. 628.

Thomas, M. and W. Harmon. 1994. Bovine Trichomoniasis: General Information, Diagnosis and Control. San Joaquin Experimental Range Research Bulletin.

Trichomoniasis: Introduction. 2011. The Merck Veterinary Manual.

Whittier, W., and J. Currin. 2009. Beef Cow/Calf Herd Health Program and Calendar. Virginia Tech Extension Publication 400-007.

Chapter 45

Stocker Management

Receiving Programs for Stocker and Feeder Cattle

An important part of both stocker and finishing programs is the receiving program. It influences subsequent performance, mortality, morbidity, and treatment expenses. Cattle received for stocker and finishing programs typically fall into one of the following groups:

1. **Freshly Weaned Calves** – These calves should be treated like highly stressed calves because the stress of weaning and shipping often results in considerable morbidity. Furthermore, they often are not accustomed to eating from bunks and exhibit lower than desired feed intakes. These cattle have not been exposed to many disease organisms and are often "naive" with respect to immunity.
2. **Preconditioned Calves** – Preconditioned calves have been weaned and exposed to feed bunks. Consequently, they usually start eating quickly. Even though they have been vaccinated before arrival, most operators vaccinate at arrival to boost immunity. Furthermore, while preconditioned calves usually exhibit less sickness than those not preconditioned, they can still exhibit some sickness and should be observed closely for the first month or so after arrival.
3. **Highly Stressed Calves** – Many lightweight (300- to 500-pound) calves are shipped out of the south and southeast into stocker programs in the High Plains such as wheat pasture grazing. These calves often exhibit high levels of morbidity and mortality. They require a high level of management with respect to both the feeding and health programs.
4. **Yearling Cattle** – These cattle have typically been through a stocker program of some type. They are older and have been previously vaccinated and exposed to the common disease organisms. They will eat readily; in fact, they will overeat if given the opportunity. Even though yearling cattle have probably been vaccinated previously, they are usually vaccinated at arrival at the feedlot.

Cattle received at feedlots for finishing are typically yearling cattle that are not highly stressed. Conversely, cattle entering stocker programs are from the first three groups and should be considered stressed. Certainly there are exceptions to these generalities. For example, in years where a lack of rainfall results in limited wheat pasture on the High Plains, calves that would have been placed on wheat pasture often are placed in commercial feedlots in a growing program before finishing. The receiving program for highly stressed calves must be different than one for a yearling animal; consequently, typical programs for these groups will be discussed separately later in this chapter.

Shrink

Both highly stressed calves and yearlings lose weight during handling and transport, which is called shrink. Some of this shrink is the loss of urine and feces, which is recovered rapidly following transport. The second type of shrink is tissue shrink, which takes much longer to regain. Shrink is expressed as a percentage of the initial payweight. The amount of shrink is a function of the time in transport, age, sex, and condition. Table 45-1 shows the effect the number of hours on a moving truck has on shrink. A general rule of thumb for shrink

Table 45-1. Effect of the Number of Hours in Transit on the Percentage of Shrink.

Hours in a Moving Truck	% Shrink
1	2
2-8	4-6
8-16	6-8
16-24	8-10
24-32	10-12

Feedlot Management Primer, Ohio State University Extension.

is 3 percent for the first 100 miles and 0.5 to 1.0 percent for each additional 100 miles. Fatter animals shrink less than thin animals. Consequently, heifers, which are typically fatter than steers, tend to shrink less than steers. Correspondingly, older animals tend to be fatter than younger animals resulting in less shrink.

Keys to Minimizing Shrink

1. Minimize the stress during gathering and loading the cattle.
2. Minimize the time between gathering and loading on the truck.
3. Minimize the time the cattle are on the truck by minimizing the stops during transit.
4. Avoid shipment during inclement weather.

Processing Incoming Cattle

Both highly stressed calves and yearlings should be processed within 24 hours of arrival. If the cattle are showing signs of sickness, some operators delay some of the more stressful operations such as dehorning and castration until the cattle are eating well and seem to be over any sickness. However, this results in another stress period when these operations are performed. Body temperature is often used as an indicator of sickness, but temperatures off the truck are poor indicators of sickness. After the cattle have had time to rest for a few hours or overnight, a temperature higher than 104 degrees Fahrenheit is considered a sign of possible sickness. Thus, cattle that are visibly sick or have a temperature of higher than 104 degrees Fahrenheit are typically treated.

Receiving Highly Stressed Calves

Highly stressed calves often do not eat well for the first few days after arrival. Furthermore, they are typically used to eating grass. Therefore, stocker operators usually offer these calves high-quality grass hay and water on arrival. Because these calves may not be familiar with bunks, this hay may need to be fed on the ground on the day of arrival. Correspondingly, these calves may not be familiar with automatic waterers, necessitating the use of stock tanks. Letting the water run over attracts the calves to the tank. Once the hay has been placed in the bunk, the receiving ration should be spread on top of the hay. When the cattle are readily consuming the receiving ration, the hay can be discontinued. If the calves are to be fed hay with a supplement, the supplement can be spread on top of the hay in the bunk.

Receiving Rations for Stressed Cattle

The receiving ration for stressed calves should be palatable. Ingredients that are not palatable, such as dry corn gluten feed and sunflower pellets, should be avoided in starter feeds. These calves typically have not been exposed to silage, and it should not be used in the initial starter program. Many feed companies offer starter feeds that contain molasses, yeast, or flavoring agents to stimulate intake. Because intake is often low, protein levels in the starter should be at least 16 percent. If intakes are extremely low, higher protein levels such as 20 percent for the first week or so may be advantageous. Recommended nutrient levels for highly stressed calves are well above those for cattle of the same weight that are not stressed because of the low intake. Nutrient levels recommended for stressed cattle by the National Research Council are shown in Table 45-2. In addition to the normal nutrients required by ruminants, stressed cattle may respond to supplemental B-vitamins because their rumens often are not functioning normally.

Roughage Levels in Receiving Rations for Stressed Calves

Research has shown a tendency for higher-roughage starter diets to reduce the level of bovine respiratory disease in stressed calves. These higher roughage diets typically produce lower gains than diets with more concentrate. It appears that a diet with

Table 45-2. Suggested Nutrient Concentrations for Stressed Calves.

Nutrient	Suggested Range	Nutrient	Suggested Range
Dry Matter, %	80-85	Copper, ppm	10-15
Crude Protein, %	12.5-14.5	Iron, ppm	100-200
NEm, Mcal/lb	0.6-0.7	Manganese, ppm	40-70
NEg, Mcal/lb	0.36-0.41	Zinc, ppm	75-100
Calcium, %	0.6-0.8	Cobalt, ppm	0.1-0.2
Phosphorus, %	0.4-0.5	Selenium, ppm	0.1-0.2
Potassium, %	1.2-1.4	Iodine, ppm	0.3-0.6
Magnesium, %	0.2-0.3	Vitamin A, IU/lb	1,800-2,700
Sodium, %	0.2-0.3	Vitamin E, IU/lb	180-225

Nutrient Requirements of Beef Cattle, 1996.

50 percent concentrate produces the best combination of health and performance.

Feed Additives in Starter Rations

1. **Antibiotics** – Many operators feed an antibiotic for the first 21 to 28 days to prevent disease outbreaks. Typical antibiotics that are used include chlortetracycline, oxytetracycline, and a combination of chlortetracycline-sulfamethazine (AS700®). Rather than feed these products daily, another common program is to feed the tetracyclines at a high level (1 g/cwt) for 3 to 5 days if a disease outbreak occurs or as a preventative.
2. **Coccidiosis Control** – Coccidiosis, which is caused by a protozoan parasite, is a common disease in growing cattle, especially in stressed calves. Commonly used products for coccidiosis control include Deccox® and the ionophores Rumensin®, and Bovatec®. All of these products control coccidiosis, but once bloody scours are observed, they are ineffective. In this situation, Amproium (Corid®), or one of the sulfa drugs should be used.
3. **Ionophores** – As noted above, the ionophores Rumensin® and Bovatec® have coccidiocidal action as well as their promotion of growth and improved feed conversion in growing programs. One of these products should be included in the starter ration.

Using Grass Traps to Receive Highly Stressed Calves

A trend that has developed in the last 20 years has been the use of grass traps of 8 to 10 acres for receiving calves as opposed to small pens. Experience shows that morbidity and mortality are lower with grass traps compared to drylot pens. Identifying and treating cattle that do become sick, however, is more difficult than when small pens are used. One approach to identifying potentially sick cattle in grass traps is to place a fence behind the feed bunk with gates that can be closed after feeding. Any cattle that do not come to the feed bunk are considered potentially sick. If they have a temperature higher than 104 degrees Fahrenheit or they exhibit signs of sickness, they are treated.

Receiving Programs for Yearling Cattle

Yearling cattle do not typically exhibit the low intakes observed with stressed calves. As noted, they often overeat. The process of starting yearlings and adapting them to the finishing ration is covered in Chapter 48.

Health Programs for Stocker/ Feeder Cattle

Vaccination Programs

The vaccination program should be developed in conjunction with a veterinarian because he/she is familiar with the disease problems on the operation and those occurring in the area. Typically, the following vaccinations are considered a basic program:

1. **Modified-live viral vaccine containing:**

   a. Infectious Rhinotracheitis Virus (IBR)

   b. Bovine Viral Diarrhea Virus – Types 1 & 2 (BVD)

   c. Parainfluenza$_3$ Virus (PI$_3$)

   d. Bovine Respiratory Syncytial Virus (BRSV)

2. **Seven- or Eight-way Clostridial Vaccine (Blackleg)**

Other vaccinations that are often given include those for protection from *Mannheimia (Pasturella)* and *Haemophilus*. The vaccinations listed above are usually given a second time 12 to 14 days after the initial vaccination to booster the level of immunity.

Vaccinations do not give 100 percent protection from disease because a high percentage (30 to 40 percent) of animals that are vaccinated fail to develop immunity for a variety of reasons. A common question is, should the animals be vaccinated if they are already showing signs of sickness? The answer is yes because vaccination tends to stimulate the immune system even if the animals might not develop immunity to the specific organism causing the disease. In fact, many feedlots will routinely revaccinate a pen if a significant number of animals start showing signs of sickness. A close working relationship with a veterinarian knowledgeable about diagnosing and treating common feedlot diseases, especially BRD, is crucial to a successful health program.

Treating Sick Animals

An important part of a good management program for receiving cattle is a good plan for identifying sick cattle and treating them.

Keys to Identifying Sick Cattle

1. **Visibly Sick** – head down, droopy ears, listless, draining from the nose.
2. **Temperature Higher than 104 degrees Fahrenheit** – Some operators take the temperature of all of the cattle or those cattle that "do not look right" and treat those with a temperature of 104 degrees Fahrenheit or higher.
3. **Not Eating** – Cattle that do not come to the feed bunk when the pen is fed are good candidates for having their temperature taken to evaluate their health status.

Importance of Good Recordkeeping in a Treatment Program

A good recordkeeping program can be of great value in a treatment program. When the animal is "pulled" from the home pen for treatment, it should receive a unique identification. The animal should be placed in a "hospital" pen to facilitate treatment and prevent the spread of disease to other animals in its home pen. However, some feedlots now place treated cattle back in their home pen because of the availability of antibiotics that produce longer sustained lung levels.

The drug and dose used to treat the animal should be recorded as well as its temperature and symptoms. This information is useful when assessing the effectiveness of the treatment. It also aids in identifying which drugs are working. It is common to change the drug used to treat the animal if its temperature does not drop below 104 degrees Fahrenheit following treatment. The number of days to wait before reassessing the animal's response may vary with the antibiotic used in the initial treatment. Most operators treat a maximum of three times then let the animal fend for itself because the drugs used to treat sick cattle are expensive and the success rate after three treatments is fairly low.

Metaphylactic (Mass Medication) Programs

It is common for highly stressed calves to be metaphylactically treated at arrival with every animal receiving an injection of antibiotic. In many cases, this program greatly reduces the level of morbidity and mortality. Many feedlots mass medicate when a significant number of animals in a pen become sick as a way of preventing more sickness. Some antibiotics are approved by the FDA for use in metaphylactic programs, while others can only be used if a certain percentage of the animals appear to be getting sick. The consulting veterinarian should provide guidance to ensure that the antibiotics are being used in accordance with FDA guidelines.

Parasite Control

- **Deworming** – Cattle should be treated with a dewomer for intestinal parasites on arrival. In addition, cattle from some locations should receive a treatment for liver flukes.
- **External Parasites** – Cattle should also receive a treatment for lice and ticks on arrival.

Profitable Management Practices in Stocker Programs

- **Implanting** – As long as stocker cattle are gaining more than 0.50 pounds per day, they will respond to implants with increased gain of 10 to 15 percent. In a drylot stocker program, feed efficiency also increases by about 10 percent. This increase in performance gives implanting the highest return on investment of any management practice in a stocker program.
- **Feeding an Ionophore** – Feeding an ionophore like Rumensin® or Bovatec® is also a profitable practice if the cattle are gaining at least 0.75 pounds per day. These ionophores increase daily gain by 0.15 to 0.20 pounds. In a drylot stocker program, feed conversion improves by about 10 percent. Furthermore, feeding either of these ionophores reduces the potential for problems with coccidia.
- **Mineral Supplementation** – Providing a free-choice mineral or adding mineral to an energy or protein supplement usually increases gain and reduces health problems. Mineral deficiencies, especially phosphorus and trace minerals, often reduce performance in stocker programs.

Supplementing Energy and/or Protein in a Stocker Program

There are many times when the basal forage will not supply enough energy and protein to provide the desired gain or most profitable gain. For example, Table 30-1 shows the typical nutrient profile of a tallgrass prairie in eastern Kansas where the forage has a high nutrient content in May and June. During this time, this forage will support gains of close to 3 pounds per day in a stocker program. However, this grass starts to mature in early July rapidly declining in both protein and energy. Gains during the late summer may only be 1 pound per day. During this period, stockers respond to energy and/or protein supplementation with additional gain. The question is, what type of supplement provides the most profitable gains?

The native grass in eastern Oklahoma has a similar growth pattern and nutrient profile to the tallgrass prairie of eastern Kansas. Oklahoma State University has conducted numerous research trials evaluating supplementation of stockers with energy and/or protein supplements. Because of this research, they developed the Oklahoma Gold program, which consists essentially of supplementing stockers in late summer with 1 pound of a 38 to 40 percent crude protein supplement. A summary of seven trials showed an increase in average daily gain of 0.37 pounds per day and a conversion of 2.7 pounds of feed per additional pound of gain. Research at other locations shows a similar response with stockers on native pasture. This excellent conversion of supplement to gain is the result of the protein in the supplement increasing digestibility and intake of the forage. Thus, for this type of supplementation to work, there must be adequate forage supply.

In the development of the Oklahoma Gold program, OSU evaluated feeding an energy supplement such as grain. Feeding 4 pounds of grain, which supplies about the same amount of protein as the 1 pound of 38 to 40 percent crude protein, actually reduced forage digestibility and didn't increase forage intake. Consequently, 1 pound of 38 to 40 percent crude protein resulted in superior performance compared to 4 pounds of grain.

This type of supplement does not have to be fed every day. The 1-pound feeding rate can be fed at 2 pounds every other day or 2.3 pounds on Monday,

Wednesday, and Friday. The supplement also can be self-fed using salt as discussed in Chapter 36.

Because there are times when more gain is desired than the 1 pound of 38 to 40 percent supplies, OSU developed the Oklahoma SUPERGOLD program, which consists of feeding 2.5 pounds of a 25 percent crude protein supplement. This program provides more protein and considerably more energy than Oklahoma Gold. OSU Extension specialists recommend that highly digestible fiber ingredients like wheat middling be used rather than grain. Oklahoma SUPERGOLD would be more effective than Oklahoma GOLD in the late fall when the protein and energy content of the native forage has declined significantly. The Oklahoma Gold and Oklahoma SUPERGOLD programs are discussed here because they have been successful and because they illustrate some basic concepts that need to be considered in developing a good supplementation program for stockers.

Considerations in Developing a Cost Effective Supplementation Program

1. **Nutrient Content of the Basal Forage** – Clearly, there is a wide range of forages used in stocker programs across the United States. These forages differ in growth patterns and nutrient composition through the year. For example, warm season native pastures, such as the tallgrass prairie of eastern Kansas, have the nutrient profile discussed previously, while cool season grasses like fescue and bromegrass have bimodal growth patterns with high forage quality in the spring and again in the fall. Having an understanding of the typical nutrient content of the basal forage is key to determining when a supplementation program should be profitable.
2. **Establish the Desired Rate of Gain** – To some extent, the most profitable rate of gain in a stocker program depends on the marketing plan for the stockers. For example, if stockers are destined to be sold as feeders at the end of the stocker program, most producers target at least 2 pounds of gain per day. If, however, ownership of the stockers is to be retained through finishing, a slightly lower rate of gain may be targeted with the idea that the stockers will make compensatory gains in the feedlot. Obviously, the relative cost of gain in the stocker and finishing phase should be considered in this decision. As another example, many producers winter calves at relatively low rates of gain (1 pound per day, for instance) if they intend to graze them during the next pasture season. The significant increase in growth rate of today's cattle compared to 30 years ago makes keeping gain low in the stocker phase to maximize pasture or finishing gains less meaningful. Research shows that today's high growth rate cattle are capable of gaining at higher rates in the stocker phase and still gaining rapidly in the finishing phase or on pasture.
3. **Determine Nutrient Requirements for the Desired Rate of Gain** – Appendix Tables 6 and 8 should be useful in determining the nutrient requirements for a desired rate of gain. Of particular interest is the dietary crude protein concentration required to support a specific rate of gain. It is important to note, however, that meeting the protein requirement may not ensure that the energy requirement will be met. It may be necessary to feed additional energy as discussed in the Oklahoma SUPERGOLD program to achieve the desired rate of gain.

Substitution of Supplement for Forage

Studies using the Oklahoma Gold program with harvested forages have shown that it increases intake and digestibility of the forage as noted previously. When the supply of the basal forage is limited, however, it may be desirable to feed enough supplement to compensate for the lack of forage. Nebraska researchers supplemented stockers on native pasture with dry distillers grains (DDG) and concluded that intake of the forage was reduced by 0.5 pound for each pound of DDG fed. It is likely that this substitution rate increases as the amount of supplement is increased.

When forage supply is limited, feeding significant quantities of hay as a supplement is often used to compensate for the lack of forage in the pasture. By-products, such as wheat middlings and corn gluten feed, also can be fed at levels high enough to reduce forage consumption.

References

Cochran, C.E. Owensby, R.T. Brandt, Jr., E.S. Vanzant, and E.M. Clary. Increasing Levels of Grain Supplementation for Intensive-Early Stocked Steers: Two-Year Summary. Kansas State University Cattlemen's Day Report of Progress.

Feedlot Management Primer Chapter 2. Shipping and Receiving Cattle. Beef Information Website. Ohio State University.

Freeman, A.S. and K.P. Coffey. 1991. Effects of Supplemental Ground Grain Sorghum During Grazing of Endophyte-Infected Tall Fescue on Grazing and Subsequent Feedlot Performance of Steers. Kansas State University Cattlemen's Day Report of Progress.

Gadberry, S. and T. Troxel. Stocker Cattle Management: Receiving Nutrition. University of Arkansas Extension Bulletin FSA3062.

Gill, D. and J. Pollreisz. 2001. Stocker Cattle Management. NCBA Cattlemen's College Proceedings.

Gill, D. and D. Lalman. Oklahoma GOLD Q&A – Late-season Supplementation Program for Stocker Cattle. Oklahoma Cooperative Extension Service ANSI-3032.

Gill, D. and D. Lalman. Oklahoma SUPERGOLD Q&A – Late-season Supplementation Program for Stocker Cattle. Oklahoma Cooperative Extension Service ANSI-3032.

Horn, G.W., D.J. Bernardo, W.E. McMurphy, M.D. Cravey, and B.G. McDaniel. 1991. High-Starch Versus High-Fiber Energy Supplements for Stocker Cattle on Wheat Pasture Cattle Performance and Economics. Oklahoma State University Animal Science Report.

Klopfenstein, T.J., L. Lomas, D. Blasi, D.C. Adams, W.H. Schacht, S.E. Morris, K. H. Gustad, M.A. Greenquist, R.N. Funston, J.C. MacDonald, and M. Epp. 2007. Summary Analysis of Grazing Yearling Response to Distillers Grains. Nebraska Beef Report.

Kuhl, G.L. 1983. Nutrition and Management of Growing and Finishing Cattle. Proceedings of Kansas Formula Feed Conference 1983.

Lalman, D. and D. Doye. Eds. Beef Cattle Manual, 6th Edition. Oklahoma Cooperative Extension Service Publication E-913.

Parish, J., J. Rhinehart, R. Hopper. Stocker Cattle Receiving Management. Mississippi State University Extension Service Publication 2506.

Powell, J. and T. Troxel. Stocker Cattle Management: Receiving Health Program. University of Arkansas Extension Bulletin FSA 3065.

Self, H. and N. Gay. 1972. Shrink During Shipment of Feeder Cattle. JAS 35:489-494.

Smith, E.F. 1981. Growing Cattle on Grass. Kansas State University Agricultural Experiment Station Bulletin 638.

Chapter 46

Balancing Rations for Growing Cattle

Developing feeding programs for growing cattle, such as replacement heifer calves, requires balancing a ration for a desired level of performance. For example, the goal might be for heifers to gain 1.5 pounds per day from weaning to breeding. Accomplishing this goal requires feeding the amount of energy and other nutrients that achieves this rate of gain. There are several computer programs that make formulating balanced rations easy; however, formulating a balanced ration by hand can be accomplished in a short time. This chapter illustrates this process and provides some tips to make it easier.

The Pearson Square

A useful tool in ration balancing is the Pearson Square, which can be used to calculate the proportion of two feedstuffs that gives a desired nutrient concentration — one nutrient at a time. Typically, only protein and energy need to be considered. Assume we have 1) meadow hay with 7 percent crude protein and 2) cottonseed meal (CSM) with 46 percent crude protein. These concentrations are on a dry matter basis. We are feeding a lactating cow requiring 11.5 percent crude protein. The Pearson Square can be used to calculate the ratio of hay and cottonseed meal that results in an 11.5 percent crude protein diet.

Using the Pearson Square – Step by Step

- **Step 1**– Draw a square or rectangle and place the desired protein level in the center. Outside the upper left hand corner place the protein content of the hay (7 percent). Outside of the lower left hand corner place the protein content of the cottonseed meal (46 percent). The square should look like this:

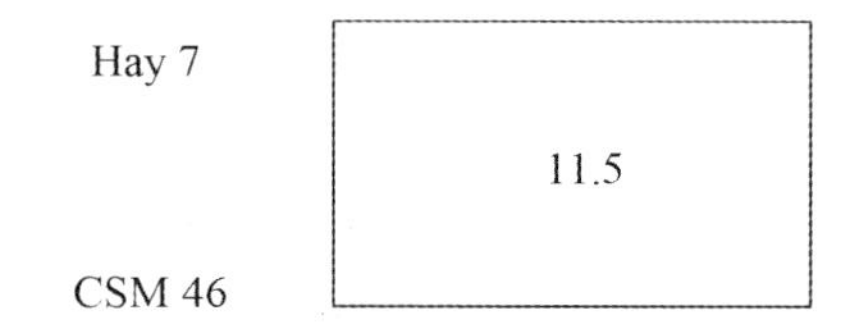

- **Step 2** – Subtract the smaller number from the larger number on the diagonal. For example, subtract 7 from 11.5 and place the value (4.5) outside the lower right corner. Correspondingly, subtract 11.5 from 46 and place the value outside the upper right corner. The box should now look like this:

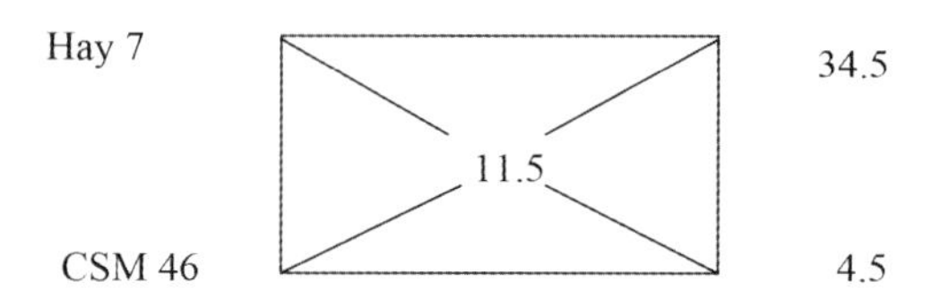

- **Step 3** – Add the two values on the right side of the box. This value (39.0) is then divided into each value on the right side to determine the percentage of each feedstuff. The box should look like this:

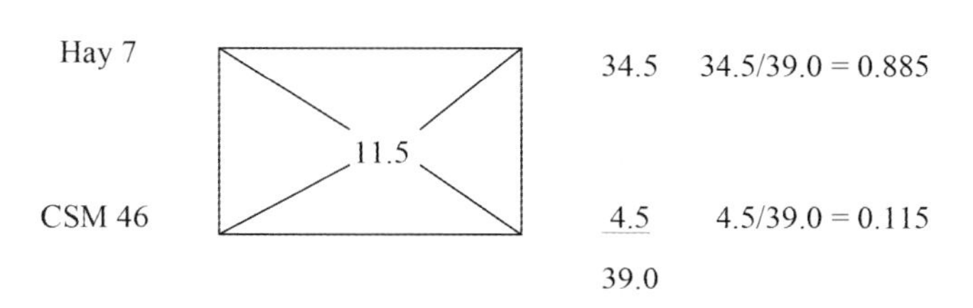

- **Step 4** – A combination of 88.5 percent meadow hay and 11.5 percent cottonseed meal would provide the 11.5 percent crude protein combination desired.

Table 46-1. Available Feedstuffs.

Feedstuff	DM, %	Energy, NEg Mcal/cwt	CP, %
Bahiagrass Hay	90	14	8
Sorghum Grain	89	59	11
Cottonseed Meal	90	53	48

Table 46-2. Heifer's Nutritional Requirement.

DMI, lbs	NEg, Mcal/cwt	CP, %
12.6	36.0	11.2

- **Step 5** – The final step is to convert these percentages into the amounts fed on an "as fed" basis. From Appendix Table 1, we find that this cow will consume about 29.2 pounds of dry matter. Assuming that both of these feedstuffs are 89 percent dry matter, the intake "as fed" should be 29.2 ÷ 0.89 = 32.8 pounds. Thus, intake of hay should be 32.8 × 0.885 = 29.0 pounds. In addition, 32.8 × 0.115 = 3.8 pounds of cottonseed meal brings the protein content up to the 11.5 percent level required. The other factor to consider is wastage, since the percent wastage for hay in many feeding situations is quite high.

Using the Pearson Square with More Than Two Ingredients

The Pearson Square works well with two ingredients; however, in many cases, more than two ingredients will be required to "balance" the ration. To illustrate, balance a ration for a 500-pound replacement heifer. We have established this heifer should gain 1.5 pounds per day from weaning to breeding. The available feedstuffs are shown in Table 46-1. The requirements of this heifer are shown in Table 46-2 (taken from Appendix Table 6).

- **Step 1** – Combine two of the ingredients into a supplement with the desired protein content. Combine the sorghum and cottonseed meal into a supplement for the Bahiagrass hay. Since it appears that energy is the key to balancing this ration, formulate a 14 percent crude protein combination as follows:

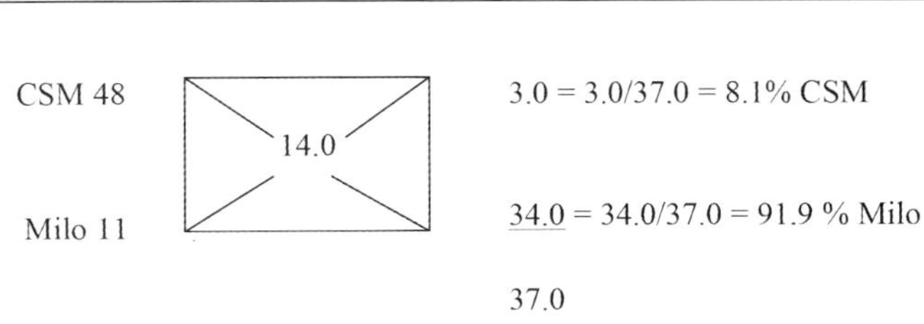

- **Step 2** – Calculate the energy level of this combination, which is (0.081 × 53) + (0.919 × 59) = 58.5 Mcal/ cwt. Now, use the Pearson Square to combine the mix of sorghum and cottonseed meal with the bahiagrass hay.
- **Step 3** – Balance the grain/protein mix with the hay for crude protein.

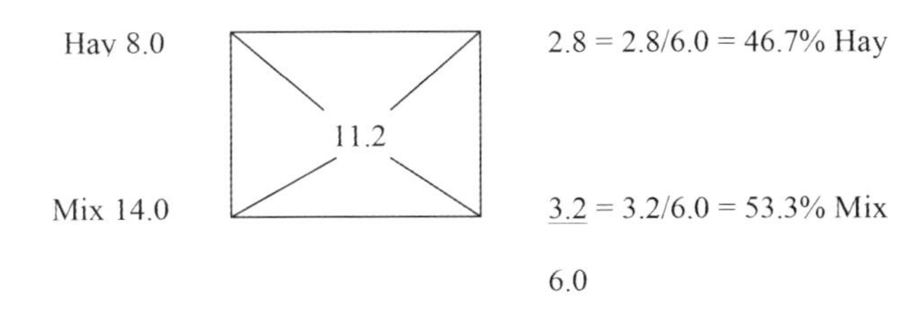

This combination would have the nutrient content shown in Table 46-3.

The energy level in the combination was calculated by taking the energy content of each ingredient times its inclusion level as follows:

(14.0 × 0.467) + (58.5 × 0.533) = 37.8 Mcal/cwt.

Table 46-3. Combined Nutrient Content.

Feedstuff	NEg, Mcal/cwt	CP, %
Bahiagrass	14.0	8.0
Grain/Protein Mix	58.5	14.0
Total	37.8	11.2

- **Step 4** – Compare the energy content of this ration to the requirement and adjust the ratio of grain to protein in the mix if necessary. The energy level is slightly higher than required but not enough to matter. With this approach,

Table 46-4. Examining a Ration for Calcium and Phosphorus.

Feedstuff	Lbs/cwt	Ca, %	Lbs Ca	P, %	Lbs P
Bahiagrass	46.7	0.48	0.22	0.20	0.09
Sorghum	49.0	0.04	0.02	0.32	0.16
CSM	4.3	0.21	0.01	1.19	0.05
Total	100.0		0.25		0.30

producers must guess at the protein content required for the mix of grain and cottonseed meal. It may take a trial-and-error approach to find the protein level that results in the right combination of protein and energy. In other words, a mix with 14 percent crude protein was selected. If the energy level of the final ration was either too high or low, the protein level of the grain/protein could be adjusted to give a mix that would be close to the energy requirement. As a rule, the lower the protein content in the roughage, the higher the protein content required in the grain/protein combination.

- **Step 5** – Determine the feeding rate for each ingredient in the final ration:

  1. Bahiagrass Hay = 46.7%
  2. Sorghum Grain = 91.9 × 0.533 = 49.0%
  3. CSM = 8.1 × 0.533 = 4.3%

- **Step 6** – Determine the adequacy of the ration for calcium and phosphorus:

  Since we calculated these values for 100 pounds of total ration, the amount of calcium and phosphorus also represents the percentage in the diet. In other words, this diet is 0.25 percent calcium and 0.30 percent phosphorus (Table 46-4). Based on Appendix Table 6, this heifer would require a diet with 0.42 percent calcium and 0.22 percent phosphorus. Thus, there is a significant deficiency of calcium, which necessitates supplementation. A free-choice mineral high in calcium could be provided, but there is no guarantee that the heifers will consume it adequately or that all of the heifers will consume it. Adding limestone (34.0 percent calcium) to the grain/protein mix would be a good way of meeting the requirement. To avoid diluting the protein and energy content of the diet significantly, the limestone should be substituted for the hay. Using a trial and error approach, 0.8 pounds of limestone per hundred pounds of ration meets the requirement. Table 46-5 shows this adjustment to the ration.

  Another option for meeting the calcium deficiency is to use a commercial mineral supplement as will be discussed later.

- **Step 7** – Determine the amounts to be fed on a daily basis.

  This ration could be fed as a total mixed ration (TMR) if the appropriate equipment is available. In most situations, the sorghum, cottonseed meal, and limestone would be fed in some type of feeder and the hay fed free-choice. Consequently, the amount of this mix to feed daily must be calculated. From Appendix Table 6, it can be determined that this heifer will consume 12.6 pounds

Table 46-5. Proportions in Adjusted Ration.

Feedstuff	Lbs/cwt	Ca, %	Ca, lbs	P, %	P, lbs
Bahiagrass	45.9	0.48	0.22	0.20	0.09
Sorghum	49.0	0.04	0.02	0.32	0.16
CSM	4.3	0.21	0.01	1.19	0.05
Limestone	0.8	22.3	0.18	---	---
Total	100.0		0.43		0.30

of dry matter per day. The grain/protein mix will constitute 54.1 percent of the diet. Thus, this heifer should be fed 12.6 × 0.541 = 6.82 pounds of dry matter from this mix. Assuming a dry matter content of 90 percent, this is 6.82 ÷ 0.90 = 7.57 pounds per day. The amount of hay the heifer should consume would be 12.6 × 0.459 = 5.78 pounds of dry matter. This would be supplied by an "as fed" level of 5.78 ÷ 0.89 (percent dry matter) = 6.50 pounds. Note: This does not allow for any wastage of the hay.

Keep in mind that these sample calculations are for a 500-pound heifer to gain 1.5 pounds per day. Consequently, as the heifer gains weight, the ration should be adjusted at 100-pound intervals. Alternatively, the amount of grain mix can be adjusted upward based on current body weight, and the hay fed free choice.

To calculate the amount of grain mix to feed at various heifer body weights, refer to the ration we just formulated above. In that example, the 500-pound heifer required 7.57 pounds of grain mix as fed per day (7.57 ÷ 500 = 1.51 percent). Thus, a producer needs to feed the grain mix at about 1.5 percent of body weight as the heifer grows. This means that a 650-pound heifer would need to be fed 650 × 0.015 = 9.75 pounds of the grain mix. In addition, free choice hay and a trace mineral supplement with vitamin A should be fed. The best option would be to add the trace mineral supplement to the limestone/grain mixture.

As discussed in a previous chapter, intake is influenced by a number of factors such as weather and body condition. The dry matter intake levels presented in the Appendix tables are good estimates, but they should not be considered absolute values. It is recommended that some estimate of the actual dry matter intake be made early in the feeding period to allow adjustment of the nutrient density to obtain the desired performance.

Using Commercial Supplements

Most feed companies have protein and mineral supplements formulated for common feeding situations such as the one in the example above. Furthermore, these supplements usually contain adequate quantities of the macro and trace minerals. In addition, they are available with the ionophores Rumensin® or Bovatec®, which are recommended in the diet of replacement heifers. Using commercial products is not typically the least expensive option, but it probably does the best job of meeting requirements in many situations.

Most feed companies have products formulated to be fed at low inclusion rates (0.25 or 0.5 pound per head per day) to compensate for nutrient deficiencies in common feeding situations. For example, most companies offer "balancer" type supplements formulated to be fed with corn gluten feed or dry distiller's grains. They can be mixed with the by-product or fed separately. These products are a good way to correct mineral imbalances common when feeding by-products and to provide the required trace minerals.

An economical way to reduce the cost of commercial supplements is to have a nutritionist formulate a set of specifications for the situation. These specifications can be submitted to several feed companies for bids. These bids should require that the company disclose the ingredient makeup of the feed and lock these ingredient specifications for the bid period. When evaluating bids, more than the price should be evaluated including service, pellet quality, and the integrity of the company.

References

Bohnert, D. and D. Chamberlain. Ration Balancing in Beef Cattle Nutrition Workbook. Oregon State University.

Church, D.C. 1972. Digestive Physiology and Nutrition of Ruminants Volume 3 - Practical Nutrition. O.S.U. Bookstores, Corvallis, OR.

Church, D.C. and W.G. Pond. 1974. Basic Animal Nutrition and Feeding. O & B Books, Corvallis, OR.

Endecott, R.L. and C.P. Mathis. Ration Balancing on the Ranch. New Mexico State University Bulletin B-125.

Nutrient Requirements of Beef Cattle. 1984. National Research Council.

Nutrient Requirements of Beef Cattle. 1996. National Research Council.

Simms, D.D. 2009. Feeding the Beef Cowherd for Maximum Profit. SMS Publishing. Amarillo, TX.

Chapter 47

Price Slides

In the direct or on-farm marketing of stocker cattle or calves, it is often necessary to use price slides to establish a fair price for both the seller and buyer because it is difficult to estimate the actual weight. This is especially true when calves or stockers are contracted in advance of the sale date. For example, many feedlots contract stockers on grass in the summer for delivery in the fall. The delivery weight will be impacted by the quality and quantity of grass during the grazing season. It also can be difficult to estimate the average weight in the pasture even at the time of sale. Consequently, a system of price slides is commonly used in the beef industry.

In the process of negotiating a sale, first estimate the expected delivery weight, which is used to establish a base price. Next, establish a price slide that goes into effect if the cattle are heavier or lighter than projected. Other items that must be negotiated include weigh-up conditions and the pencil shrink to be applied to the final weight. The pencil shrink usually is from 2 to 4 percent, depending on how much the cattle will be handled before being weighed. For example, if the cattle are to be rounded up, loaded on a truck, hauled a short distance, and weighed on the truck, the buyer might want a 4 percent pencil shrink. Whereas, if the cattle go through the same process but are unloaded at a sale barn to be weighed, a more appropriate pencil shrink might be 2 percent. The distance that the cattle are hauled before weigh-up should be considered with smaller pencil shrinks used for longer hauls before weigh-up.

Establishing Price Slides

Price slides are based on the relative price differentials of cattle of different weights at the time the contract is negotiated. These price differentials are a function of the expected cost of gain in the next phase of production. For example, high costs of gain in feedlots tend to reduce price slides for feeder cattle, while low costs of gain increase price slides. In addition, lighter feeder cattle tend to be more efficient than heavier feeder cattle. Tables 47-1 and 47-2 show average price slides for steers and heifers from 2000 to 2010 taken from the USDA Agricultural Marketing Service seven-market combined weighted average feeder cattle market report.

A review of these tables indicates that there is a tendency for the slides to be higher for lighter-weight steers and heifers, and the slides tend to be larger for steers than heifers. Lighter-weight animals tend to have more value per pound because they are more efficient and the opportunity exists to put on more pounds before they are finished at a lower cost per pound. The slides are higher for steers than heifers because steers have better feed conversions than heifers resulting in lower cost-of-gain.

Tables 47-1 and 47-2 also indicate that price slides vary from year to year. This variability is primarily a function of differences in feed prices. Typically, when corn prices are high resulting in high costs-of-gain, heavier weight feeders are discounted less than when corn is relatively cheap.

Example of a Price Slide

To illustrate a price slide, assume that 500-pound steers are selling for $155 per hundredweight and 600-pound steers are selling for $146 per hundredweight. This results in a price slide of $9.00 per hundredweight or $0.09 per pound. As part of the contract, it is typical that there is a "window" around the estimated base weight where a price slide will not be applied.

For example, if the base price was established for 575 pounds, it may be agreed that no price slide will be applied if the cattle weigh between 565 and 585 pounds. In addition, the slide may only be applied if the cattle are heavier than the projected base weight. Obviously, this agreement is in favor of the buyer, who benefits greatly if the cattle are significantly lighter than the weight used to establish the base price.

To be fair to both the buyer and seller, the price slide should reflect current market conditions.

Table 47-1. Annual Average Feeder Steer Price Slides, by Weight Category.

Year	250-350 lb Slide ($/lb)	350-450 lb Slide ($/lb)	450-550 lb Slide ($/lb)	550-650 lb Slide ($/lb)	650-750 lb Slide ($/lb)	750-850 lb Slide ($/lb)	850-950 lb Slide ($/lb)
2000	0.09	0.08	0.08	0.10	0.07	0.05	0.05
2001	0.07	0.08	0.09	0.09	0.07	0.04	0.04
2002	0.04	0.10	0.09	0.07	0.05	0.04	0.05
2003	0.09	0.01	0.18	0.08	0.05	0.04	0.03
2004	0.12	0.11	0.10	0.10	0.07	0.05	0.05
2005	0.17	0.15	0.12	0.12	0.09	0.06	0.06
2006	0.18	0.15	0.13	0.11	0.08	0.05	0.06
2007	0.08	0.10	0.09	0.09	0.06	0.04	0.04
2008	0.07	0.07	0.07	0.08	0.06	0.03	0.04
2009	0.09	0.05	0.08	0.08	0.06	0.04	0.04
2010	0.03	0.10	0.07	0.08	0.07	0.05	0.04

University of Nebraska NebGuide G2088.

Table 47-2. Annual Average Feeder Heifer Price Slides, by Weight Category.

Year	250-350 lb Slide ($/lb)	350-450 lb Slide ($/lb)	450-550 lb Slide ($/lb)	550-650 lb Slide ($/lb)	650-750 lb Slide ($/lb)	750-850 lb Slide ($/lb)	850-950 lb Slide ($/lb)
2000	0.05	0.08	0.08	0.06	0.04	0.04	0.03
2001	0.09	0.09	0.09	0.07	0.04	0.04	---
2002	0.05	0.08	-0.09	0.22	0.04	0.03	---
2003	0.05	0.08	0.08	-0.03	0.11	0.03	0.04
2004	0.13	0.10	0.09	0.08	-0.01	0.11	0.06
2005	0.15	0.11	0.12	0.09	0.07	0.06	0.04
2006	0.09	0.13	0.11	0.09	0.07	0.05	0.04
2007	0.05	0.09	0.08	0.06	0.04	0.02	0.04
2008	0.03	0.06	0.06	0.05	0.03	0.02	0.04
2009	0.04	0.05	0.06	0.05	0.04	0.03	0.04
2010	0.02	0.05	0.06	0.07	0.05	0.03	0.05

University of Nebraska NebGuide G2088.

Example: Actual weight is higher than projected

Assumptions:

Projected base weight = 600 lbs

Base price = $146/cwt

Pencil shrink = 2%

Slide = $8.00/cwt with a 10 lb window above and below the base weight

Actual weight = 640 lbs at delivery

The price slide will be used because the actual weight exceeds the projected weight plus 10 lbs.

Calculations:

Pay weight = 640 less 2% = 640 – 13 = 627 lbs

Weight subject to the price slide = 627 – 610 = 17 lbs = 0.17cwt

Price slide adjustment = 0.17 cwt × $8.00/cwt = $1.36/cwt

Adjusted final price = $146 – $1.36 = $144.64/cwt

Price received per animal = 6.27 cwt × $144.64 = $906.89

Example: Actual weight is below the projected weight

Assumptions:

Same as the example above except the actual weight = 550

Calculations:

Pay weight = 550 less 2% = 550 – 11 = 539 lbs

Weight subject to price slide = 590 – 539 = 51 lbs = 0.51 cwt

Price slide adjustment = 0.51 cwt × $8.00/cwt = $4.08

Adjusted final price = $146.00 + $4.08 = $150.08

Price received per head = 5.39 cwt × $150.08 = $808.93

References

Feedlot Management Primer Chapter 2. Shipping and Receiving Cattle. Beef Information Website. Ohio State University.

Rasby, R. and D. Mark. Feeder Cattle Price Slides. University of Nebraska NebGuide. G2088.

Self, H. and N. Gay. 1972. Shrink During Shipment of Feeder Cattle. JAS 35:489-494.

Chapter 48

Finishing Cattle Management

Feeding Programs

Chapter 45 covered the handling and feeding of highly stressed cattle. The following recommendations refer to backgrounded and yearling cattle. Most cattle entering a feedlot to be finished have been on a roughage-based ration with little grain or other concentrate. Cattle that have been on a drylot growing program that have received 1.0 to 1.5 percent of their body weight in grain or by-products per day are the exception. The final finishing ration in a feedlot typically has 90 to 92 percent concentrate, if the protein supplement is considered concentrate. The microorganisms that digest fiber are different than the microorganisms that digest starch, which means that a major shift in the microbial population must occur as the cattle are placed on the finishing program. Overloading the rumen with concentrate results in high populations of lactic-acid producing bacteria. The high levels of lactic acid damage the rumen and other tissues when absorbed into the blood stream resulting in a condition known as acidosis characterized by bloat, going off feed, and in severe cases, death. To avoid acidosis, the cattle must be "worked up" or adapted to the final ration. On average, feedlots take about 21 days to make this transition to the finishing diet.

Common Adaptation Programs

Currently, there are two ways cattle are adapted to the finishing ration:

1. **Step-Up Rations** – This is the traditional way of adapting cattle to the finishing diet. It is the most common method used today. Typically, three to five rations are fed, each slightly higher in energy. Normally, the cattle are on each energy level for 5 to 7 days. Table 48-1 shows the energy and protein levels that might be in typical step-up rations.

   A comparison of the crude protein levels in typical rations and animal requirements indicates that the protein levels typically fed are well above the NRC requirements. It is common for feedlot nutritionists to feed high levels of protein to ensure protein is adequate for all the cattle in the feedlot. Nutritionists do not want protein to be a performance-limiting nutrient.

2. **Two-Ration Blending Program** – In this system, only two rations are made by the feed mill – the starter and the finisher. The cattle receive only the starter for the first few days, then the percentage of finisher is increased and the starter decreased until all the starter is replaced after 21 to 28 days. The two rations

Table 48-1. Nutrient Levels in Typical Step-up Rations.

			Ration		
Nutrient	**Starter**	**1**	**2**	**3**	**Finisher**
CP, %	14-16[a]	13.5	13.5	13.5	13.5
NEg, Mcal/Cwt	47	52	57	62	67

[a] Level depends on body weight and level of stress at arrival.

Figure 48-1. A Traditional Step-Up Program.

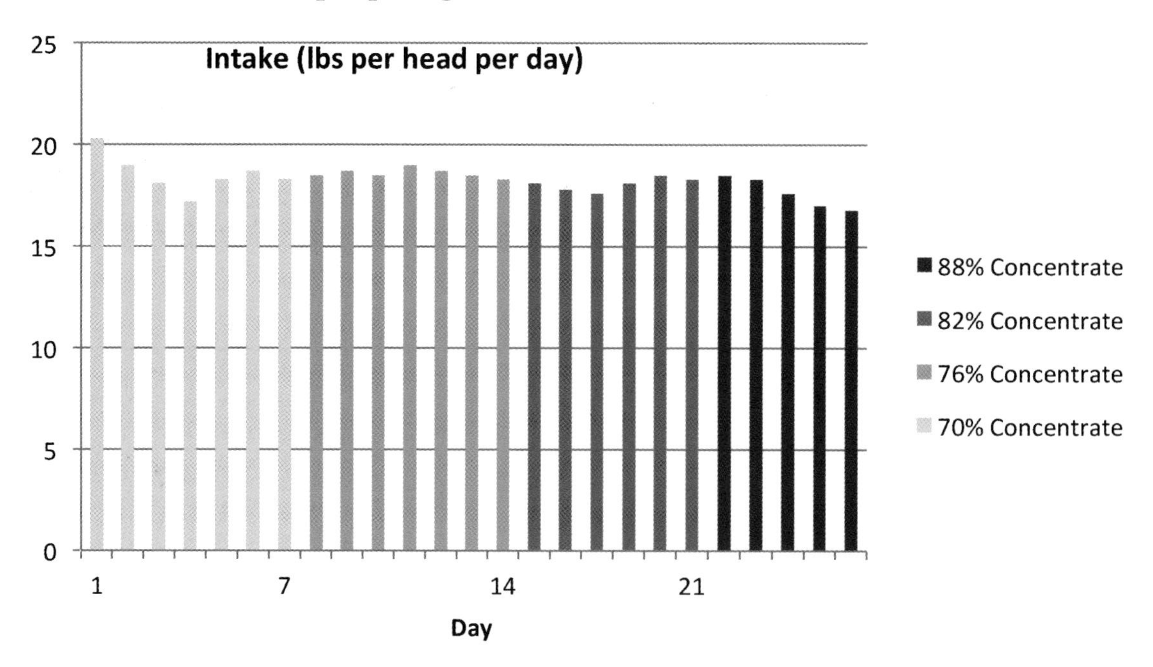

Oklahoma State University

Figure 48-2. Two Ration Blending Program

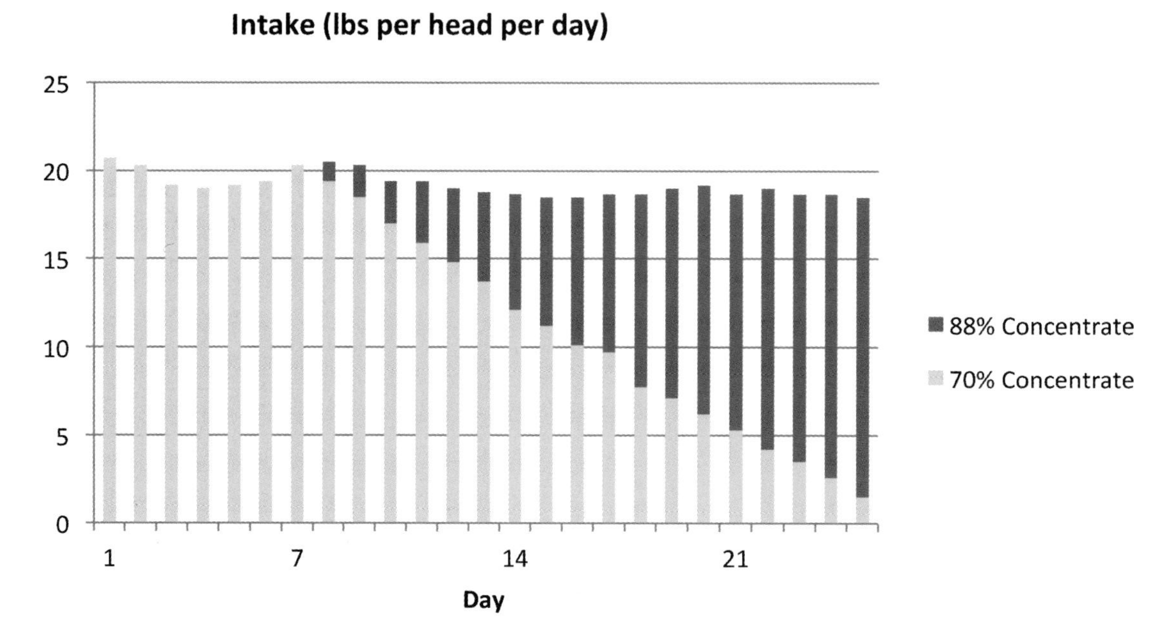

Oklahoma State University

Table 48-2. Comparison of Feed Calls (Delivery to the Bunk) with a Traditional Step-up Program and a Two Ration Blending Program

	Traditional Step-Up				Two Ration Blending			
	AM Call		PM Call		AM Call		PM Call	
Day	% of Call[a]	% Con[b]	% of Call[a]	% Con[b]	% of Call[a]	% Con[b]	% of Call[a]	% Con[b]
0-7	50.0	70	50.0	70	50.0	70	50.0	70
8	50.0	76	50.0	76	95.5	70	4.5	94
9	50.0	76	50.0	76	90.9	70	9.1	94
10	50.0	76	50.0	76	86.4	70	13.6	94
11	50.0	76	50.0	76	81.8	70	18.2	94
12	50.0	76	50.0	76	77.3	70	22.7	94
13	50.0	76	50.0	76	72.7	70	27.3	94
14	50.0	76	50.0	76	68.2	70	31.8	94
15	50.0	82	50.0	82	63.6	70	36.4	94
16	50.0	82	50.0	82	59.1	70	40.9	94
17	50.0	82	50.0	82	54.6	70	45.4	94
18	50.0	82	50.0	82	50.0	70	50.0	94
19	50.0	82	50.0	82	45.5	70	54.5	94
20	50.0	82	50.0	82	40.9	70	59.1	94
21	50.0	82	50.0	82	36.4	70	63.6	94
22	50.0	88	50.0	88	31.8	70	68.2	94
23	50.0	88	50.0	88	27.3	70	72.7	94
24	50.0	88	50.0	88	22.7	70	77.3	94
25	50.0	88	50.0	88	18.2	70	81.8	94
26	50.0	88	50.0	88	13.6	70	86.4	94
27	50.0	88	50.0	88	9.1	70	90.9	94
28	50.0	88	50.0	88	4.6	70	95.4	94
29-end	50.0	94	50.0	94	50.0	94	50.0	94

[a]% of Call = percent of daily feed call at that feeding. [b]% Con = percent concentrate ration fed at that feeding.
Oklahoma State University

are not mixed on a truck before delivery to the bunk, but delivered separately. For example, when the ration consists of 25 percent of the starter ration and 75 percent of the finisher ration, deliveries of the desired amounts of each ration will be made by different trucks and likely at different times. Table 48-2 shows the delivery protocol used in an Oklahoma State University research trial to illustrate the concept. The advantage of the two-ration blend approach is that it allows a feed truck containing starter, for example, to feed a high percentage of pens in an alley. The net result is fewer miles driven by each truck and lower fuel consumption. Oklahoma State University researchers concluded that, compared to the traditional approach, this approach may require increased management of feed calls, feed timing, and feed distribution.

Managing a Traditional Step-Up Program

Getting the cattle on full feed has a tremendous influence on their performance through the entire finishing program. It is important to manage intake and energy levels to avoid acidosis and erratic intake. Typically, cattle are fed good-quality grass hay on arrival. Then the starter ration is fed on top of the hay at 1.5 to 2.0 percent of body weight (dry matter basis). After a few days, the hay is no longer fed, and the amount of starter is gradually increased by about 1 pound per day until the cattle are eating approximately 2.5 percent of their body weight. The cattle are then fed the same amount of the next step-up ration. The amount of this ration is increased only if the cattle clean up the current amount for 3 days, and the amount fed increased by a maximum of 0.5 pound per day. After 5 to 7 days, the cattle are switched to the next step-up ration with the same amount being fed. The amount fed is never increased when the switch is made to the next step-up ration. The goal is to bring the cattle up on feed gradually avoiding days of high intakes followed by days with low intakes — a condition referred to as a "yo-yo" or "roller coaster" pattern of feed intake. Most feedlots feed twice per day, while some feed three times per day. So the amounts delivered are spread between two or three deliveries.

Many feedlots use silage as their primary roughage, which complicates the starting program to some extent because calves and yearlings that have never been exposed to silage are often reluctant to eat it. Consequently, feedlots use another roughage such as alfalfa hay in the starter and gradually replace it with silage in the step-up rations. This changes the moisture content in the rations and delivery amounts. These changes must be accounted for in the feeding program since the goal is to manage dry matter consumption on a consistent basis.

Starting Yearling Cattle with High Levels of Corn Gluten Feed

Recently, many feedlots have been starting yearling cattle with RAMP®, a complete starter based on Sweet Bran® corn gluten feed. This product consists of Sweet Bran®, roughage, and mineral. Performance using RAMP® has been at least as good and in some research trials better than with traditional starter rations. Starting cattle with RAMP® has several advantages for a feedlot. For example, the yard can usually reduce the number of ingredients it has to handle such as the starter supplement and roughages used in the traditional starter program. Moreover, since RAMP® is delivered as a complete feed, its usage reduces the volume of feed going through the feed mill.

Managing Cattle on Full Feed

Once the cattle are on the finisher ration, increases in the delivered amount are typically made when the cattle have consumed their entire ration for 3 consecutive days. The increase in feed delivered at any time is usually limited to no more than 0.5 pound of dry matter per day. It is important to not let the cattle get too hungry because the more aggressive cattle tend to overeat resulting in acidosis. Feedlots use "bunk readers" that score the bunks for feed availability at least once and possibly three times per day. As a minimum, the bunks are read before the first feeding in the morning, and the amount delivered at the first feeding is reduced if feed remains in the bunk. If the bunk has been clean before the first delivery for 3 days, the amount delivered is typically increased. The bunk readers may make an evaluation of the bunks before the afternoon feeding and make adjustment if necessary. In addition, many feedlots read the bunks at midnight with increases in delivery made to pens that have clean bunks.

There are several events that can disrupt the feeding program making it difficult to maintain consistent intake:

1. Rain gets the feed wet and produces mud, both of which create problems. Wet feed should be consumed quickly or removed from the bunk. Fresh feed should never be placed on top of wet feed. Severe mud greatly reduces intakes. Cattle are reluctant to fight their way through heavy mud to get to the bunks.

2. Storms also create other issues with intake. Cattle tend to eat more in response to a pending storm and less after the storm passes. Feedlots typically have a "storm" ration that is higher in roughage that is fed before a storm. This allows cattle to eat more without causing acidosis. After the storm, the normal finisher is often fed at a slightly reduced level slowly

working the cattle back up to their intake level before the storm.

3. Cattle tend to eat the most during the time of day that they are most comfortable, which is during the day in winter and at night in the summer. The percentage of the ration given at each feeding may be adjusted to compensate for these changes in eating pattern.
4. It is important to spread the feed evenly down the entire length of the bunk to accommodate as many cattle at once that want to eat.
5. Bunks should be cleaned frequently to remove any clumps of spoiled feed. Water tanks should be cleaned regularly, especially during the summer. Dirty water reduces water consumption, which reduces feed intake.

Typical Feedlot Nutrition Programs

Protein Levels in Finishing Rations

As noted previously, the protein level that feedlot nutritionists use is higher than the NRC requirement. In a survey conducted by Texas Tech University, the average crude protein level in finishing diets was 13.3 percent. Nutritionists commonly feed 1.0 percent urea in the ration to supply 2.83 percent crude protein.

Use of Supplemental Fat in Finishing Rations

In the Texas Tech University survey, 71 percent of the feedlots were adding fat to their rations with an average added level of 3.1 percent. It is likely that both the number of feedlots adding fat and the level added has declined since this survey because of the increase in the price of fat.

Mineral Levels in Finishing Diets

Table 48-3 shows the mineral feeding levels recommended by consulting nutritionists. Again, in several cases these levels are significantly higher than NRC recommendations. For example, the levels for copper and zinc are well above NRC recommendations.

Feed Additives

- **Ionophores** – Ionophores are polyether antibiotics originally developed as coccidiostats in poultry. Ionophores effect the rumen microflora, increasing the production of propionate, which increases the energy available from the ration. They also reduce protein degradation in the rumen increasing post-ruminal digestion, and they reduce methane production. The net effect is an increase in feed efficiency. Recent research indicates an improvement of about 3 to 4 percent.

  Commonly used ionophores include monensin (Rumensin®), lasalocid (Bovatec®), and laidlomycin (Cattlyst®). Given the importance of feed conversion on profitability in cattle feeding and the relatively low cost of ionophores, they should be included in the ration unless the cattle are being fed for a natural or branded beef program that does not allow ionophores.

Table 48-3. Average Mineral Levels Recommended by Feedlot Consulting Nutritionists as Cited in a Survey by Texas Tech University in 2007.

Major Minerals		Trace Minerals	
Mineral	%	Mineral	ppm
Calcium	0.70	Copper	17.6
Phosphorus	0.30	Zinc	93.0
Magnesium	0.22	Manganese	47.9
Potassium	0.70	Iron	51.7
Sulfur	0.22	Selenium	0.24
Salt	0.30	Iodine	0.75
		Cobalt	0.38

Vasconcelos, et al., 2007.

- **Beta-Agonists** – Beta-agonists are relatively new feed additives for finishing cattle. They are termed "repartitioning agents" shifting nutrients from fat deposition to lean tissue deposition. Currently, two beta-agonists are approved for use in cattle.

Optaflexx™ (Elanco Animal Health) – Approved for feeding during the last 28 to 42 days of the finishing period. Fed as recommended, Optaflexx™ increases gain, especially carcass gain with no increase in intake resulting in an improvement in feed conversion.

Zilmax™ (Merck) – Approved for feeding during the last 20 to 40 days on feed. It must be withdrawn from the ration for at least 3 days before harvest. Zilmax™ increases weight gain, carcass gain, and feed efficiency

Table 48-4 shows a comparison of Optaflexx™ and Zilmax™. As shown, Zilmax™ provides for more improvement in average daily gain, feed conversion, dressing percent, and ribeye area. Managing cattle fed Zilmax™ to an equal body composition mitigates some of the reduction in marbling score. However, it has shown a significant reduction in marbling score in some studies; and it has significantly increased Warner-Bratzler shear force in some studies. Studies using taste panels have demonstrated no difference in flavor, juiciness, and tenderness when compared to cattle not fed Zilmax™. In general, Optaflexx™ is used when improved performance with minimal influence on carcass quality is desired. Conversely, Zilmax™ is used when maximum performance and lean meat yield is desired.

Status of Zilmax™

In August of 2013, Merck temporarily removed Zilmax™ from the market in response to reports that Zilmax™-fed cattle were arriving at packing plants with severe lameness. It was unclear whether Zilmax™ was the cause of the problem or even a contributing factor; however, because of animal welfare concerns, Merck removed Zilmax™ from the market. Merck plans to study the problem and conduct research trials if necessary before reintroducing Zilmax™ to the market.

Melengesterol Acetate (MGA)

This synthetic progestin is routinely fed to heifers to prevent estrus and subsequent riding activity. It improves gain and feed conversion slightly.

Table 48-4. A Comparison of Optaflexx™ with Zilmax™.

Item Active ingredient	Optaflexx™ Ractopamine hydrochloride	Zilmax™ Zilpaterol hydrochloride
Duration of feeding, days	28-42	20-40
Optimal days	28-35	20
Withdrawal time, days	None	3
ADG, lbs/day	Increase 9-21%	Increase from 14-40%
Feed conversion	Improve 9-21%	Improve 14-40%
Dressing percent	Minimal impact	1 to 2 percentage unit increase
Hot carcass weight, lbs	10-25 lbs increase	13 to 28 lbs increase
Ribeye area, sq. in.	Up to 0.5 sq. in. increase	0.3 to 1.0 sq. in. increase
Marbling	Minimal impact	0 to 25 degree reduction in marbling score
Tenderness	Minimal impact	None up a 1.7 kg increase in Warner-Bratzler shear force

Source: UW Extension Wisconsin Beef Information Center.

Other Common Food Additives

- **Antibiotics** – In addition to the ionophores, there are other antibiotics that are often fed to finishing cattle. Commonly fed antibiotics include Tylan® and the tetracyclines.

Tylan® (Elanco Animal Health) – Tylan is commonly fed to finishing cattle in combination with an ionophore to prevent liver abscesses.

Table 48-5. Growth Promoting Implants Approved for Use in Beef Cattle.

Implant Trade Name	Marketing Company	Active Ingredient	Estimated Payout in Days
Low Potency Estrogen or Estrogen-like			
Ralgro	Merck	36 mg Zeranol	70-100
Synovex-C Component E-C	Pfizer Elanco	10 mg Estradiol benzoate 100 mg Progesterone	100-140
Medium Potency Estrogen or Estrogen-like			
Synovex-S Component E-S	Pfizer Elanco	20 mg Estradiol benzoate 200 mg Progesterone	100-140
Synovex-H Component E-H	Pfizer Elanco	20 mg Estradiol benzoate 200 mg Testosterone propionate	100-140
Compudose	Elanco	25.7 mg Estradiol	180-200
Encore	Elanco	43.9 mg Estradiol	300+
Androgen Implants			
Finaplex-H Component T-H	Merck Elanco	200 mg TBA 200 mg TBA	100-120 100-120
Lower Potency Combination Implants			
Revalor-G Component TEG	Merck Elanco	8 mg Estradiol 40 mg TBA	100-140
Revalor-IS Component TEIS	Merck Elanco	16 mg Estradiol 80 mg TBA	120-160
Revalor-IH Component TEIH	Merck Elanco	8 mg Estradiol 80 mg TBA	120-160
Synovex Choice	Pfizer	14 mg Estradiol benzoate 100 mg of TBA	120-160
High Potency Combination Implants			
Revalor-S Component TE-S	Merck Elanco	24 mg Estradiol 120 mg TBA	120-140
Revalor-H Component TEH	Merck Elanco	14 mg Estradiol 140 mg TBA	120-140
Synovex Plus	Pfizer	28 mg Estradiol benzoate 200 mg TBA	140-160
Revalor-XS	Merck	40 mg Estradiol 200mg TBA	170-200

Implants with an S or H in their name are for use in steers or heifers, respectively. Component implants are generic versions of the original implants.

- **Chlotetracycline (CTC) and Oxytetracycline (OTC)** – These antibiotics are not typically fed to finishing cattle at low levels on a continuous basis. Typically, they are fed at high levels (1 gram/cwt/day) for 3 to 5 days to treat respiratory diseases.

Types of Supplements Used in Finishing Cattle Diets

A supplement containing protein, minerals, and additives is normally included in a finishing ration at 5 to 6 percent of the total ration. The Texas Tech University survey indicated that 45.4 percent of the feedlots used liquid supplements, 38.6 percent used pelleted, dry supplements, and 10.4 percent used loose, dry supplements. Micro-ingredient machines were used by 49.0 percent of the feedlots.

Implants

Without a doubt, implanting is one of the most profitable management practices in a finishing program. It increases gain and efficiency by roughly 10 percent, and thus it greatly reduces cost of gain. Consequently, finishing cattle should be implanted unless they are being fed for a "natural" or organic beef program. There are a multitude of different implants (Table 48-5) available to cattle feeders making the choice of implant and implant program rather complicated.

Developing an Implant Strategy

While implants significantly increase rate of gain and feed efficiency, they also tend to reduce marbling and thus quality grade. Consequently, there is a trade-off between performance and carcass quality that must be considered in designing an implant program. The following factors are key to developing an effective implant strategy:

1. **Anticipated Length of the Finishing Program** – A review of Table 48-5 indicates a wide range in potency and payout period for the currently available implants. Most feedlots use a combination implant (containing both estrogen and testosterone analogues) as the final implant because the final implant has the greatest effect on performance and cost of gain. If the cattle are fed for only 120 to 140 days, a single combination implant is typically used. For cattle fed for 140 to 160 days, a low-potency or medium-potency implant is often given at the start of the finishing period. The cattle are re-implanted with a medium- or high-potency implant a minimum of 80 days before anticipated slaughter. Another option for cattle fed more than 140 days is to use one of the implants with a long payout period. Obviously, developing a good implant strategy requires the ability to forecast the number of days the cattle will be fed. Unfortunately, adverse weather can lengthen the finishing period. Correspondingly, better-than-average feeding conditions can allow the cattle to finish sooner than anticipated.
2. **Marketing Program** – Another consideration in developing the implant program is the anticipated marketing program. For example, cattle feeders selling on a grid that favors retail product yield tend to be aggressive in their implant program, often using the high-potency combination implants. Conversely, cattle feeders targeting the high-quality, branded-beef programs tend to use a much less aggressive program to avoid reducing carcass quality. The moderate potency combination implants appear to provide good performance enhancement with minimal reduction in carcass quality.

Common Disease Problems in Feedlots

Shipping fever/pneumonia is easily the most common disease problem in feedlots. It significantly increases death loss and reduces performance. Research shows animals that get shipping fever perform poorer than those that do not. Footrot is fairly common in feedlots, especially in muddy conditions. In addition, clostridial diseases occur occasionally.

Common Infectious Diseases in Feedlots

Footrot

Footrot is characterized by lameness and swelling between the claws. It is an infectious disease caused by bacteria. It is most common when the cattle have been standing in mud for long periods. It frequently occurs when the cattle have been walking on frozen, rough ground. Several systemic antibiotics are effective in treating footrot, and footbaths using copper or zinc sulfate reduce the number of cases.

Pneumonia/Shipping Fever/Bovine Respiratory Disease (BRD) Complex

A number of viral and bacterial organisms that infect the lung cause this disease complex. Any stress like weaning or shipping will increase the incidence of bovine respiratory disease. An affected animal has an increased rate and depth of respiration, has a cough, and breathes with its mouth open. In addition, they often have a nasal discharge. Key to treating bovine respiratory disease successfully is early detection and treatment with an antibiotic.

Nutritional Diseases in Feedlots

Acidosis

Acidosis is the most common nutritional disorder in feedlots. A typical feedlot diet results in the production of large quantities of volatile fatty acids, which lowers the pH in the rumen. In some cases, lactic acid producing bacteria proliferate resulting in an even lower pH. When this occurs, the lactic acid enters the blood stream, lowering blood pH and causing damage to the liver and other tissues in the body. Acidosis is defined as being either acute or subacute depending on the severity.

- **Acute Acidosis** – Cattle suffering from acute acidosis go off feed, avoid movement, and become dehydrated. They may appear to be blind and kick at their belly. The rumen may be distended, and they will have a foul smelling diarrhea. In severe cases, the animals will lie down, unable to move. These animals generally die with their head tucked to the side. Survivors of acute acidosis often have suffered damage to their rumen leading to poor performance. They often exhibit overgrown hooves 30 to 60 days after an acute acidosis episode.

- **Subacute Acidosis** – Cattle experiencing a milder form of acidosis consume less feed, resulting in poor performance. Their stool often is a flat gray color. In addition, these cattle experience some reduction in rumen function and tend to develop liver abscesses.

Common causes of acidosis include:

1. Increasing energy intake too rapidly either by increasing intake or energy density of the ration too rapidly.
2. Allowing cattle to overconsume the ration. This often occurs when cattle eat more before or after a storm.
3. Feeding the wrong ration to a pen. This often happens when a feed truck driver delivers the finisher ration to a pen that should receive the starter or one of the step-up rations.

Bloat

In feedlot cattle, the typical bloat differs from that normally observed in a cow herd situation. It is usually free-gas bloat rather than frothy bloat. The effect on the animal, however, is similar with both types of bloat. Inserting a tube into the rumen via the esophagus usually relieves the pressure. Some animals become chronic bloaters with repeated incidences of bloat. These cattle should be sent to slaughter.

Polioencephalomalacia (PEM)

Cattle with PEM exhibit blindness, have seizures, and are observed "swimming" on their side. Before it became common to feed by-products with high levels of sulfur like distillers grains and corn gluten feed, most cases of PEM were caused by a thiamine deficiency. Currently, the most common cause is sulfur toxicity. The increased feeding of by-products from corn milling and ethanol production, which are high in sulfur, has made this a serious problem. When sulfur is fed in excess, the rumen microorganisms produce high levels of hydrogen sulfide gas, which is eructated and inhaled. It enters the blood stream and damages the brain causing the symptoms of PEM. Many nutritionists feed additional thiamine to prevent sulfur induced PEM. It does reduce the severity of the symptoms.

Sudden Death Syndrome

A number of disease problems can cause death late in the feeding period. These deaths are grouped into what is referred to as "sudden deads." Causes include bloat, enterotoxemia (overeating disease), acidosis, and pneumonia. Sudden deads should be necropsied to identify the cause of death so a prevention plan can be implemented.

Urinary Calculi

Urinary calculi are mineral deposits in the bladder or urethra of male cattle. If the urethra is blocked for a sufficient time, the bladder or urethra ruptures, resulting in the condition known as "water belly." Affected cattle strain to urinate and often kick at their belly. If the blockage is incomplete, urine may drip from the sheath. There are two primary types of urinary calculi:

1. **Phosphatic Urinary Calculi** – A high phosphorus level or an imbalance between calcium and phosphorus are the main causes of this type of calculi. The Ca:P ratio should be between 2:1 and 1.2:1. This type of urinary calculi occurs more often in the winter probably because of lower water consumption.
2. **Siliceous Urinary Calculi** – These calculi normally occur in cattle grazing range plants in areas with mineral imbalances. In many cases these calculi cause problems shortly after the cattle enter a feedlot.

Typical Feedlot Vaccination Program

A basic vaccination schedule includes:

1. IBR, BVD, PI_3, BRSV – Modified Live Vaccine.
2. 7-way or 8-way Clostridial Vaccine – Prevents blackleg, enterotoxemia, and other diseases caused by the various clostridial organisms.

References

Burken, D.B, J.L. Wahrmund, B.P. Holland, C.R. Krehbiel, and C.J. Richards. Two Rations Blending vs. Traditional Step-Up Adaptation to Finishing Diets: Performance and Carcass Characteristics. Proceedings of the 2010 Plains Nutrition Conference.

Feedlot Management Primer Chapter 2. Shipping and Receiving Cattle. 2010. Beef Information Website. Ohio State University.

Hicks, B. 2010. Beef Cattle Research Update – Adaptation Programs for Starting Cattle on Feed. Oklahoma Panhandle Research & Extension Center.

Lawrence, J., S. Shouse, W. Edwards, D. Loy, J. Lally, R. Martin. Beef Feedlot Systems Manual. Iowa Beef Center Website. Iowa state University.

Prince, S. 2003. Ionophores. Student Research Summary. Texas A&M University.

Profit Tip: Feed Additives. University of Nebraska Beef Cattle Production Website.

Radunz, A. 2011. Use of Beta Agonists as a Growth Promoting Feed Additive for Finishing Beef Cattle. UW Extension Wisconsin Beef Information Center.

Symposium: Intake by Feedlot Cattle. 1995. Oklahoma State University Publication P-942.

Symposium: Impact of Implants on Performance and Carcass Value of Beef Cattle. 1997. Oklahoma State University Publication P-957.

Vasconcelos, J. and M. Galyean. 2007. Nutritional Recommendations of Feedlot Consulting Nutritionists: The 2007 Texas Tech University Survey. JAS 85:2772-2781.

Weichenthal, B., I. Rush, and B. Van Pelt. 1999. Dietary Management for Starting Finishing Yearling Steers on Feed. 1999 Nebraska Beef Report. Pg 44-46.

Chapter 49

Grain Sources and Grain Processing for Finishing Cattle

Grain Sources

From time to time, all major cereal grains are used in cattle finishing rations including wheat when it is priced competitively. The two major feed grains, however, are corn and grain sorghum (milo). Table 49-1 shows the relative performance and energy levels of the major feed grains.

Corn

Corn is the predominant feed grain used to finish cattle. It is palatable and can be fed whole, yet it responds to processing, especially steam flaking, with improved performance. Before development of the ethanol industry, corn often comprised 85 percent of typical finishing rations on a dry matter basis. It is still the major ingredient in most feedlot rations.

Grain Sorghum

Grain sorghum can be grown on nonirrigated land in areas where rainfall will not support corn production in most years. For example, milo is grown as a dryland crop in western Kansas and the panhandles of Texas and Oklahoma. Furthermore, in areas, such as western Kansas, where water levels for irrigation have declined in recent years, some producers have shifted from corn to milo because it requires less water. Milo has about 85 percent of the energy value of corn when each grain is rolled or ground; however, steam flaking has a greater influence on digestibility with milo than with corn. The seed coat of milo is essentially indigestible meaning that milo must be rolled or ground as a minimum level of processing. Milo is much more variable in starch and protein content than corn.

Barley

Barley is used as a feed grain in the northern states, the northwest, and in Canada where it is difficult to get corn to mature. Barley has a tough outer hull that necessitates some processing. The feeding value of barley is more variable than corn with test weight a good indicator of its value — the heavier the test weight, the higher the energy value. Barley is less palatable than corn and tends to cause more bloat than corn, especially when fed with alfalfa.

Table 49-1. A Comparison of the Performance of Cattle Fed Different Feed Grains and the Energy in the Grains.

Item	Barley	Corn	Milo	Oats	Wheat
ADG, lbs	3.26	3.30	3.04	3.31	2.94
DMI, lbs/day	20.22	20.02	21.01	20.13	18.81
Feed/Gain	6.22	6.45	6.95	6.12	6.39
ME of grain DM, Mcal/lb					
Observed	1.52	1.48	1.40	1.46	1.52
NRC (1984)	1.38	1.47	1.38	1.26	1.44

Adapted from Owens, et al.

Danny D. Simms

Oats

Oats are used in some finishing rations, especially in the northern states. Oats have a fibrous hull and relatively small kernel, which means that they should be processed for finishing diets. They are usually fed as a part of the grain portion of the ration rather than as the sole grain. The hull can provide significant fiber in a finishing diet. They can be fed whole to calves because they will chew them better than a finishing animal. As with barley, test weight is a good indicator of the starch content and energy value.

Wheat

Wheat often is price competitive with the other feed grains based on its high energy content, which is similar to corn. Its high protein content compared to corn also makes it an attractive ingredient in finishing diets. Wheat is rapidly fermented in the rumen, which limits the amount that should be included in the ration. Typically, wheat is fed at no more than 50 percent of the grain portion of the ration. This rapid fermentability places a premium on good bunk management and constant intake levels to avoid acidosis.

Grain Processing

Most of the energy in grains is in the form of starch, which is held within a protein matrix. This protein matrix reduces the digestibility of the starch to some extent. The degree to which the protein matrix reduces digestibility varies between grains. For example, the protein matrix in milo has a much greater effect on digestion than the protein matrix in corn. To reduce the influence of the protein matrix on digestibility, grains used in finishing diets are typically processed in some fashion. The most notable exception is the feeding of whole shell corn, which will be discussed later. The most common processing methods used in the industry are:

1. **Dry Roll** – The grain is forced between rollers that break it into several pieces.
2. **High Moisture** – The grain is harvested at higher moisture than normal (usually 26 to 32 percent) and stored whole in an oxygen limiting structure or ground and stored in a bunker. The grain stored whole is ground before feeding. The grain goes through fermentation, which breaks down the protein matrix to some extent.
3. **Steam-Flaked** – The grain is exposed to steam before being rolled into a flake. The combination of heat from the steam and rolling breaks up much of the protein matrix.

Table 49-2 shows a comparison of corn processing methods on gain, intake, feed efficiency, and metabolizable energy. Surprisingly, dry rolling the corn decreased performance relative to whole corn. Whole corn is typically fed with lower roughage levels than used with dry rolled corn, which has some influence on this comparison. Correspondingly, high-moisture corn resulted in poorer performance than whole corn. Steam flaking improved conversions slightly and increased the metabolizable energy compared to whole corn.

Table 49-3 shows the effect of processing milo on gain, dry matter intake, feed/gain, and metabolizable energy. Since the digestibility of whole milo is low, it was not included in the comparisons. In general, the finer the particle size, the better the performance when feeding dry rolled milo. As shown, feed intake

Table 49-2. Effect of Corn Processing on Gain, Intake, Feed Efficiency, and Metabolizable Energy.

Item	Whole	Dry Rolled	High Moisture	Steam Flaked
Gain, lb/day	3.20	3.15	3.04	3.13
Dry Matter Intake lb/day	19.22	21.01	19.91	18.21
Feed/Gain	6.07	6.68	6.69	5.82
Metabolizable Energy (ME), Mcal/lb	1.52	1.40	1.47	1.66
% Improvement in F/G compared to Whole Corn	---	-10.0	-10.2	4.1

Adapted from Owens, et al. 1995

Table 49-3. Effect of Milo Processing on Gain, Intake, Feed/Gain, and Metabolizable Energy.

Item	Dry Rolled	High Moisture	Steam Flaked
Gain, lb/day	3.17	2.89	3.02
Dry Matter Intake lb/day	23.21	19.82	19.11
Feed/Gain	7.39	6.94	6.46
Metabolizable Energy (ME), Mcal/lb	1.32	1.35	1.54

Adapted from Owens, et al. 1995

Table 49-4. Effect of Barley Processing on Gain, Intake, Feed/Gain, and Metabolizable Energy.

Item	Whole	Dry Rolled	Steam Flaked
Gain, lb/day	3.26	3.37	3.13
Dry Matter Intake lb/day	22.93	20.02	18.92
Feed/Gain	6.79	6.07	5.99
Metabolizable Energy (ME), Mcal/lb	1.41	1.51	1.60

Adapted from Owens, et al. 1995

Table 49-5. Effect of Oat Processing on Gain, Intake, Feed/Gain, and Metabolizable Energy.

Item	Dry Rolled	Steam Flaked
Gain, lb/day	3.37	3.26
Dry Matter Intake lb/day	20.33	20.13
Feed/Gain	6.01	6.18
Metabolizable Energy (ME), Mcal/lb	1.47	1.46

Adapted from Owens, et al. 1995

Table 49-6. Effect of Wheat Processing on Gain, Intake, Feed/Gain, and Metabolizable Energy.

Item	Dry Rolled	Steam Flaked
Gain, lb/day	2.91	3.04
Dry Matter Intake lb/day	19.31	17.92
Feed/Gain	6.68	5.95
Metabolizable Energy (ME), Mcal/lb	1.51	1.54

Adapted from Owens, et al. 1995

with high moisture milo is lower than with dry rolled, but conversions are better. Clearly, steam flaking has a much greater effect on feed conversion and energy level with milo than corn.

Table 49-4 shows the effect of barley processing on animal performance and observed energy level. Dry rolling improved conversions and energy level compared to whole barley, and steam flaking only improved conversions and energy slightly over dry rolling. Given the cost of steam flaking, these slight improvements make steam flaking barley questionable from an economic standpoint.

Table 49-5 shows the effect of processing oats on animal performance and energy level. As opposed to other grains, steam flaking did not improve performance with oats. However, oats should be rolled for finishing diets.

Table 49-6 shows the effect of processing wheat on animal performance and energy level. As with milo,

Table 49-7. Performance of Cattle Fed Corn Flaked to Different Thicknesses.

	Pounds Per Bushel		
Item	<22	23 to 29	>29
ADG, lb	3.20	3.06	3.06
DMI, lb	17.3	16.1	16.3
Feed/Gain	5.4	5.2	5.3

Table 49-8. Performance of Cattle Fed Milo Flaked to Different Thicknesses.

	Pounds Per Bushel		
Item	<22	23 to 29	>29
ADG, lb	3.04	3.31	3.04
DMI, lb	18.4	19.4	19.3
Feed/Gain	6.0	5.9	6.0

the smaller the particle size of rolled wheat, the better the performance. However, the smaller the particle size the greater the potential for acidosis. Steam flaking wheat significantly reduced intake and improved conversions compared to dry rolling.

As shown, there are differences in how processing effects the digestibility, intake, and performance parameters of the different feed grains. Additionally, there is a considerable difference in the cost of the processing methods. However, given the high cost of grains, processing seems to be warranted because even slight increases in feed conversion result in greater profitability.

Feeding Whole Corn

Based on Table 49-2, whole corn appears to have potential, especially considering the energy costs associated with the various processing options. A typical whole corn ration is low in roughage (5 to 7 percent). This low roughage level stimulates the cattle to eructate the corn and chew it, effectively processing it themselves. In addition, whole corn stimulates the rumen wall. Whole corn is digested slower than processed corn, and more of the digestion occurs in the small intestine. Many cattle feeders do not like feeding whole corn because they observe what appear to be undigested kernels in the feces; however, the performance on whole corn is good as shown in Table 49-2.

Feeding High-moisture Corn

From a review of Table 49-2, high-moisture corn does not appear to be a good processing system. However, the research comparing high-moisture corn with other processing methods shows considerable variability with high-moisture corn performing well in some trials and poorly in others. The variability in results with high-moisture corn is not surprising when considering that getting high-moisture corn put up at the right moisture level is difficult. Research shows high-moisture corn put up at less than 24 percent moisture results in much poorer feed conversions than obtained with higher moisture levels. The cattle will eat more of this below optimum moisture level, high moisture corn, but the gains are the same resulting in poorer conversions. Ideally, high-moisture corn should be put in the pit at 26 to 30 percent moisture, but as harvest proceeds, the corn becomes dryer often resulting in moisture levels falling below 24 percent. Water can be added to the corn if it is in the 24 to 25 percent moisture range to raise the moisture level to the desired level. However, when the corn is in the 22 to 23 percent ranged it should not be put in a pit.

High-moisture corn that was put up at the right moisture level is rapidly digested in the rumen, which means that it is prone to cause acidosis. Many feedlots feed high-moisture corn in combination with steam-flaked corn or dry-rolled corn to reduce the potential for acidosis.

Steam-flaked Corn

Steam flaking consists of passing the corn through a steam chest where it is exposed to steam for approximately 20 minutes. During this time, the moisture level of the corn is increased to about 20 percent. The corn is then rolled to make a flake similar to a breakfast cereal. The distance between the rolls dictates the thickness of the flake. As the flakes become thinner, the pounds of steam-flaked corn per bushel decreases. Consequently, pounds per bushel is often used to describe the thickness of the flakes. Table 49-7 shows a comparison of animal performance with different flake thicknesses. Flake thicknesses in the intermediate range (23 to 29 pounds per bushel) resulted in the best feed conversions.

Steam-flaked Milo

As shown in Table 49-3, steam flaking improves the energy level of milo significantly. Table 49-8 shows a comparison of the performance of cattle fed milo of different flake densities. The intermediate densities resulted in the best gains and feed conversions.

References

Brethour, J. 1992. Very Finely Rolled Milo for Finishing Steers. Fort Hays Experiment Station Roundup Report. Kansas State University.

Eck, T.P., R.S. Swingle, M.H. Poore, J.A. Moore, and C.B. Theurer. 1991. Effect of Sorghum Grain Flake Density on Digestibility of Dry Matter and Starch by Steers under Feedlot Conditions. JAS 69(Suppl 1):542.

Hale, W. 1973. Influence of Processing on the Utilization of Grains (Starch) by Ruminants. JAS 37:1075.

Krehbiel, C., B. Holland, and C. Milton. Grain Processing Effects on Management: Adaptation Diets. Oklahoma State University.

Kuhl, G. 1987. Feeding Value of Sorghum Grain and Forage. Proceedings KSU Sorghum Field Day.

Mader, T. Optimizing Feedlot Diet Utilization. University of Nebraska – Northeast Research and Extension Center.

Owens, F., D. Secrist, and D. Gill. 1995. Impact of Grain Sources and Grain Processing on Feed Intake by and Performance of Feedlot Cattle. In Symposium: Intake by Feedlot Cattle. Oklahoma State University Cooperative Extension Publication P-942.

Robbins, M. and R. Pritchard. 1994. Effect of Corn Processing and Reconstitution in High Grain Diets on Feedlot Performance and Carcass Characteristics of Steers and Heifers. South Dakota Beef Report 1994.

Stock, R. and T. Mader. Grain Sorghum Processing for Beef Cattle. Great Plains Beef Cattle Handbook. GPE-2010.

Swingle, R.S., T.P. Eck, M. DeLallata, C.B. Theuer, M.H. Poore, and J.A. Moore. 1991. Density of Steam Flaked Sorghum Grain and Performance of Growing/Finishing Steers. JAS 69(Suppl 1):542.

Wagner, D. and C. Hibberd. 1983. Grain Processing. Oklahoma State University.

Wagner, D., D. Gill, and R. Totusek. Sorghum Processing. Great Plains Beef Cattle Feeding Handbook. GPE 2001.

Zinn, R.A. 1990. Influence of Flake Density on the Comparative Feeding Value of Steam Flaked Corn for Feedlot Cattle. JAS 68:767.

Chapter 50

By-Products and Roughages in Finishing Rations

Traditionally, finishing rations contained about 80 to 85 percent grain, 10 percent roughage, and a protein/mineral supplement. Small amounts of various by-products were included when priced competitively. Over the past 10 to 20 years, as the percentage of the U.S. corn crop being used for purposes other than livestock feed has increased, cattle feeders have had to substitute by-products for part of the grain portion in finishing rations. For example, approximately 40 percent of the U.S. corn crop was used to make ethanol in 2011. A significant portion of the corn crop is used to make high-fructose corn syrup and starch. This shift in the use of corn has dramatically increased the price of corn and resulted in a large increase in by-products available for cattle feeding. Currently, finishing cattle rations typically contain 10 to 40 percent of by-products, and in some areas, the percentage is even higher. Consequently, an understanding of how by-products fit into finishing rations is crucial to profitably finishing cattle.

While there are numerous by-products that are used in finishing rations at low levels, several are used at high levels. These include distiller grains (wet and dry), corn gluten feed (wet and dry), and potato by-products.

Two major milling processes produce large quantities of by-products used as feedstuffs in finishing diets. Wet milling of corn produces starch and high fructose corn syrup as the major products and corn gluten feed as the major by-product as shown in Figure 50-1. The dry milling process uses primarily corn and sorghum to produce ethanol as its main product with distillers grains as the major by-product as shown in Figure 50-2.

Corn Gluten Feed (Wet)

As shown in Figure 50-1, corn gluten feed is the primary by-product of making starch and high-fructose corn syrup out of corn. The starch produced is used as a substrate in a number of fermentation processes and as human food. The other major by-product of wet milling is steep liquor. In most plants, it is added back to the corn gluten feed; however, in some cases the steep liquor is sold separately. It is a major ingredient in liquid feeds and liquid supplements. It also is used as a conditioner in feedlot rations.

For many years, the wet version of corn gluten feed has been used as an ingredient in finishing rations in areas close to corn milling plants. The moisture content in typical wet corn gluten feed varies significantly making delivery of a consistent level of dry matter a problem. The dry matter level may be as low as 32 percent or as high as 50 percent in conventional wet corn gluten feed. To overcome the problems caused by this variability, Cargill has been producing a consistent version of wet corn gluten feed (Sweet Bran®). This product is a consistent 60 percent dry matter. They have been selling it out of eastern Nebraska for many years and shipping it by rail to the Texas panhandle from Iowa for more than 10 years.

On average, wet corn gluten feed is about 20 percent crude protein with an NEg of 65 Mcal/cwt. The response to the addition of wet corn gluten feed to finishing diets depends on grain processing. For instance, wet corn gluten feed is usually fed at 30 to 40 percent of the dry matter when the grain portion is rolled corn. In these diets, inclusion of wet corn gluten feed increases intake and gain and improves feed conversion slightly. In diets based on steam-flaked corn, wet corn gluten feed is typically fed at no more than 20 percent of the diet dry matter. In these diets, it typically increases intake and gain, but results in slightly poorer conversions. Even with

Figure 50-1. Wet Milling of Corn to Produce Starch and High Fructose Corn Syrup.

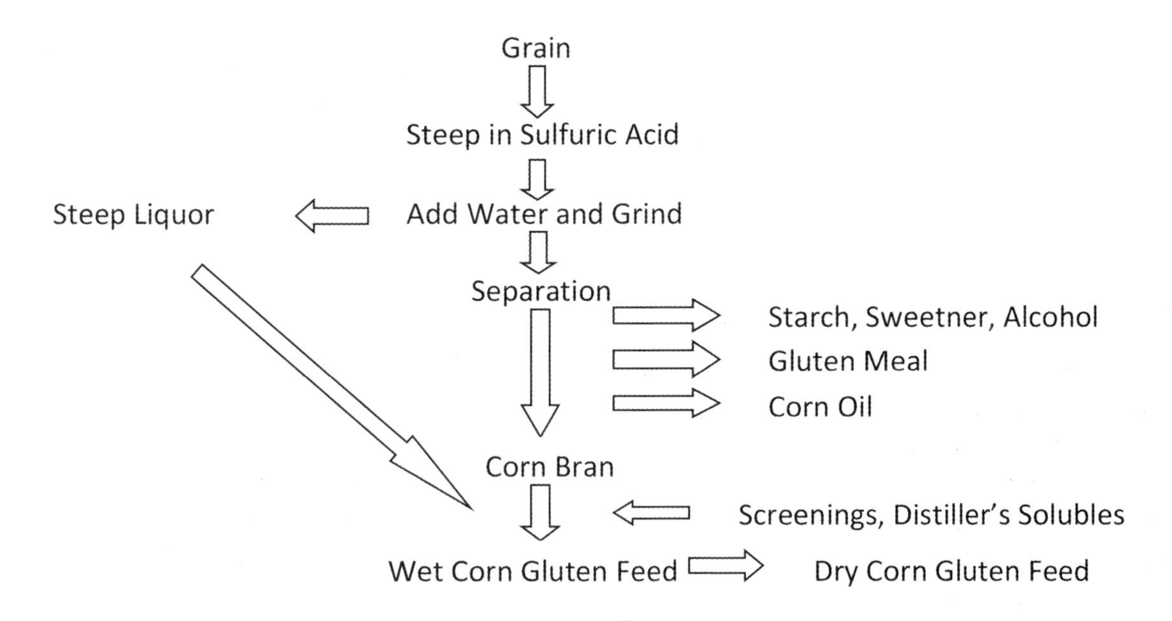

this slight reduction in conversion, wet corn gluten feed inclusion in steamed-flaked corn diets usually reduces the cost of gain because it reduces ration costs and increases gain. Feeding a significant level of corn gluten feed reduces the incidence of acidosis.

Feeding corn gluten feed will allow a reduction of roughage in the diet. For example, it appears that little if any roughage is required in diets where dry rolled corn is the grain source. However, roughage levels can only be reduced slightly in diets where steam-flaked corn is the grain source. For example, in a steam-flaked corn diet, the level of roughage might be reduced by a couple of percentage points. Even this slight reduction in roughage level could have a significant effect on cost-of-gain given the high cost of roughages over the past few years.

One of the issues with corn gluten feed, both wet and dry, is the high level of sulfur that it often contains. Not only is the level often high, it is variable from load to load. The NRC Publication, *Mineral Tolerance of Animals,* lists a maximum tolerable concentration of 0.30 percent for high concentrate diets. It is easy to exceed this level if a high level of corn gluten feed is included in the diet. There appears to be an interaction of roughage level and sulfur level with the tolerable sulfur level increasing as the roughage level is increased. For instance, in high roughage diets the maximum tolerable level is 0.50 percent.

High sulfur levels decrease intake and gain, and they can result in polioencephalomalacia (PEM) termed sulfur induced PEM. The sulfur ingested is reduced to hydrogen sulfide (H_2S) by ruminal bacteria. This gas is eructated and inhaled. It enters the blood stream and results in lesions on the brain. It should be noted that this form of PEM is different from the PEM caused by thiamine deficiency.

Corn Gluten Feed (Dry)

As with distillers grains, some plants cannot sell all of their corn gluten feed wet and are forced to dry it. This dry version is shipped all over the United States; at times, a considerable amount is exported. Nebraska research indicates that the dry version has about 88 percent of the feed value of dry rolled corn whereas the wet version is at least as good as dry rolled corn. The dry version does provide some of the same benefits in terms of increasing feed intake and gain and reducing acidosis if fed at low levels. If fed at significant levels (over 15 percent), feed conversions will be poorer. Thus, the price of dry corn gluten feed relative to grains is a key factor in its use.

Distillers Grains with Solubles (Wet)

In the process of making ethanol, two major by-products are produced: distillers grains and solubles as shown in Figure 50-2. The distillers grains consists of essentially the grain excluding the starch, which was converted into ethanol. Thus, distillers grains consist of the bran and the germ. The solubles contain the portion

Figure 50-2. Schematic of the Dry Milling Process Used to Produce Alcohol.

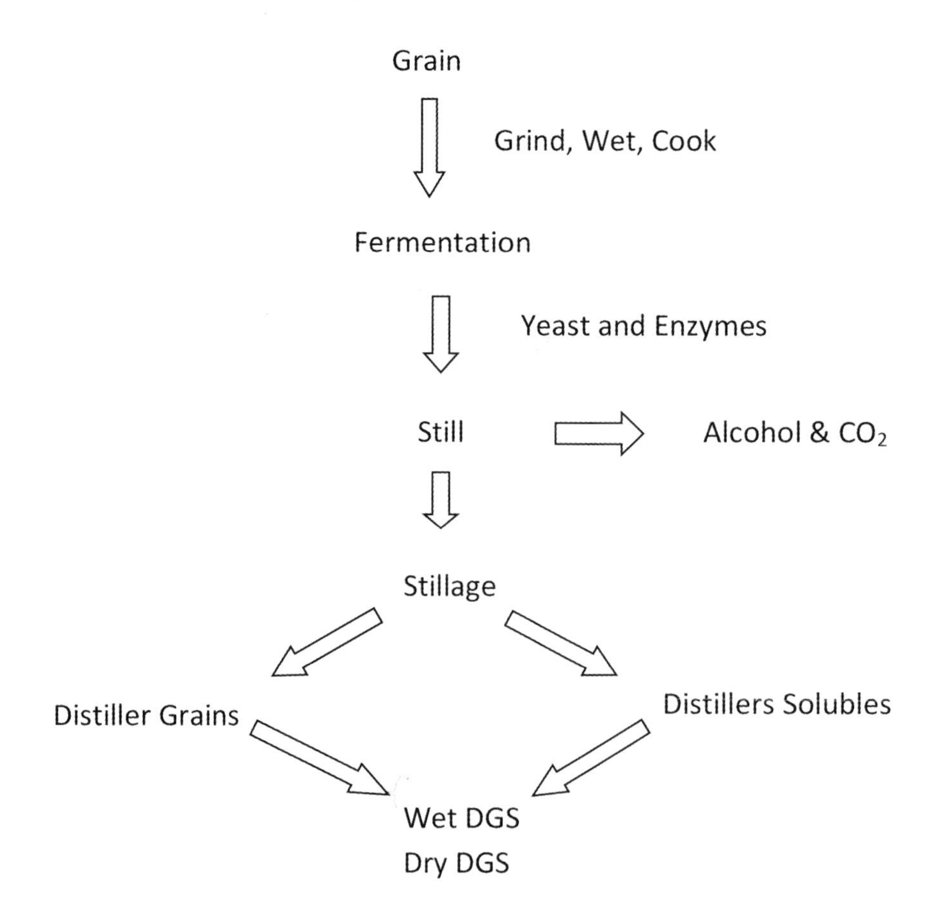

of the kernel that went into solution during the fermentation process that was not converted to ethanol. In most cases, the solubles are added back to the distillers grains resulting in distillers grain with solubles, which is used extensively as an ingredient in finishing cattle rations.

Wet corn distillers grains with solubles (WDGS) is typically 32 to 36 percent dry matter, 30 percent crude protein, and has an NEg of 77 Mcal/cwt. Thus, DGS have more energy than steam-flaked corn (NEg = 71 Mcal/cwt) and about three times the protein on a dry matter basis. Traditional DGS has about 10 percent fat, which contributes greatly to its high energy content. Milo distillers grains with solubles averages 35 percent dry matter, 32 percent crude protein and an NEg of 64 Mcal/cwt. Thus, milo DGS is considerably lower in energy than corn distillers grains with solubles. It also is less palatable than distillers grains from corn. The protein in distillers grains is more than 50 percent bypass protein, which makes it especially valuable in growing cattle rations.

In addition to providing energy and protein to the ration in finishing diets, distillers grains with solubles tend to increase intake resulting in higher daily gains. It also improves feed conversion, especially when it replaces dry rolled corn in the diet. When it replaces steam-flaked corn, it increases intake and gains, and improves conversions slightly. It also has the added benefit of reducing acidosis. Finally, its moisture content makes it a good ration conditioner. On the negative side, it spoils rather rapidly in the summer, which can be a problem for smaller cattle feeders. Distillers grains also have a high sulfur level with the same potential for causing sulfur-induced PEM as corn gluten feed.

Distillers Grains with Solubles (Dry)

Many ethanol plants cannot sell all their distillers grains wet and are forced to dry them. This allows them to be shipped to distant locations for feeding. In general, the dry version is lower in energy than the wet version probably because the heating that occurs during drying ties up some nutrients. For example, dry corn distillers grains with solubles has an NEg of 75 Mcal/cwt. Even with slightly lower energy content, dry distillers grains provides considerable protein and energy, and it stimulates additional intake.

In some cases, ethanol plants will not add the solubles to the grains because of the added cost of drying them. This doesn't appear to have a negative influence on the feeding value. To complicate feeding distillers grains, there are some plants that extract the fat from the grains before fermentation, which increases the protein concentration in the distillers grains but reduces the energy level. Because it is likely that there will be changes in the composition of distillers grains in the future, an analysis is warranted.

Dry distillers grains result in the same increases in intake produced by the wet version. In addition, feeding it at a significant level (greater than 10 percent) reduces the potential for acidosis.

Potatoes and Potato Waste

Cull potatoes and other by-products from potato processing have been used in finishing cattle diets for years, especially in the northwest. These by-products tend to be high in moisture, but they are equivalent to barley in energy value on a dry basis. The upper limit on inclusion is probably 50 percent because intakes decline when the inclusion rate is higher. Their high moisture content and variability in moisture content make getting consistent dry matter intakes difficult, but their low cost as an energy source offsets this problem to some extent. Cattle should be adapted to potatoes slowly because the starch they contain is rapidly fermented.

While the cost of potato by-products is low, the high moisture content of some of the by-products should be considered in comparing them to other energy sources. The cost of hauling must be factored into price comparisons because some potato by-products are 80 to 90 percent water. Storage losses are also an important factor because the starch in potato waste ferments rapidly with significant losses occurring in only a few days. Potato waste can be ensiled, but a dry material like hay or wheat middlings should be added to promote good fermentation. Whole potatoes should be ground before they are ensiled.

Rejected fried potato products are also included in finishing diets in the northwest. These by-products have a high level of fat, and thus they have a high energy level.

Because choking can be a problem when feeding whole potatoes, grinding or chopping them is preferable. Sprouted potatoes should not be fed. They contain a toxic alkaloid. Ensiling sprouted potatoes alleviates this toxicity problem.

Roughages in Finishing Rations

Roughage is an essential component of finishing rations because it has important effects on saliva production and gastrointestinal tract buffering. It influences the nature of fermentation in the rumen, digesta flow rates, the site and extent of nutrient digestion, and the supply of nutrients to the animal. Because roughages are the most expensive ingredient in finishing rations on a per unit of energy basis, feedlot nutritionists try to obtain the benefits of roughage while keeping the percentage in the diet as low as possible. The roughage level in the final finishing ration is normally between 8 to 10 percent on a dry matter basis. Corn silage is the most commonly used roughage followed by alfalfa hay. Research has shown that the digestibility of the fiber in roughages in high concentrate diets is low. Thus, the primary role of the roughage portion of the diet is to provide effective fiber to maintain rumen function and health.

Evaluating Roughage Sources

Neutral detergent fiber (NDF) level is the best indicator of the effective fiber potential of a roughage. Consequently, dietary neutral detergent fiber concentration can be used to balance diets to provide equal dry matter and energy intake. In addition to neutral detergent fiber concentration, particle size influences the effectiveness of a roughage with larger particle sizes providing more effective fiber. While neutral detergent fiber concentration in roughages can be used to estimate their effectiveness as fiber sources, the neutral detergent fiber in by-product feeds like corn gluten feed and distillers grains appear to provide little effective fiber. Consequently, the neutral detergent fiber in these by-products should not be considered when balancing rations for neutral detergent fiber content.

Common Roughages Used in Finishing Rations

- **Corn Silage** – In good corn silage, approximately 50 percent of the dry matter is actually in the grain portion. Thus, it is surprising that it is such an effective roughage. It has the added benefit of increasing the moisture content, which conditions the ration and reduces fines and separation of ingredients.
- **Alfalfa Hay** – This roughage typically adds natural protein as well as providing effective fiber. It is palatable, which stimulates intake. There is some tendency for more bloats with alfalfa hay than other roughages, especially if it is high quality hay. In most cases, a ration conditioner like corn steep liquor or molasses is used in alfalfa hay diets to reduce dustiness. The movement of the dairy industry into the High Plains has created considerable competition for alfalfa hay forcing some feedlots to shift to other roughages.
- **Sorghum Silage** – Reduced water availability for irrigation has forced some feedlots to use sorghum silage rather than corn silage because it requires less water. It is an effective fiber source, but does not provide as much energy as corn silage. The use of forage sorghum silage will probably increase across the High Plains as water for irrigation becomes more limited.
- **Cottonseed Hulls** – At one time, cottonseed hulls were one of the primary roughages used in the Texas panhandle, but in recent years, they have not been priced competitively as a feedstuff because of competition for other uses. Cottonseed hulls are an excellent roughage providing effective fiber.
- **Cotton Burrs** – As the use of cottonseed hulls has decreased, the use of cotton burrs as roughage has increased. They are an effective fiber source providing coarse material. They can contain considerable dirt, but at the levels they are typically included in the diet it does not cause problems in finishing cattle.
- **Sudangrass Hay** – Many of the forage sorghums and sudan hays are baled and used as roughages in finishing rations. They must be ground, but they do provide effective fiber. These hays have much lower protein levels than alfalfa, but they are often priced very competitively as roughage sources.
- **Corn Stalks** – Because drought has reduced the production of common roughages in recent times, ground corn stalks have been used as roughage in feedlot diets. They are high in neutral detergent fiber and are an effective roughage.

References

DeFoor, P.J., M.L. Galyean, G.B. Salyer, G.A. Nunnery, and C.H. Parsons. 2002. Effects of Roughage Source and Concentration on Intake and Performance by Finishing Heifers. JAS 80:1395.

Drewnoski, S. Hansen, D. Loy and S. Hoyer. 2011. How Much Distiller Grains Can I Include in My Feedlot Diets? Iowa Beef Center Publication 46.

Erickson, G.E., V.R. Bremer, T.J. Klopfenstein, A. Stalker, and R. Rasby. Feeding of Corn Milling Co-Products to Beef Cattle. Nebraska Corn Board. *www.nebraskacorn.org*.

Gaylyean, M.L. and P.J. DeFoor. Effects of Roughage Source and Level on Intake by Feedlot Cattle. JAS 81:E8-E16.

Galyean, M.L. and M.E. Hubbert. 2012. Traditional and Alternative Sources of Fiber – Roughage Values, Effectiveness, and Concentrations in Starting and Finishing Diets. 2012 Plains Nutrition Council Spring Conference.

Mader, T.L., J.M. Dahlquist, and L.D. Schmidt. 1991. Roughage Sources in Beef Cattle Finishing Rations. JAS 69:462.

National Research Council. 1996. Nutrient Requirements of Beef Cattle. 7th Revised Edition. The National Academies Press.

National Research Council. 2005. Mineral Tolerances of Animals. 2nd Ed. The National Academies Press.

Simms, D. D. 2009. Feeding the Beef Cowherd for Maximum Profit. SMS Publishing.

Uwituze, S., G.L. Parsons, M.K. Shelor, B.E. Depenbusch, K.K. Karges, M.L. Gibson, C.D. Reinhardt, J.J. Higgins, and J.S. Drouillard. 2010. Evaluation of Dried Distillers Grains and Roughage Source in Steam-Flaked Corn Finishing Diets. JAS 80:258.

Chapter 51

Natural and Organic Beef

There are essentially four types of beef marketed in the United States: Conventional, natural, organic, and grass-fed. The last three types currently account for less than 2 percent of cattle slaughtered, but the market for these types appears to be growing.

Conventional

In the typical conventional program, cattle are raised on pasture or range and placed on a high-concentrate diet for the last 120 to 200 days before harvest. This finishing period, primarily using grains, increases the tenderness and palatability of the final product. Producers can feed grains and hays produced in a conventional manner, and they can feed growth promotants and antibiotics, and use implants.

Natural Beef

The official USDA definition of natural refers only to meat products — it does not refer to how the animal was raised. The meaning of the term natural, as used in the industry, is vague. Most of the industry defines natural as being raised without antibiotics, growth promotants, ionophores, or implants. Since the guidelines for claiming the beef was produced "naturally" are vague, some producers use these technologies and still apply the "natural" claim.

Organic Beef

Whereas the guidelines for natural beef are vague, organic beef must meet rigid guidelines established by the USDA National Organic Program. Key requirements include:

1. Animals cannot receive any antibiotics or growth promotants.
2. Cattle must be fed 100 percent organic feed.
3. Cattle can be fed mineral and vitamin supplements, but these supplements must meet rigid standards with respect to their ingredients.
4. Cattle must have access to pasture.

Table 51-1. A Comparison of Four Types of Production Systems.

	Conventional Beef	Natural[a] Beef	Organic Beef	Grass-fed Beef
Antibiotics (therapeutic)	Yes	No	No	Optional
Antibiotics (subtherapeutic)	Optional	No	No	No
Hormones (implants)	Yes	No	No	No
Ionophores	Yes	No	No	No
Pesticides	Yes	Optional	No	Optional
Vaccinations	Yes	Yes	Yes	Yes
Feed Grains and Grain By-products	Yes	Yes	Optional	No

Adapted from a paper by Acevedo, et al. Iowa State University. [a]Requirements for the "natural" label are not well defined, but most producers follow these guidelines.

5. A government-approved certifier must inspect the premises and certify that the producer is following all of the requirements for organic production.

Grass-fed Beef

The USDA defines grass-fed beef as an animal whose sole diet during its lifetime was forage except for milk consumed before weaning. The animals cannot have been fed grain or grain by-products, and they must have access to pasture during the growing season. Grass-fed beef is typically leaner than conventional beef, and it is higher in vitamin A, conjugated linoleic acid, and Omega-3 fatty acids, which some producers claim makes it healthier than conventionally produced beef. Table 51-1 shows a comparison of these production systems.

Economic Comparison of Conventional and Natural Beef Production

With the growth of "natural" beef programs, many producers have been tempted to produce cattle for these programs because in many cases cattle that meet the requirements of these programs command a premium. Table 51-2 shows the results of a study conducted at the Dixon Springs Agricultural Center, University of Illinois. In this research, they compared a conventional program on pasture and in confinement

Table 51-2 Results of a University of Illinois Study Comparing Conventional to Natural Beef Production.

Item	Conventional Confinement	Natural Confinement	Conventional Pasture	Natural Pasture
Weight @ 13 mo.	1,210	1,099	1,175	1,117
Average Daily Gain, lbs	3.1	2.8	3.0	2.7
Carcass Weight, lbs	738	666	715	668
Back Fat, in.	.46	.48	.37	.37
Yield Grade	2.9	3.1	2.5	2.7
Marbling	552	643	519	584
Feed Conversion lb/lb	6.1	6.8	6.1	6.7
Value Above Production Costs	$733	$634	$709	$634

Adapted from a presentation by Frank A. Ireland.

Table 51-3. A Comparison of a Conventional Finishing Program to a Natural Program.

Item	Natural	Conventional	P Value(<)
Average Carcass Weight, lbs	822	896	0.0001
Average Dressing Percent	64.4	65.2	0.0240
Average Daily Gain, lbs	2.70	3.71	0.0001
Median Marbling Score	634	576	0.0020
Percent Choice	93	82	0.0770
Percent Ave. Choice of Higher	58	38	0.0370
Average Backfat, in	0.50	0.54	0.0770
Average Ribeye Area, sq in	13.3	14.6	0.0003
Calculated Yield Grade	3.64	3.48	0.0450
Dry Matter Intake, lbs/day	26.1	27.90	0.0010
Lbs Gain/lbs of Dry Matter	0.1031	0.1326	0.0001

KSU Agricultural Research Center – Hays Roundup Report 2005.

and a natural program on pasture and in confinement. The natural cattle were never fed any ionophores or antibiotics, and never implanted. The cattle on pasture were fed a high concentrate diet late in the finishing period with the pasture providing the roughage. All cattle were harvested at the same time at approximately 13 months of age.

The conventionally produced cattle were heavier at 13 months of age because of higher average daily gain. As would be expected, they produced heavier carcasses. Confinement-produced cattle had more backfat than the pasture-produced cattle; however, there was not a difference in backfat between conventional and natural treatments. Yield grades were higher for the confinement-produced cattle but again there was not a difference in yield grades between conventional and natural treatments. Marbling scores were higher for naturally produced cattle illustrating the negative effect that implants have on marbling. Marbling scores were lower for the cattle on pasture. Feed conversions were much better for conventionally produced cattle, but there was not a difference between confinement and pasture treatments. Value of the animals above production costs greatly favored the conventional system because of heavier carcasses and better feed conversions. Based on this analysis the cattle on the natural program would have required a $75 to $100 per head premium to just break even with the conventionally produced cattle.

Table 51-3 shows a summary of two trials conducted at the Kansas State University Agricultural Research Center – Hays comparing a conventional feeding program to a natural program. There were significant differences in performance and carcass characteristics. As occurred in the Illinois study, the cattle on the natural program graded better than the conventional program cattle, but their much lower rate of gain resulted in much lighter carcasses. One might think the cattle on the natural program should have been fed longer to increase carcass weight; however, at the end of the feeding period their average yield grade was 3.64, which means that had they been fed longer they would have received significant discounts for yield grade 4s. The researchers calculated that the cattle on the natural program would have to receive a premium of $7.50 and $10.08 per hundredweight of carcass to breakeven in trials 1 and 2, respectively.

Both the Illinois study and the Kansas study indicate:

1. Producing beef in a natural program costs much more than in a conventional program meaning significant premiums are required to make it profitable.
2. The importance of implants and ionophores in reducing the cost of producing beef.
3. The adverse effect that implants have on marbling.
4. The significant influence of implants and ionophores on rate of gain, feed conversion, and cost of gain.

References

Acevedo, N, J. Lawrence, and M. Smith. 2006. Organic, Natural and Grass-fed Beef: Profitability and Constraints to Production in the Midwestern U.S. Value-added Agriculture Program, Iowa State University.

Boland, M., E. Boyle, and C. Lusk. 1999. Economic Issues with Natural and Organic Beef. Bulletin MF2432 Kansas State University Cooperative Extension Service.

Brethour, J.R. and B. J. Bock. 2005. Costs of Producing Beef in a "Natural" Program Without Implants or Antibiotics. KSU Agricultural Research Center – Hays Roundup Report.

Cox, B and J. Turner. 2007. USDA Establishes Grass (Forage) Fed Marketing Claim Standard. USDA Agricultural Marketing Service AMS No. 178-07.

Ireland, F. Natural vs Traditional Beef Production. Dixon Springs Agriculture Center, University of Illinois. Available on Website: dsac.ansci.uiuc.edu/pods/dsac_070507-extra.pdf.

Melroe, T. and E. Loe. 2007. Conventional, Natural, and Organic Beef Production and Consumption. Bulletin Ex2059 South Dakota State University Cooperative Extension Service.

Smith, M. 2007. Resources for Organic, Natural and Grassfed Beef Producers. Iowa State University Value Added Agriculture Extension.

Stovall, T. and J. McCaffery. 2005. Natural Beef in the Feedlot: Risk & Return to Feeder Calf Premiums. Proceedings, The Range Beef Cow Symposium XIX.

Chapter 52

Marketing Fed Cattle

Fed cattle marketing has changed significantly in the past 30 years. Historically, cattle were sold on a live basis with a pencil shrink (usually 4 percent) or on a carcass basis. The feedlot manager negotiated the price with a packer-buyer. Marketing under this system resulted in little price differentiation. The difference in price for the best pen of cattle in a feedlot might have been only $1.00 per hundredweight higher than the worst pen of cattle. This system sent bad signals back to the cow/calf and stocker segments of production. For instance, cow/calf producers were frustrated that their efforts to produce quality cattle were not rewarded because all pens of cattle, good or bad, sold for the same money. Cattle feeders also realized that there was more profit potential in poorer-quality cattle than in good-quality cattle because the poorer-quality cattle could be purchased for less money and sold for essentially the same price. Fortunately, there have been changes in marketing options for fed cattle that offer potential for premiums for higher-quality cattle.

Currently, cattle feeders have three major options for marketing their cattle: live pricing, dress-weight pricing, or grid pricing. Table 52-1 shows the comparative attributes and risk factors for these systems.

Live Pricing

Packers typically establish the live price by determining the average carcass value and multiplying it by the average dressing percentage. For example, if

Table 52-1. Attributes and Risk Factors of Fed Cattle Marketing Systems.

	Fed Cattle Pricing System		
Attribute or Risk Factor	**Live**	**Dressed**	**Grid**
Value based	No	No	Yes
Pricing unit	Pen	Pen	Individual Carcass
Price variability among animals	None	None	High
Price variability among packers	Little	Moderate	High
Trucking costs paid by	Packer	Producer	Producer[a]
Pricing location	Feedlot	Plant	Plant
Prices reflect meat yield	Estimated	Dressed	Yield Grade
Prices reflect quality grade	Estimated	Estimated	Detailed actual
Base price determination	Live	Dressed	Varies
Discounts for out cattle	No	Some	Yes
Carcass performance risk burden	Packer	Packer	Producer
Need to know carcass quality	No	No	Yes

Adapted from Marketing Cattle via the Price Grid by Orlen Grunewald, Kansas State University.
[a]Some packers pay a set amount for freight, which may or may not cover all of the bill.

the average carcass is worth $192 per hundredweight and the average dressing percentage is 64 percent, the live price would be $192 × 0.64 = $122.88 per hundredweight usually with a 4 percent pencil shrink. Establishing a live price by a packer is a little more complicated than simply multiplying the average carcass value times the average dressing percentage. Other factors such as supply and demand, by-product value, and cost of processing must be considered. Many feedlot managers like selling cattle on a live basis because they feel more in control of the outcome compared to relying on the grading in the plant, which they feel is totally out of their control. In addition, they do not have to be as knowledgeable about the relative merit of their cattle. Typically, when they have cattle to market, they keep in touch with other feeders or an organization like Cattle Fax to know the current bids that packer-buyers are offering.

Dressed-Weight Pricing

With dressed-weight pricing, the cattle feeder is offered a price per hundredweight of hot carcass weight. The packer starts with the current average carcass value and adjusts it for by-product value, processing costs, etc. As noted in Table 52-1, the feeder typically pays the transportation cost to the packing plant. In the example above, the cattle feeder would receive $192 per hundredweight of hot carcass regardless of how the cattle graded or yielded.

Grid Pricing

With grid based pricing, each animal receives a price based on its quality and yield grade and defects. The cattle feeder receives the sum of the values of all of the animals in the lot. A grid is essentially a list of specifications with premiums and discounts for various carcass characteristics. There are many grids available to cattle feeders. All of the major packers offer grids, and the branded-beef programs also offer grids. In some cases, packers offer several grids representing the types discussed below. Grids can be categorized as:

1. **High Quality** – These grids provide significant premiums for cattle that grade in the top two-thirds of choice or higher. The target market is the high end restaurant trade and consumers willing to pay a premium for a quality steak.
2. **Commodity** – These grids offer premiums and discounts for both quality and yield grades but neither type of grade is over emphasized. Most of the major packers have a grid that fits into this category.
3. **Lean Yield** – These grids emphasize premiums for carcasses that yield a high percentage of lean meat. Significant premiums are provided for yield grade 1 and 2 carcasses and heavy discounts for yield grade 4 and 5 cattle.

Matching Cattle to Pricing Grids

Grid pricing offers cattle feeders the opportunity to receive premiums for better than average cattle, but receiving a higher net return requires a thorough understanding of both the premiums and discounts contained in a grid. Discounts can be especially severe, and only a small percentage of cattle receiving discounts can more than offset any premiums for other cattle. The other key item required to effectively use grids is the ability to predict carcass characteristics, especially with respect to quality and yield grade. For instance, heavily muscled cattle should produce a high percentage of yield grade 1 & 2 carcasses making them appropriate for a lean grid. Correspondingly, cattle with the genetic ability to marble should be targeted for a high-quality grid. To some extent, the ability to grade and yield is set by genetics, but there are several management practices that can influence carcass characteristics. These practices are discussed later in this chapter.

Base Price

In addition to the premiums and discount specifications in grids, they often differ in how the base price is established. Some grids use the average live price for the week, while others use the plant average for the past week or the last 4 weeks. In Nebraska, the dressed price is typically used to set the base. The base is usually set using the plant average dressing percentage. Thus, cattle that dress higher than the plant average usually receive a premium. In comparing grids compare the base price because two grids may have exactly the same premiums and discounts but differ significantly in base price.

Target or Threshold Levels

Many grids use target or threshold levels as the basis for premiums or discounts. For example, a grid might have a target of 50 percent for choice or better carcasses. Pens of cattle with more than 50 percent receive a premium, while pens below 50 percent receive a discount. In many cases, these target levels are the plant averages for the last week or 4 weeks. In this case, the targets change over time influencing the actual price received for the cattle. Average levels might be quite different between two plants, which means that the price received could be different. For instance, if two packing plants use their 4 week average for choice as a target and they differ markedly in the percent grading choice, it could result in higher premiums for choice carcasses in the plant with the lower average.

An Example Grid – U.S. Premium Beef, LLC "Market" Grid

This grid was taken from the U.S. Premium Beef website (*www.uspremiumbeef.com*) in early 2012. This website also contains grid specifications for other grids including one for natural beef.

Specifications – U.S. Premium Beef "Market" Grid

Category	Threshold	Premium/Discount (+/-)
Prime	4-week plant avg*	+/- Blend of Prime/Choice spread* and $8/cwt floor × weight above (premium) or below (discount) plant avg*
Certified Angus Beef®	4-week plant avg*	+/- Blend of CAB/Choice spread* and $3/cwt floor × weight above (premium) or below (discount) plant avg*
Choice or higher	4-week plant avg*	+ Choice/Select spread* × weight difference above plant avg* (premium)
Select	4-week plant avg*	+ Choice/Select spread* × weight difference below plant avg* (discount)
Ungraded	None	- $8/cwt (includes all grades below Select, but not Hardbone
Hardbone	None	- $25/cwt
Over 30 months	None	- $10/cwt (determined by dentition; do not get Prime or CAB premiums)
YG 1	4-week plant avg*	+/- $4/cwt × weight difference above (premium) or below (discount) plant avg*
YG 2	4-week plant avg*	+/- $2/cwt × weight difference above (premium) or below (discount) plant avg*
YG 3	4-week plant avg*	None
YG 4	4-week plant avg*	-/+ $12/cwt × weight difference above (premium) or below (discount) plant avg*
YG 5	4-week plant avg*	-/+ $17/cwt × weight difference above (premium) or below (discount) plant avg*
550 lbs and under	None	- $25/cwt
1,000 lbs and over	None	- $15/cwt
Steer premium	None	+ $5/head for all cattle in the lot
Heifer premium	None	+ $2/head for all cattle in the lot

**4-week rolling average. **Out weight carcass are included in determining the averages for Quality and Yield grades. To qualify for steer premium, the lot must be 95% steers*

Cattle Type

The program is non-breed specific; however, the USPB Market grid does not allow heiferettes, Holsteins and other dairy breeds, intact bulls, late-cut bulls, Corrientes, or high-percentage Brahman influenced cattle (over 50 percent Brahman as measured by visual appraisal). USPB reserves the right to refuse delivery of individual cattle or entire pens based on visual appraisal.

Base Price

The base live price will be the weighted average price from USDA Price Reporting data from cattle sold in Kansas, Texas, and Oklahoma during the week before delivery plus $0.25 per hundredweight. The base carcass price is determined by dividing the base live price by the plant average yield. The plant average yield is the average hot carcass yield for all non-formula (cattle purchased on a live or carcass basis) cattle purchased in Kansas, Texas, and Oklahoma by National Beef Packing Company, LLC (NBP) and processed during the same delivery week. No USPB cattle are included in the plant average yield.

Premiums and Discounts

Actual performance of the cattle is compared to a 4-week rolling average performance of cattle that NBP purchased on the spot market (non-formula). Plant

Cattle Settlement Worksheet – Example (USPB)

Feedyard: Example Feedlot
NBP Lot # 235 USPB Lot #92876
Method: USPB Market Grid
Harvest: 1/2/09 Plant: Dodge City
FY Lot # 160 Pen # 26

Lot Statistics

Live Wt: 124,000	Hot Yield: 64.04%	Avg Wt: 1240	Net Live Price: $93.19	Net Prem/Disc: $7.28/hd
Hot Wt: 79,410		Avg Wt: 794	Net Hot Price: $145.61	Net Hot Difference: $0.59/cwt
				Gross Prem/Disc: $12.25/Hd

Base Price

USDA KS/TX/Ok Weekly Avg	92.60
Formula Allowance	0.25
Grid Allowance	0.00
Base Live Price	92.85
Hot Yield Threshold	63.65%
Base Hot Price	145.86

Prem Summary

Choice	$7.79
Prime	$17.89
CAB	$3.87

Heifers	Head	Live Wt	Hot Wt	Base Price	Amount	$/cwt hot	$/Hd
Base Lot Amount	100	124,000	79,410	145.86	115,826.84	145.86	1158.27
Quality Grade Adjustments					-326.81	-0.41	-3.27
Yield Grade Adjustments					521.52	0.66	5.22
Weight Adjustments					-150.3	-0.19	-1.50
Other Adjustments					175.55	0.22	1.76
Gross Cattle Proceeds	100	124,000	79,410		116,046.80	146.14	1160.47
Excess Freight					-497.26	-0.63	-4.97
Net Cattle Proceeds					115,549.54	145.51	1155.5

averages do NOT include ASPB cattle or any cattle purchased on a grid or formula basis by NBP. Cattle grading above plant average for **Choice or higher,** Prime and Certified Angus Beef® (CAB) receive a premium for the total carcass weight that is above plant average or a discount for the total carcass weight below plant average. The Choice premium/discount is calculated using a 4-week rolling average of the

Adjustments to Net Returns

Quality Grade Adj.		Head	Hot Wt	Actual	Threshold	Price	Wt Diff	Total $	$/Hd
Quality Grade Adj.									
	Prime	1	782	1.00%	0.86%	17.89	112	20.03	
	Choice or higher	54	43,034	55.00%	54.32%	7.79	542	42.22	
	Select	42	33,234	41.85		0.00	33,234	0.00	
	Ungraded	3	2,347	3.00%		-8.00	2,382	-190.56	
	Hard Bone	1	794	1.00%		-25.00	794	-198.50	
Total Quality Adj.								**-326.81**	**-3.27**
Carcass Adj.									
	CAB	8	6,353	8.00%	8.79%	3.87	-631	-24.45	
	Over 30 months	0	0	0.00%		-10	0	0.00	
Total Carcass Adj.								**-24.45**	**-0.24**
Yield Grade Adj.									
	YG1	10	7,824	10.00%	10.20%	4.00	-162	-6.48	
	YG2	41	32,080	41.00%	36.99%	2.00	3,188	63.76	
	YG3	42	32,862	42.00%	41.45%	0.00	436	0.00	
	YG4	7	5,477	7.00%	10.13%	-12.00	-2,486	298.32	
	YG5	0	0	0.00%	1.23%	-17.00	-976	165.92	
Total Yield Grade Adj.								**521.52**	**5.22**
Weight Adjustments									
	550 and under	0	0	0.00%		-25.00	0	0.00	
	1,000 and up	1	1,002	1.26%		-15.00	1,002	-150.30	
Total Weight Adj.								**-150.30**	**-1.50**
Lot Adjustments									
	Age and Source Verified, $/Head					35.00		0.00	
	Steer Premium, $/Head					5.00		0.00	
	Heifer Premium, $/Head					2.00		200.00	
Total Lot Adj.								**200.00**	**2.00**
Total Adj. to Net								**219.96**	**2.20**

USDA-reported Choice/Select cutout spread. **Prime and Certified Angus Beef®** (CAB) premiums are calculated using Weekly USDA Prime (LM_XB456) and CAB (LM_XB452) beef cut prices minus $0.075 per pound. Respective weekly Prime and CAB cutout values are calculated as a premium for USDA Choice for that particular week using adjusted Prime and CAB beef cut prices, 2004 USDA primal/carcass standards, and NBP average carcass weight. Prime and CAB premiums are determined by blending the 4-week rolling average spread of USDA-reported boxed beef cutout values for Prime and CAB versus Choice together with a floor value of $8 per hundredweight for Prime or $3 per hundredweight for CAB. Prime and CAB premiums will not drop below the floor values. Yield Grade premium/discount calculations are based on the 4-week rolling plant average performance on non-formula cattle purchased in Kansas, Texas, and Oklahoma by NBP during the preceding 4 weeks. Yield Grade 1 or 2 will receive a premium for the total carcass weight heavier than plant average, but will be discounted if below plant average. Yield Grade 4 and 5 will be discounted for the total carcass weight that is greater than the plant average, but will receive a premium for being below plant average.

Cattle Settlement Worksheet – Example

This example shows how just a few cattle that fail to meet specifications result in severe discounts. For, example, the three head that failed to reach the Select grade reduced the per-head returns for the entire pen by $1.91 per head. In addition, the one hard bone carcass reduced returns by $1.99 per head. The one overweight carcass reduced the returns by $1.50 per head. Thus, while the pen received a significant premium for exceeding the yield grade specifications, these outliers significantly reduced the overall premiums for the pen.

Environmental Factors Influencing Carcass Composition

There are several environmental factors that influence the composition of a group of cattle at slaughter. For example, weather can influence the rate of gain and subsequently the amount of finish at slaughter. Cattle feeders typically extend the length of the feeding period to compensate for adverse weather. A more serious problem is the effect of adverse weather just before slaughter on the degree of marbling. A good example of this is the so-called "summer slump" in July and August when the percentage of cattle grading choice is typically low because of the heat. There is little cattle feeders can do to avoid this environmental effect.

Management Factors Influencing Carcass Composition

There are several management practices that influence carcass composition. Consider a few of these in more detail.

- **Age at Slaughter** – Over the past 25 years there has been a trend to place cattle on feed at a younger age. To some extent, the increase in growth potential has made growing programs unnecessary with some cattle capable of entering a finishing program at weaning. While research comparing calf-feds with cattle placed on feed as yearlings has not produced consistent results, it appears calf-feds produce approximately 6 to 8 percent fewer choice cattle than yearlings. Additionally, calf-fed cattle reach the same external fat endpoint at lighter weights than they would if grown longer and fed as yearlings.

- **Implanting** – An aggressive implanting program, especially one using multiple implantation with implants containing trenbolone acetate (a testosterone analogue) and estrogen, has been shown to reduce the percentage of cattle grading choice. Conversely, these implants increase muscle deposition resulting in higher dressing percentages and lower yield grades.

- **Beta-agonists** – Feeding either Optaflexx® or Zilmax® at the end of the feeding period results in increased muscle deposition significantly increasing dressing percentage and lowering yield grade. As noted in Chapter 48, there are some differences in performance effects and impact on carcass characteristics between these two beta-agonists (Table 48-4).

- **Length of Feeding Period** – The number of days that cattle are fed a high-energy ration has a tremendous influence on carcass composition. Overall, increased time on feed results in more fat deposition with correlated responses being increased dressing percentage, decreased

Table 52-2. Effects of Time-on-Feed on Gain and Dressing Percentage.

Days-on-Feed	Slaughter Wt., lbs	Dressing Percentage, %	Cumulative ADG, lbs
0	763	56.8	
28	949	57.3	5.14
56	992	61.1	3.35
84	1,105	61.6	3.51
112	1,163	64.6	3.22
140	1,250	64.8	3.09
168	1,295	64.6	2.80
196	1,430	67.0	3.11

Oklahoma State University, 1987.

Table 52-3. Effect of Time-on-Feed on Carcass Characteristics.

Days-on-Feed	Marbling Score[a]	Fat, in.	Ribeye, Sq. in.	Carcass Wt., lbs	Yield Grade
0	254	0.12	9.8	433	1.4
28	299	0.16	10.8	522	1.7
56	336	0.27	12.2	581	1.7
84	373	0.39	11.8	652	2.4
112	472	0.57	12.8	721	2.9
140	428	0.59	13.3	778	3.2
168	472	0.72	13.1	804	3.7
196	464	0.83	14.5	920	4.0

[a]small = 400–499 (choice). Oklahoma State University, 1987.

Table 52-4. Effect of Time-on-Feed on Beef Palatability.

Days-On-Feed	Juiciness[a]	Flavor Intensity[b]	Tenderness[c]	Shear Force, lbs
0	4.7	4.6	3.5	18.1
28	5.0	4.9	4.2	14.7
56	5.0	4.8	5.3	11.2
84	5.1	4.8	5.9	9.6
112	5.5	5.0	6.4	8.4
140	4.9	5.0	6.4	9.4
168	5.5	5.0	6.4	9.4
196	4.8	5.0	5.3	11.9

[a]1 = extremely dry, 8 = extremely juicy. [b]1 = extremely bland, 8 = extremely intense. [c]1 = extremely tough, 8 = extremely tender. Oklahoma State University, 1987.

percentage of retail meat yield, increased yield grade, increased percentage grading choice, increased carcass weight, decreased daily gain, and decreased feed conversion. To illustrate these changes in carcass composition, examine the results of a study conducted at Oklahoma State University in which 48 Angus X Hereford steers approximately 16 months of age were slaughtered at 28-day intervals. Six steers were slaughtered at each time, including at the start of the feeding trial. Results of this research are shown in Tables 52-2, 52-3, and 52-4.

This research indicates that the rate of change in many carcass traits slows or plateaus after a certain number of days on feed (112 days in this case). Other research evaluating the effect of days on feed supports this conclusion. Of particular importance is marbling score where there was little change after 112 days on feed even though external fat increased substantially. It appears that the maximum level of marbling that a group of cattle will exhibit is governed to some extent by their genetic potential to marble, which establishes the point where marbling will plateau no matter how long the cattle are fed. This is an important point to keep in mind when feeding cattle for a specific grid, especially a grid where there is a significant premium placed on quality grade.

Another important finding in this research was that dressing percentage tended to plateau as well, even though external fat cover increased linearly. Yield grade, which is a significant component of all pricing grids, increased in a linear fashion in this research.

Table 52-5. Effect of Additional Days on Feed on Finishing Steer Performance and Carcass Traits.

	Additional Days-on-Feed		
Item	**0**	**14**	**28**
Initial Weight, lbs	701	702	703
111-day Weight, lbs	1,093	1,094	1,094
Final Weight, lbs[a]	1,093	1,124	1,158
0-111 Days			
Daily Gain, lbs	3.56	3.57	3.56
Daily Feed, lbs of DM	21.3	21.4	21.5
Feed/Gain	5.98	5.98	5.98
Additional Days			
Daily Gain, lbs		2.07	2.28
Feed/Gain		10.07	9.33
Entire Feeding Period			
Daily Gain, lbs	3.56	3.40	3.30
Feed/Gain	5.98	6.25	6.47
Cost of Gain, $/lbs[b]	0.487	0.507	0.522
Carcass Traits			
Hot Carcass Weight, lbs	683	719	744
Dressing Percentage	62.5	63.9	64.2
Ribeye Area, sq. in.	12.4	13.0	13.4
Adjusted Backfat, in.	0.48	0.53	0.55
Yield Grade	2.80	2.90	2.88
Marbling Score	5.07	5.35	5.27

[a] Weights pencil shrunk 4%. [b] Using ration costs of $95/ton (as fed), $.35/hd/d yardage, interest, medicine, etc.
[c] Marbling score: Sl^{50} = 4.5; Sm^{00} = 5.0; Sm^{50} = 5.5, etc. Kansas State University, 1992.

Table 52-6. Effect of Additional Days on Feed on Carcass Quality and Yield Grade.

Item	Additional Days-on-Feed 0	14	28
Choice and Prime, %	67	71	75
Yield Grade:			
Yield Grade 1, %	15	11	12
Yield Grade 2, %	45	40	38
Yield Grade 3, %	38	47	43
Yield Grade 4, %	2	2	7

Kansas State University, 1992

The effect of additional days on feed on beef palatability was also interesting. There did not appear to be an effect on juiciness or flavor intensity, but tenderness improved up to 168 days on feed. The final 28 days on feed, however, appeared to reduce palatability significantly based on both test panel scores and Warner-Bratzler shear force. This reduction in palatability probably resulted from increased maturity.

In a large research trial on the effect of days on feed conducted at Kansas State University, the effect of adding an additional 14 or 28 days was evaluated as shown in Table 52-5. In general, the results of this study agree with the OSU study. In this research, feed conversion was also measured. Daily gain of the steers was only slightly over 2.0 pounds during the additional 14-day periods compared to more than 3.5 pounds for the first 111 days on feed. Feed conversion was much poorer with conversions in the 9- to 10-pound range, much below the 6.0-pound average for the first 111 days. One of the most important points illustrated in Table 52-5 is that feeding for an additional 28 days resulted in an increase of $3.50 per hundredweight in cost of gain over the entire feeding period. And this increase occurred when feed costs were much lower than in recent times.

Table 52-6 shows the effect of additional days on feed on both quality and yield grade. As would be expected, the percentage of Choice and Prime carcasses increased slightly while the percentage of yield grade 4 carcasses increased significantly. Under formula selling, this represents a trade-off with the premium for choice increasing while the discount for yield grade 4 carcasses was also increasing.

A review of the OSU and KSU studies and other research clearly shows that there are numerous factors to consider in determining the optimum length of the feeding period as it relates to matching cattle to a formula. Feeding for more days tends to increase the cost of gain, increase the premiums for higher quality carcasses, increase the discounts for yield grade 4 and 5 carcasses, and increase the potential for discounts resulting from heavy weight carcasses.

As discussed in Chapter 8, the genetics for growth has changed significantly over the past 30 years, and both the Oklahoma State University and Kansas State University studies were conducted with cattle with lower growth rates than prevalent in the industry today. This means that the results might be different if these studies were repeated with current genetics. However, the results of large pen serial harvest studies discussed in the next section tends to indicate that the findings in the OSU and KSU studies are still relevant.

Large Pen Serial Harvest Studies

In 2012, a summary of several large pen serial harvest studies conducted by Merck Animal Health was published in the Plains Nutrition Council Proceedings. This summary included four studies totaling 129 pens and 12,310 head of steers. Correspondingly, three studies including 84 pens and 7,607 head of heifers were summarized. Both steers and heifers were slaughtered in 21-day increments. Live performance of both the steers and heifers is shown in Table 52-7. As the days on feed increased, the average daily gain decreased while feed conversion became poorer in both steers and heifers. Dry matter intake was not reduced with the increased days on feed in either the steers or heifers.

Table 52-7. Summary of Live Performance in Large Pen Serial Harvest Studies.

	Days on Feed (Day of Serial Harvest)		
Item	153 (0)	174 (21)	195 (42)
Steers			
Initial Weight, lbs	712	712	712
Final Weight, lbs	1,272	1,318	1,369
ADG, lbs	3.73	3.47	3.35
Feed:gain	5.57	5.97	6.15
DMI, lbs	20.8	20.7	20.6
	Days on Feed (Day of Serial Harvest)		
	146 (0)	167 (21)	188 (42)
Heifers			
Initial Weight, lbs	621	620	622
Final Weight, lbs	1,115	1,168	1,217
ADG, lbs	3.43	3.34	3.20
Feed:Gain	5.10	5.30	5.49
DMI, lbs	17.47	17.74	17.69

Table 52-8. Summary of Hot Carcass Weights, Dressing Percentages, and Carcass Weight Gain as a Percentage of Live Weight Gain.

	(Day of Serial Harvest)		
Item	0	21	42
Steers			
Hot Carcass Weight, lbs	809	847	891
Dressing Percentage, %	63.6	64.2	65.0
Carcass Gain, % of Live Gain		82.6	87.3
Heifers			
Hot Carcass Weight, lbs	707	751	783
Dressing Percentage, %	63.6	64.2	64.4
Carcass Gain, % of Live Gain		86.6	65.8

Table 52-9. Summary of Carcass Characteristics in Large Pen Serial Slaughter Studies.

	(Day of Serial Harvest)		
Item	0	21	42
Steers			
Choice or Greater, %	41.5	50.6	54.0
Select, %	54.2	47.5	44.7
Ribeye Area, Sq. In.	14.1	14.3	14.5
Yield Grade 1, %	21.8	18.1	13.9
Yield Grade 1 & 2, %	64.1	56.2	49.8
Yield Grade 4, %	5.4	8.8	12.8
Yield Grade 4 & 5, %	5.9	10.7	15.7
Heifers			
Choice or Greater, %	43.9	46.2	51.2
Select, %	52.7	50.7	46.3
Ribeye Area, Sq. In.	13.7	14.0	14.1
Yield Grade 1, %	27.1	20.9	19.2
Yield Grade 1 & 2, %	75.2	66.1	59.5
Yield Grade 4, %	1.9	4.4	8.3
Yield Grade 4 & 5, %	2.2	5.0	9.6

Carcass data are shown in Table 52-8. A comparison of Tables 52-7 and 52-8 allows the computation of the percentage of gain that is carcass. For example, live gain in the steers between 0 and 21 days was (1,318 – 1,272) = 46 pounds while the gain in hot carcass weight was (847 – 809) = 38 pounds. Thus, carcass gain was (38/46) = 82.6 percent of live gain. Because carcass gain is worth much more than live gain on a per pound basis, the high percentage of gain that is carcass rather than hide, digestive tract, etc., greatly influences the profitability of different marketing options discussed later in this chapter.

Table 52-9 shows a more detailed summary of the carcass characteristics. The percentage of choice or better carcasses in the steers followed the trend in other research with a 9.1 percent increase from day 0 to day 21 and only a 3.4 percent increase from day 21 to day 42. The heifers did not show this same trend with essentially the same increase in percent choice for each 21-day period. The changes in yield grade composition are also relevant if the cattle are sold on a grid since there are typically premiums for yield grade 1 and 2 carcasses and discounts for yield grade 4 and 5 carcasses.

As part of this summary, the authors conducted an economic analysis evaluating the profitability of marketing options as impacted by days on feed. The assumptions used in the analysis of the steer data are shown in Table 52-10, and the corresponding assumptions for the heifers are shown in Table 52-11. It should be noted that this economic analysis represents a snapshot in time based on the assumptions for feed costs and premiums and discounts for carcass characteristics. Obviously, these variables are constantly changing, which means that the relative profitability of marketing options and days on feed will also change.

Table 52-12 shows the live and carcass cost of gain based on the assumptions for rations costs and the animal performance in the summary. As would be expected, the cost of gain increased with increased days on feed because of poorer feed conversion. The incremental cost of gain on a live basis was much higher at $146.67 per hundredweight than the live sale price of $126.00 meaning that on a live basis, the most profitable days-on-feed was day

Table 52-10. Assumptions Used in Economic Analysis for Steers.

		(Day of Serial Harvest)		
Item	Value	0	21	42
Purchase Weight, lbs	$143.50/cwt	712	712	712
ADG, lbs		3.73		
Incremental ADG, lbs			2.45	2.45
Feed Efficiency		5.57		
Incremental Efficiency			8.41	8.41
DMI, lbs	$300/ton	20.6	20.6	20.6
Final Weight, lbs	$126/cwt	1,270	1,321	1,373
Hot Carcass Weight, lbs	$202/cwt	808	852	897
Heavy Wt. Carcasses, %	<$21.54/cwt>	2.8	7.0	15.1
Prime Premium	$15.11/cwt			
Premium Choice	$3.00/cwt			
Choice-Select Spread	$2.00/cwt			
Yield Grade 1 Premium %	$3.38/cwt			
Yield Grade 2 Premium	$1.46/cwt			
Yield Grade 4 Discount	<$11.38/cwt>			
Yield Grade 5 Discount	<$17.46/cwt>			

Table 52-11. Assumptions Used in the Economic Analysis for Heifers.

		(Day of Serial Harvest)		
Item	Value	0	21	42
Purchase Weight, lbs	$146/cwt	621	621	621
ADG, lbs		3.43		
Incremental ADG, lbs			2.64	2.30
Feed Efficiency		5.10		
Incremental Efficiency			6.67	7.66
DMI, lbs	$300/ton	17.6	17.6	17.6
Final Weight, lbs	$126/cwt	1,114	1,169	1,218
Hot Carcass Weight, lbs	$202/cwt	709	748	787
Heavy Wt. Carcasses, %	<$21.54/cwt>	0.2	0.6	1.6

Note: Premiums and discounts were the same used in the steer analysis.

Table 52-12. Live and Carcass Cost of Gain and Incremental Cost of Gain.

	(Day of Serial Harvest)		
Item	0	21	42
Steers – Live			
Cost of gain, $/cwt	102.83	106.53	109.64
Incremental cost of gain, $/cwt		146.67	146.39
Breakeven, $cwt	125.63	126.45	127.20
Steers – Carcass			
Cost of gain, $/cwt	140.10	143.07	145.51
Incremental cost of gain, $/cwt		170.56	170.56
Breakeven, $/cwt	197.53	196.96	195.69
Heifers - Live			
Cost of gain, $/cwt	100.32	101.96	104.54
Incremental cost of gain, $/cwt		116.55	133.96
Breakeven, $cwt	125.78	125.34	125.68
Heifers - Carcass			
Cost of gain, $/cwt	134.48	137.74	140.21
Incremental cost of gain, $/cwt		165.56	165.99
Breakeven, $/cwt	197.83	195.38	195.19

0 (actually 153 days on feed for the steers). On a carcass basis, however, the incremental cost of gain for the steers was $170.56 per hundredweight compared to an assumed carcass value of $202.00 per cwt. Thus on a carcass basis, the most profitable days on feed was day 42. This shows the significance of the fact that a high percentage of gain late in the finishing period is carcass gain. It also indicates why cattle sold either in the beef or on a grid can and should be fed for more days.

Figure 52-1 shows a graphic representation of the relative profitability of different marketing systems for steers as influenced by days on feed. Of particular note is the reduction in profitability for the steers fed to day 42 compared to day 21 and sold on a grid. This reduction resulted from the dramatic increase in heavy weight carcasses with their large discount. When the heavy weights were removed, selling on a grid and feeding for more days was much more profitable. Figure 52-2 shows the same profit analysis for the heifers. In this case, heavy-weight carcasses were not a problem making selling on a grid and feeding for more days more profitable. With heifers, yield grade 4 and 5 carcasses are more of a problem because of their significant discounts.

Conclusions from Serial Slaughter Studies

1. Rate of gain declines significantly late in the finishing period.
2. Feed conversion also declines significantly.
3. Cost of gain increases dramatically late in the finishing period.
4. A high percentage of live gain late in the finishing period is carcass gain, which greatly influences the most profitable days on feed for a specific marketing option.
5. Cattle marketed in the beef or on a grid can typically be fed for more days than cattle sold live.
6. Heavy weight carcasses limit the number of days on feed that is most profitable for steers sold on a grid unless the heavy-weight steers are sorted off and sold before they get large enough to produce a carcass more than 1,000 pounds.
7. The percentage of yield grade 4 and 5 carcasses increases rapidly late in the finishing period, which can markedly influence profitability if the cattle are sold on a grid. This is especially true for heifers.

Figure 52-1. Net Economic Returns for Steers Marketed Live, as Carcasses in the Beef, or on an Average USDA Grid.

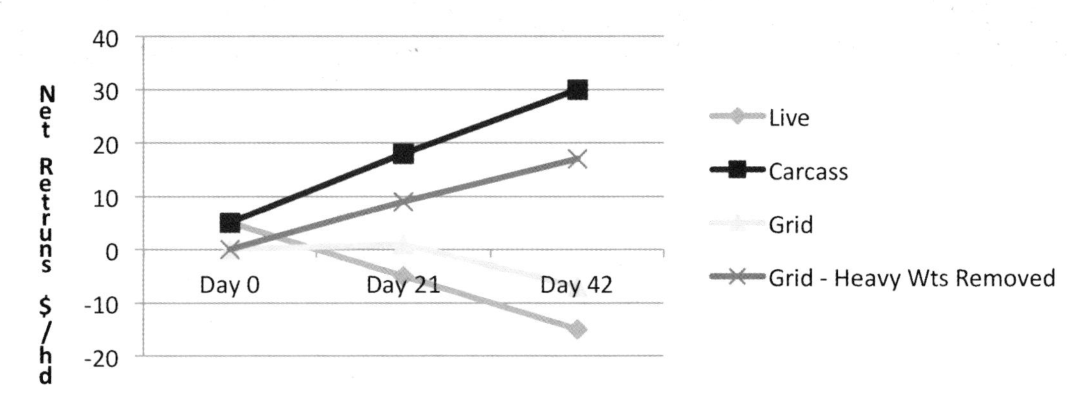

Figure 52-2. Net Economic Returns for Heifers Marketed Live, as Carcasses in the Beef, or on an Average USDA Grid.

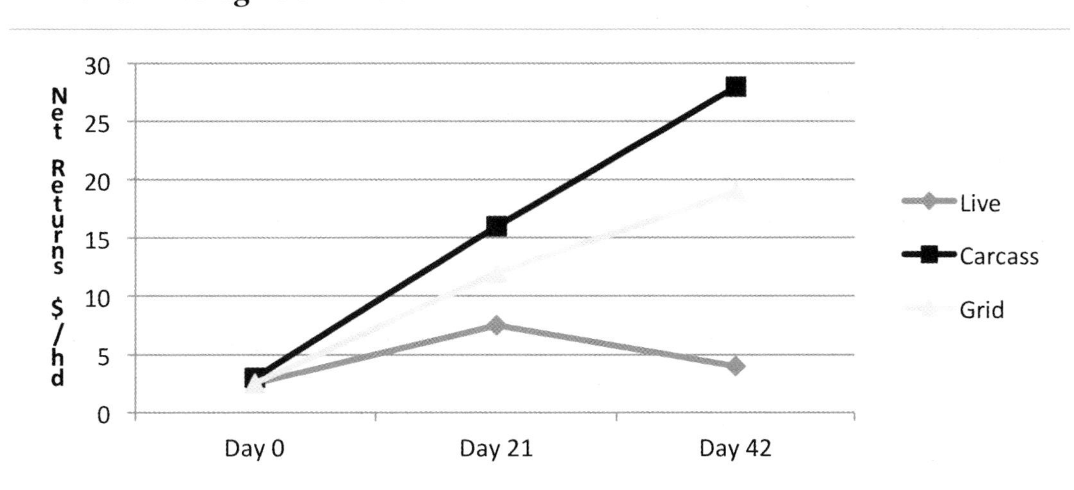

References

Brandt, R.T., Jr., M.E. Dikeman and S. Stroda. 1992. Effects of Estradiol or an Estradiol-Trenbolone Acetate Reimplant Scheme and Time-on-Feed on Performance and Carcass Traits of Finishing Steers. Kansas state University Cattlemen's Day Report of Progress 651.

Fuez, D.M. 1997. Pricing/Formula Grids: Which Fit and Which Don't Fit. 1997 Range Beef Cow Symposium Proceedings.

Fuez, D.M. 1999. Issues to Consider When Selling Cattle on a Grid or Formula. University of Nebraska NebGuide G98-1352-A.

Grunewald, Orlen. 2004. Marketing Cattle Via the Price Grid. Risk and Profit Conference Proceedings.

Hicks, R.B., F.N. Owens, D.R. Gill, J.J. Martin, H.G. Dolezal, F.K. Ray, V.S. Hays, and C.A. Strasia. 1987. The Effect of Slaughter Date on Carcass Gain and Carcass Characteristics of Feedlot Steers. Oklahoma State University Animal Science Research Report.

Parrish, J.A., J.D. Rhinehart, and J.D. Anderson. 2009. Marketing Fed Cattle. Mississippi Cooperative Extension Publication 2556.

Schroeder, T.C., J. Mintert, C. E. Ward, and T.L. Wheeler. 2003. Fed-Cattle Grid-Pricing Valuation: Recommendations for Improvement. Kansas State University Cooperative Extension Service Bulletin MF-2637.

Streeter, M.N., J.P. Hutcheson, W.T. Nichols, D.A. Yates, J.M. Hodgen, K.J. Vander Pol, and B.P. Holland. 2012. Review of Large Pen Serial Slaughter Trials – Growth, Carcass Characteristics, Feeding Economics. 2012 Plains Nutrition Council Spring Conference Proceedings.

Taylor, J., S.C. Cates, S.A. Karns, S.R. Koontz, J.D. Lawrence, and M.K. Muth. 2007. Alternative Marketing Arrangements in the Beef Industry: Definition, Use, and Motives. Livestock Marketing Information Center LM-2.

Ward, C.E., D.M Fuez, and T.C. Schroeder. 1999. Formula Pricing and Grid Pricing Fed Cattle: Implications for Price Discovery and Variability. Research Bulletin 1-99 Research Institute on Livestock Pricing Agricultural and Applied Economics, Virginia Tech University.

Ward, C.E. 2006. Grid Pricing Usage by Cattle Feeders. Oklahoma Cooperative Extension Service AGEC-604.

Chapter 53

Custom Cattle Feeding

Custom cattle feeding is a program where the owner of the cattle sends them to a commercial feedlot to be fed. The owner of the cattle may be a cow/calf or stocker producer who retains ownership, or simply an investor who purchases cattle and places them in a commercial feedlot. As noted previously, many feedlots specialize in custom feeding with few of the cattle in the lot owned by the feedlot or its owners. However, since cattle feeding has not been extremely profitable over the last 10 to 12 years, there are fewer people willing to place cattle in commercial feedlots, which has forced many feedlots to own a higher percentage of the cattle in the feedlot to maintain an adequate number of cattle for efficient use of the facility.

Many cow/calf producers place at least part of their calf crop in custom feedlots to evaluate the performance and carcass characteristics of their cattle. In addition, taking full advantage of quality genetics may require retaining ownership through finishing.

While cattle feeding has not been profitable over the past 10 to 12 years, the cattle industry has historically been cyclical with the cow/calf segment profitable for a few years followed by the feeding segment being profitable for a few years. Consequently, there likely will come a time when cow/calf producers will be forced to retain ownership of their cattle through finishing to realize a profit. Obviously, custom feedlots will play an important role when that time comes.

Considerations in Retaining Ownership

The primary consideration in retaining ownership is the potential for making a profit. Thus, placing a value on the cattle and performing a breakeven analysis are essential tasks. Another consideration is determining the number of similar cattle that can be retained. In most feedlots, the smallest pens are designed to hold 100 head; however, there are a few custom feedlots that have smaller pens to accommodate cow/calf producers with small herds. One of the major considerations for a cow/calf producer is the potential impact of retaining ownership on taxes because retaining ownership usually means shifting the year the cattle will be sold. In addition, retaining ownership usually requires the assistance of a lender because of the shift in income and the capital required to pay feed bills.

Costs of Retaining Ownership

- **Freight Costs** – One of the expenses that some cow/calf producers fail to consider is the freight cost to the feedlot. Depending on the distance to the feedlot, this can be a significant expense.
- **Feedlot Costs** – Obviously, custom feedlots are in business to make money. They typically charge at least $0.45 per head per day for their services. They may mark up the feed to make this $0.45 or charge a yardage fee or both. This overhead expense tends to be similar across the industry because custom feedlots are in competition with each other for customers (cow/calf producers who want to retain ownership or investors).
- **Feed** – Feedlots typically bill cattle owners every 2 weeks. Feed is normally charged at its cost or market value. In some cases, these two values may be quite different. For example, if the feedlot contracted corn at $5.00 per bushel and it later went to $8.00 per bushel, the cost and market value when it is fed are quite different. Feedlots differ in how they price corn to a

customer in this situation. This price differential actually occurred in 2012 resulting in significant differences in feed costs between feedlots. Obviously, comparing how custom feedlots price their feed is a key question when considering placing cattle in a custom feedlot.

- **Yardage** – The most common yardage charge is $0.05 per head per day with the remainder of the $0.45 coming from feed markup. Some feedlots do not have a yardage fee; they simply mark the feed up to get the $0.45 per head per day in income.
- **Medical Costs** – Custom feedlots charge for medicine used to treat any sick cattle. These medicines may be charged at cost or with a markup. The high cost of some of the most effective medicines means that this can be a significant expense, which makes a good preconditioning program essential for cow/calf producers retaining ownership. Most custom feedlots also have a chute charge for each time an animal is handled.
- **Death Loss** – Since most death loss in custom feedlots is the responsibility of the cattle owner, it can become a significant expense. Typical death loss in yearling cattle is about 1 percent. It will be slightly higher for cattle fed through the winter. Death loss in calves placed in feedlots is higher and quite variable. Cattle that die because of catastrophe such as a blizzard or lightning strike are usually covered by insurance. Otherwise, death loss is the responsibility of the cattle owner.
- **Miscellaneous Costs** – Other costs that need to be considered include a charge for processing, insurance, taxes, and Beef Check-off.

Custom feedlots can provide a projected cost-of-gain based on current feed costs and projected performance. While these cost-of-gain figures are averages, they are useful in preparing a breakeven analysis before placing cattle in a custom feedlot. A key consideration is how the cattle will perform relative to the average of cattle being fed in the feedlot.

Services Provided by Custom Feedlots

- **Cattle Feeding and Management** – Successful custom feedlots typically do a good job of feeding and managing cattle because their success depends on repeat customers. Essentially all custom feedlots have both a consulting nutritionist and veterinarian. In addition, custom feedlots are knowledgeable on how to maximize the benefits of feed additives and implants.
- **Cattle and Feed Financing** – Most custom feedlots offer financing for both the cattle and feed.
- **Cattle Partnering** – Custom feedlots often partner on cattle placed in the yard. This can be valuable for a cow/calf producer who needs to sell some of his/her calves to provide income but would like to retain ownership of some of the calves. The feedlot may retain ownership of part of the cattle or sell part of the cattle to another individual.
- **Risk Management** – Custom feedlots typically have either an "in house" risk management specialist or they have a close relationship with a firm specializing in risk management. In either case, they can provide assistance in using futures or options to reduce the risk associated with cattle feeding.
- **Marketing Finished Cattle** – Custom feedlots usually market cattle using the option that they feel will result in the highest net return whether it is selling live, in the beef, or on a grid. In addition, many feedlots have access to special grids that fit certain types of cattle. For example, a number of feedlots in Kansas purchased shares in U.S. Premium Beef and can sell cattle using its grid.
- **Prepaid Feed** – Most custom feedlots allow a producer placing cattle in their feedlot to prepay for grain and other commodities at the time of placement. This essentially fixes the cost of feed and allows for a more accurate breakeven analysis. It is illegal to prepay a feed bill, but the IRS does allow the purchase of commodities that will be used in the feeding program.
- **Cattle Performance Records** – Custom feedlots provide a closeout record showing the performance of the cattle, and they will arrange to have carcass data collected.

Figure 53-1: Typical Closeout Summary

P E R F O R M A N C E   R E C O R D

PEN #: J04

HEAD IN :	51 Steers	AVG DATE IN: 01/16/12	ORIGIN: RAYMONDVILLE	
			HEAD DAYS	
HEAD OUT:	49	CLOSE DATE: 09/10/12	TOTAL:	11,604
DEATHS :	2	DEATH LOSS: 3.92%	AVG:	228

	--DEADS OUT--- TOTAL	AVG	---DEADS IN--- TOTAL	AVG	AVG WEIGHT
WEIGHT OUT:	66,355	1,354	66,355	1,301	954 DEADS IN
WEIGHT IN :	29,694	606	30,905	606	980 DEADS OUT
TOTAL GAIN:	36,661	748	35,450	695	
GAIN/HEAD/DAY:	3.16		3.05		

F E E D I N G   S U M M A R Y

	AS FED	DRY		DEADS IN	DEADS OUT
TOTAL LBS. FED	302,712	196,944	CONVERSION FACTOR		
			AS FED ---:	8.54	8.26
AVERAGE			DRY ------:	5.56	5.37
LBS. PER HEAD	5,936	3,862	PERCENT CONSUMPTION TO AVG WEIGHT		
LBS. / HEAD / DAY	26.087	16.972			
RATION COST / TON	237.70	365.36	AS FED ---:	2.73	2.66
COST / HEAD / DAY	3.30		DRY ------:	1.78	1.73

F I N A N C I A L   S U M M A R Y

	TOTAL COST	- AVG COST PER HEAD - DEADS IN	DEADS OUT	-- AVG/CWT --- DEADS IN	DEADS OUT
DELIVERED COST	39,956.00	783.45	815.43	129.29	134.56
FEEDING EXPENSES					
FEED & RATIONS	35,978.68	705.46	734.26	101.49	98.14
MEDICINE	336.42	6.60	6.87	.95	.92
PROCESSING	1,927.33	37.79	39.33	5.44	5.26
COST OF GAIN	38,242.43	749.85	780.46	107.88	104.31
				= $/CWT	SOLD ==
TOTAL INVESTMENT	78,198.43	1,533.30	1,595.89	117.85	117.85
SALES	79,247.78	1,553.88	1,617.30	119.43	119.43
BIC CHARGES	49.00	.96	1.00	.07	.07
OTHER SALES EXPENSES	48.00	.94	.98	.07	.07
NET SALES	79,150.78	1,551.98	1,615.32	119.28	119.28
GROSS PROFIT	952.35	18.67	19.44	1.44	1.44
CUSTOMER'S COST					
OTHER CHARGES	117.70	2.31	2.40	.18	.18
NET PROFIT	834.65	16.37	17.03	1.26	1.26

C A T T L E   I N

Date	# Hd	Pay Wt.	Off-Truck	Shrink	Amount	Prc/CWT	Buyer
01/16/12	51	30,905	30,000	2.93	39,956.00	129.29	
		Avg: 606	588			Origin:	RAYMONDVILLE

C A T T L E   D E A T H S

Date	# Hd	Cause	$ Salvaged	Wt.	DOF
04/04/12	1	BLOAT PH		744	76
05/05/12	1	RESPIRATORY 75% OLD		832	106
Total	2			1,576	

Death %: 3.92

C A T T L E   S H I P M E N T S

Date	# Hd	Pay Wt.	Scale Wt.	Shrink	Amount	Prc/CWT	Buyer/Dest
/30/12	1	902	940	4.04	704.18	78.07	H & B PACKI DoF: 135
		CHRONIC RESP RATILER					
09/07/12	48	65,453	68,180	4.00	78,543.60	120.00	CARGILL DoF: 235
		Avg: 1,364 CASH	1,420				
Total	49	66,355	69,120	4.00	79,247.78	119.43	
		Avg: 1,354	1,411				

Other Sales Expenses:	
Total BIC Expenses	49.00
Total CBP FEES Expenses	48.00
Net Proceeds from Shipments:	79,150.78

Feedlot Closeout Summary

Feedlots prepare a closeout summary for each lot of cattle, whether the cattle are being custom fed or are owned by the feedlot. These summaries are useful in evaluating the performance of the cattle and are often used to evaluate cattle buyers. For example, feedlots will stop using buyers who procure cattle that consistently perform poorly or exhibit high levels of sickness. Conversely, they will order more cattle from buyers who consistently procure cattle that perform above average.

Closeouts also allow managers and nutritionists to compare performance and costs information to industry standards. This type of comparison can be extremely useful in "fine tuning" the management and nutrition programs. Figure 53-1 shows a typical closeout summary. It shows that 51 head of steers were purchased weighing an average of 606 pounds. They weighed an average of 588 pounds when they arrived at the feedlot resulting in an average shrink of 2.93 percent. The purchase price was $129.29 per hundredweight. This price was well below the average price for 600-pound steers at the time indicating that these cattle were probably below average in quality or exhibiting some sickness.

As shown, the analysis is performed on both a deads in and deads out basis. The deads in analysis is especially valuable in determining the economic performance of the lot since death loss has a significant influence on profitability. The deads out analysis is useful in assessing the feeding performance of the cattle. For example, it may be more meaningful to compare the average daily gain and conversion on a deads out basis to other lots of cattle.

The average daily gain of 3.16 pounds was probably below average. Feed conversions should always be compared on a dry matter basis because rations vary in dry matter so much. In this case, the ration was 65 percent dry matter. The conversion of 5.37 pounds of dry matter per pound of gain is much better than the industry average. The consumption of 1.73 percent of body weight is also lower than industry average.

The financial summary section shows that the cost of gain was $107.88 per hundredweight with the deads in and $104.31 per hundredweight with the deads out. In this case, the cost of gain with the deads in is a better indicator of the actual cost. These steers made $16.37 profit per head. This profit was largely a function of the low purchase price and better than average feed conversion. This profit points out the importance of the buy-sell margin and feed conversion on profitability in cattle feeding.

The cattle deaths section shows the cause of death, date they died, and estimated weight at the time of death for each animal that died. The cattle shipments section shows that a chronic was removed from the lot and sold separately. This is typically done to avoid selling an animal that will produce a carcass that is too small resulting in a severe weight discount. These animals usually bring more when sold to a small local packer than to a large packer because of carcass weight discounts.

The sales section shows that 48 head were sold live with a 4 percent pencil shrink. Sales expenses included the Beef Check-off and marketing group membership expenses.

References

Boyles, S., D. Frobose, and B. Roe. Ownership Options for Feeding Cattle. Ohio State University Extension Fact Sheet AS-15-02.

Gill, D., K. Barnes, and D. Lalman. Ranchers' Guide to Custom Cattle Feeding. Oklahoma Cooperative Extension Service Bulletin ANSI-3022.

Lardy, G. 1999. A Cow-Calf Producer's Guide to Custom Feeding. North Dakota State University Cooperative Extension Service Bulletin AS-1162.

Chapter 54

Feeding Cull Cows

Cull cows account for 15 to 25 percent of the gross revenue in cow/calf operations. Most producers simply sell their cull cows in the fall after the calves are weaned. These cull cows should be viewed as a potential source of additional income. While most cull cows go directly to slaughter, some cattle feeders purchase cull cows when their calculations indicate that there is a good chance to make a profit. Key factors to consider in whether cull cows should be sold in the fall or fed include:

1. Seasonality of cull cow prices.
2. Price differences between slaughter grades.
3. Cost of feeding the cull cows.

Seasonality of Cull Cow Prices

Probably one of the most consistent trends in the beef business is the seasonality of cull cow prices with prices lowest in November and December and highest in the summer months. Obviously, this trend is based on supply with the highest number of cows being culled in November. Figure 54-1 shows the seasonal price index averaged over a number of markets. This index is based on dividing the average for each month by the average for the year and expressing the difference as a percentage. Thus, the average increase in the price received for a cull cow in May averages about 12 percent (107-95) more than in January.

The seasonality of prices for cull cows is essentially the same for all market classes. The data represented in Figure 54-1 helps predict the cull cow market, at least based on historical price relationships. To illustrate, assume cull cows are bringing $62 per hundredweight in November. Based on historical relationship, these

Figure 54-1. Average Monthly Cull Cow Prices Expressed as an Index Value, 2005-2009.

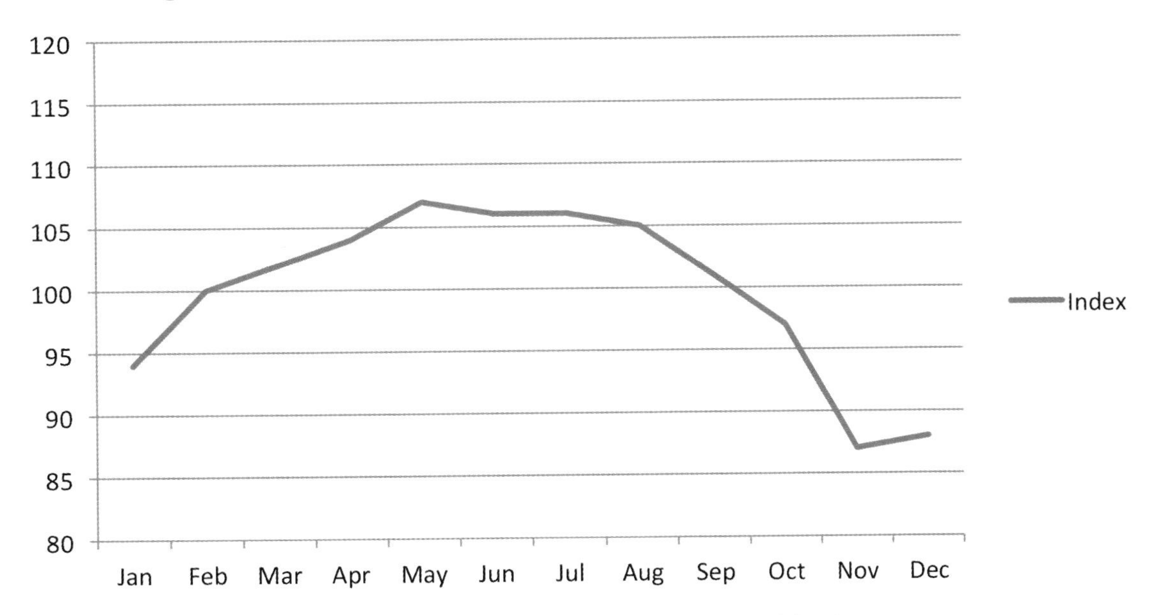

Adapted from Marketing and Feeding Cull Cows by Fuez, D.M and J.P. Hewlett

Table 54-1. Average Prices for Cull Cows.

Market Class	Price, $/lb	% Change Relative to Lean Class	Body Condition Score
Breaker	0.513	3.8	7–9
Boner	0.521	5.3	5.5–7
Lean	0.494	-	3–5.5
Light	0.416	-15.9	1–3

Adapted from Livestock Market Information Center, January 2004-July 2007.

cows will bring 15 percent (102-87) more in March with an expected price of $71.30 per hundredweight.

Price Differences between Slaughter Grades

In general, fatter cows receive a higher price. Table 54-1 shows this relationship. Thus, feeding cull cows purchased in the fall takes advantage of the seasonality of prices and the trend for fatter cows to bring more than thin cows, which means that there is often a positive buy-sell margin.

Selecting Cull Cows for Feeding

As with any feeding program, selecting the right animals to feed and getting them bought at a good price are key factors in profitability. Cull cows should be healthy and relatively thin. Cows in a body condition of 3 to 4 probably offer the greatest potential for good performance. In addition, cull cows should show the genetic potential for heavy muscling. Extremely thin cows are usually avoided because of concerns about their health even though they can be purchased at a much lower price as shown in Table 54-1. Finally, it may be wise to have purchased, "open" cull cows pregnancy checked. In many cases cows diagnosed as open are actually pregnant, and pregnant cows sold for slaughter will be severely discounted.

Typical Cull Cow Feeding Programs

- **Grazing Programs** – Some producers place thin, cull cows on crop residues or dry, winter grass with some supplement. Gains on these programs are minimal, and the impact on the quality of the carcass is minimal. The goal of this type of program is to take advantage of the seasonality of cull cow prices.
- **Quality Roughage Programs** – Some producers drylot cull cows and feed them quality roughage and possibly some grain. Gains on these programs will usually average 2.0 pounds per day depending on the amount of grain included in the ration. The goal in this type of program is to add weight and take advantage of the seasonality of cull cow prices. This program allows producers to market homegrown roughage. However, it does not improve carcass quality as much as the high-energy program. For example, cull cows fed corn silage would still have yellow fat at the end of the feeding period.
- **High Energy, High Concentrate** – When the goal is to add weight and markedly improve carcass quality, the cows are typically placed on a diet similar to that used to finish steers and heifers. Gains and feed conversions are usually good on this type of program at least for the first 60 days of the feeding period.

Impact of Feeding Cull Cows High Energy, High Concentrate Diets on Carcass Characteristics

As discussed in Chapter 5, beef from cull cows is tougher, leaner, and less juicy than beef from grain-fed steers and heifers. In addition, the fat in the meat from cull cows is often yellow from the deposition of carotenoids (precursor to vitamin A) while on pasture. Moreover, it has a less desirable taste than that of conventional beef. One way to improve the palatability of the beef from cull cows is to place them on a high-energy, high-concentrate diet before harvest. Feeding cull cows in this manner increases red meat yield, increases marbling, improves lean color (brighter), improves fat color (from yellow to white), and improves palatability.

Table 54-2. The Effect of Management Strategy (Fed vs. Non-Fed) on the Carcass Characteristics of Cull Cows.

Item	Preharvest Management	
	Non-Fed	Fed
Number of animals	104	108
Hot carcass weight, lbs	553[a]	813[b]
Adj. fat thickness, in	0.07[a]	0.51[b]
Ribeye area, sq in	8.7[a]	12.4[b]
Skeletal maturity score[1]	562	566
Lean maturity score[1]	540[a]	340[b]
Marbling score[2]	215[a]	362[b]
Ribeye intramuscular fat, %[3]	3.56[a]	7.02[b]
Ribeye lean color score[4]	4.4[a]	3.5[b]
Fat color score[5]	5.3[a]	2.2[b]
Ribeye shear force, lbs	62.4[a]	46.7[b]
Ribeye collagen, mg/g[6]	5.09[a]	3.60[b]

[1]100 to 199 = A^{00} to A^{99}; 200 to 299 = B^{00} to B^{99}; 300 to 399 = C^{00} to C^{99}; 400 to 499 = D^{00} to D^{99}; 500 to 599 = E^{00} to E^{99}. [2]200 to 299 = $Traces^{00}$ to $Traces^{99}$; 300 to 399 = $Slight^{00}$ to $Slight^{99}$. [3]% reported on a wet sample basis. [4]1 = light, pinkish-red, 2 = cherry red, and 6 = dark, purplish-red. [5]0 = bright white; 5 = pale yellow, and 8 = dark yellowish-orange. [6]SF measurements were computed across postmortem aging period (14,21,28, & 35 days) [a, b]Means in the same row without a common superscript differ ($P<0.05$). Adapted from Woerner 2010.

In addition, high-concentrate feeding before harvest increases tenderness. While feeding periods of 56 days elicit many of these positive changes in palatability, feeding for more days results in even greater palatability. Table 54-2 shows a comparison of carcass characteristics of cull cows fed a high-energy diet for approximately 95 days to those of cull cows that were not fed before harvest.

The research summarized in Table 54-2 shows the significant effect that high-concentrate feeding has on palatability characteristics. Lean maturity score, marbling score, intramuscular fat, lean color score, fat color score, slice force, and collagen content were all improved significantly by feeding for 95 days. In addition, feeding cull cows has been shown to reduce the chance of off-flavors. This improvement in palatability is especially meaningful given that steaks and other cuts from cull cows are being merchandised into the low-end restaurant trade.

Feeding and Managing Cull Cows

Cull cows to be fed high-concentrate diets should be "worked-up" to the final ration in a fashion similar to the techniques used with younger cattle to avoid digestive problems. Adequate bunk space (probably at least 24 inches per cow) should be available. Rations should be at least 80 percent grain and 11.5 percent crude protein. The diet should contain Rumensin® or Bovatec®. Research on feeding beta-agonists has shown that these products increase carcass weight and the yield from the carcass. However, since cull cows are typically sold on a live basis, the feeder probably will not receive enough additional income to pay for these products. The cows should be implanted with a combination implant.

Cull Cow Performance on High Concentrate Diets

To illustrate the performance of cull cows on a high-concentrate diet, consider the results of a research project at Kansas State University shown in Table 54-3. Gains were good, even out to 56 days on feed. In addition, feed conversions were good, especially in the implant treatments. Intakes increased from day 28 to day 56, probably because the cows adapted to the high-energy diet. These data clearly show the value of an implant in a cull cow-feeding program. While there was not a significant difference between implant

Table 54-3. Feedlot Performance of Implanted and Nonimplanted Cull Beef Cows Fed for 28 or 56 Days.

Item	Control	Finaplex- H	Synovex-H	Both Implants
		28 Days on Feed		
Daily gain, lbs	3.9[a]	5.1[b]	5.3[b]	5.8[b]
Feed/gain	7.2[c]	5.5[d]	5.4[d]	5.0[d]
Daily intake, lbs of DM	26.0	26.8	27.5	26.8
		56 Days on Feed		
Daily gain, lbs	3.5[c]	4.8[d]	4.5[cd]	5.3[d]
Feed/gain	9.0[c]	6.5[d]	6.8[d]	5.5[d]
Daily intake, lbs DM	29.3	29.5	30.1	29.2

[ab]Values in the same row without a common superscript are different (P<.10). [cd]Values in the same row without a common superscript are different (P<.05). The gains in the Kansas State University trial are higher than those reported in other research trials on average. The key factors influencing gains of cull cows are the condition of the cows at the start of the trial, the genetic potential of the cows for growth and muscle deposition, and the length of the feeding program.

programs, the combination implant program tended to produce better gains and efficiency.

Risk Associated with Feeding Cull Cows

Obviously, there is a certain level of risk in cattle feeding; however, in cull cow feeding the buy-sell margin can be a significant factor in profitability because a high percentage of the final weight is purchased. For example, when purchasing 1,100-pound thin cows to feed and selling them at 1,300 pounds, 85 percent (1,100 ÷ 1,300) of the final weight was purchased. If the buy-sell margin is even slightly negative, it can result in a loss even if the cattle perform well. Consequently, market swings can be a significant factor in the profitability of feeding cull cows.

References

Cranwell, C.D., J.A. Unruh, D.D. Simms, and J.R. Brethour. Performance and Carcass Characteristics of Cull Beef Cows Implanted with Growth Promotants and Fed a High Concentrate Ration. Cattlemen's Day Report of Progress, Kansas State University.

Dijkhuis, D.D. Johnson, and J.N. Carter. 2008. Feeding Ractopamine Hydrochloride to Cull Cows: Effects on Carcass Composition, Warner-Bratzler Shear Force, and Yield. Professional Animal Scientist 24:634-638

Fuez, D.M. and J.P. Hewlett. 2012. Marketing and Feeding Cull Cows. Right Risk™ Information Sheet #IS-12-01.

Peel, D.S. and D. Doye. Cull Cow Grazing and Marketing Opportunities. Oklahoma Cooperative Extension Service Bulletin AGEC-613.

Simms. D.D. 1997. The Effect of Implanting Cull Cows on Gain, Intake, Feed Conversion, and Carcass Characteristics. Oklahoma State University Implant Symposium.

Woerner, D.R. 2010. Beef from Market Cows. National Cattlemen's Beef Association White Paper. *www.beefuse.org* Website.

Wright, C.L. 2011. Managing and Marketing Cull Cows. Proceedings, The Range Beef Cow Symposium XIX.

Chapter 55

Risk Management

There are many risks associated with stocker production and cattle feeding. The cattle might die or not perform up to expectations. While some risk is unavoidable, there are some tools available to reduce price risk. For instance, in many cases it is possible to forward contract cattle when concerned about price declines. For example, many ranchers contract their calf crop in the summer if they think the price of calves is going to go down. Cattle feeders may forward contract cattle with packers when they think the market might go lower. In addition to forward contracting, many cattle feeders use the futures market to minimizing price risk. The futures market offers stocker operators and cattle feeders the opportunity to establish a sales price at some date in the future.

Futures Market

Futures Contract

A futures contract is an agreement to buy or sell a commodity at a date in the future. The contract specifies the commodity such as live cattle, the quantity (40,000 pounds in the case of live cattle, 50,000 pounds for feeder cattle), product quality, delivery points (for live cattle), and the delivery date.

Futures Trading Terminology

- **Short** – A seller of a futures contract, which is a commitment to sell the commodity in the future.
- **Long** – A buyer of a futures contract, which is a commitment to buy the commodity in the future.
- **Bull** – A person who expected a commodity to increase in price.
- **Bear** – A person who expects a commodity to decrease in price.

What is Hedging?

Hedging is using a futures contract to protect against unfavorable price changes. In true hedging, the commodity must be currently owned or purchased in the future. A short hedge is used when planing to sell the commodity in the future. A long hedge is when one plans to purchase the commodity such as grain for use in cattle feeding. In the case of a long hedge, the goal is to protect against an increase in the price of corn or feeder cattle purchased in the future. For each contract, there is an opposite transaction. In other words, if a cattle feeder sells a contract of live cattle (i.e., short hedge), someone else must assume the long position (i.e., buy the contract). When the contract to sell the cattle comes due, the feeder can deliver the cattle or buy back the contract at the current price. This is called "offsetting" the original contract.

Basis

As mentioned, commodities are clearly defined in futures contracts. In many cases, producers do not produce exactly what the contract specifies or there may be a difference between where the product is located and a delivery point. In these cases, there may be a difference between the futures market price and the local cash price on the contract due date. This difference is called the basis. The formula for basis is:

Basis = Cash Price – Futures Price

Because the cash price is what is of interest, the formula can be arranged as the following:

Cash price = Futures Price + Basis

Table 55-1. Kansas Fed Cattle Basis Relative to Chicago Mercantile Exchange October Live Cattle Futures ($/cwt).

Week	2009	2010	2011	3-Year Ave.
		$/cwt		
10/2	-1.39	-0.38	-1.26	-1.01
10/9	-0.29	0.04	-1.94	-0.73
10/16	-0.96	0.34	-0.75	-0.46
10/23	2.23	-0.43	-1.51	0.10
10/30	1.33	-0.18	-3.51	-0.79

Kansas State University Department of Agricultural Economics Publication AM-GTT-2012.3.

Basis is important because it influences how close the estimated hedge price is to the actual cash price received. To illustrate, assume that the December live cattle contract is at $121 per hundredweight and historically the basis for the contract has been -$1.00. The expected price in December will be $120 ($121 – $1.00). Estimating basis accurately is key to predicting the expected price. Unfortunately, the basis changes, resulting in discrepancies between the expected price and the realized price. Table 55-1 shows the historical basis for western Kansas during the weeks in October in 2009, 2010, and 2011. There was considerable variability in the basis for each week in October across years. Research shows that using a 3-year historic average to predict basis is a relatively accurate approach. Typically, it is recommended that the historical basis should be kept on a weekly basis.

Example of a Short Hedge

A cattle feeder places a pen of 100 steers on feed in November with an expected finishing date of April at an expected weight of 1,300 pounds. Total pounds to be sold equals 130,000. Since the futures contract for live cattle is for 40,000, the closest fit is to sell three futures contracts for a total of 120,000 pounds. The current price of the April live cattle contract is $127 per hundredweight and the historic basis is -$1.00. This means the expected cash price is $126 per hundredweight. This expected price is appealing because the feeder is concerned that the cash market may be lower in April and this price offers some profit. In April, the cash price for the steers is $129 per hundredweight and the futures price is $130 per hundredweight.

Cash Sales: 130,000 × $1.29 = $167,700

Futures Market: Buy back the April Futures Contract @ $130

Futures results: -$3.00/cwt × 120,000 lbs = -$3,600

Net = $167,700 – 3,600 = $164,100 = $126.23/cwt ($164,100 ÷ 130,000)

Thus, the price received was close to the expected price with the difference resulting from the fact that the 10,000 pounds that were not hedged sold for $129 per hundredweight. In this example, the basis was exactly at the historical average. In the real world, this rarely happens — the basis is usually higher or lower, which results in higher or lower returns than expected. For example, if the basis at the time of sale was -$2.00, the cash price would be $128 with sales of $166,000. Net returns would be (166,000 – 3,600) $162,800. The net return per hundredweight would be $162,800 ÷ 130,000 = $125.23. Since basis is constantly changing, it can be a significant factor in how close the actual returns are compared to the expected returns.

This example also shows that it is difficult to match the number of pounds contracted to the number produced, which can cause some inaccuracy in obtaining the expected price. Another consideration is the match between the date of the contract and the date the cattle will be sold. For instance, the cattle in the example will be fed over winter, and if the weather is severe, it will probably move their sale date to May or even June. This type of delay can cause differences between the expected price and the actual price received because the April futures contract has to be settled by the last trading day in April.

Advantages of a Short Hedge

1. Protects against the risk of a price decline.
2. Because of the protection it offers, it can make obtaining credit easier.
3. Easier to implement and cancel than a typical forward contract.

Disadvantages of a Short Hedge

1. Cannot take advantage of an increase in price.
2. Accuracy in forecasting basis is crucial to the success of the hedge.
3. Sometimes difficult to match quantities produced to the contract quantities.
4. Margin requirements and margin calls can be significant factors. Margin requirements are discussed later in this chapter.

Using Long Hedges to Protect Against an Increase in the Price of an Input

Cattle feeders often use long hedges to protect against an increase in the price of an input. To illustrate, assume that it is May and corn is at $5.00 per bushel. A cattle feeder thinks that corn is going higher. The December futures price of corn is $5.10. He decides to cover a portion of his corn needs for the coming year by purchasing futures contracts (i.e., long hedge). He buys 20 December corn contracts at $5.10. Each corn contract is 5,000 bushel (280,000 pounds) for a total of 100,000 bushel (5,600,000 pounds). The historic corn basis at this location is +$0.20 per bushel. Consequently, his expected price for corn in December is $5.30 per bushel (futures price + basis: $5.10 + $.20).

In December, corn is selling locally for $7.20 and the December futures contract is at $7.05 per bushel.

Cash purchases of corn 100,000 bushel @ $7.20 = $720,000

Futures Market: Sell back the December contract @ $7.05

Futures results: $1.95 profit per bushel = $195,000

Net paid for corn = $720,000 - $195,000 = $525,000 = $5.25 per bushel.

Again, the result is close to the expected price for corn differing only because the basis was lower than the historical value.

Changes in a Futures Contract Value

In the short hedging example, the three April contracts sold for $127 per hundredweight, and each live cattle contract is for 40,000 pounds. Thus the value of the contracts at the time of the trade was 120,000 × $1.27 = $152,400. If the April contract moved to $130 per hundredweight the contracts would be valued at 120,000 × $1.30 = $156,000.

Margin Money

When the futures contract is bought or sold, an initial margin must be placed in a trading account. This is essentially earnest money guaranteeing that funds will be available if the market moves against the position. Since this account must be kept at a certain level, more money may have to be deposited. These requests for additional money are termed "margin calls." If one is on the wrong side of a move in the market, margin calls can be large. Consequently, it is crucial that funds are available to cover potential margin calls; otherwise, the broker may close the account canceling the price protection position.

To illustrate a margin call, consider the producer who short hedged the cattle in the previous example. Initially, he would have had to deposit 5 percent of the value of the contracts (152,400 × 0.05 = $7,620) in his margin account. When the April contract went to $130 per hundredweight, he would have had to deposit an additional $3,600 ($156,000 – 152,400). When the contract increased in value to $130 per hundredweight, $3,600 was moved from his margin account to the account of the buyer of the contract.

Functions of a Broker

All futures transactions are handled through brokers who deal with the appropriate futures market. The Chicago Mercantile Exchange handles live cattle, feeder cattle, and grain contracts. A good broker can be a real asset in using the futures market because the broker can assist in choosing the right contract. In addition, good brokers follow the markets and have a good feel for the expected trends.

Using Feeder Cattle Futures in Stocker Operations

While the examples presented are all using live cattle futures, stocker producers can use feeder cattle contracts to reduce price risk in a similar fashion. A couple of differences are that the contracts are for 50,000 pounds and there are no delivery points as feeder cattle contracts are cash settled.

Options

While futures contracts have been around since the 1800s, another risk management tool, options, was made available to cattle producers in the 1980s. They are based on futures contracts, but they offer a more flexible pricing tool. They offer a type of insurance against an adverse price move while also allowing buyers to take advantage of favorable price moves.

What is an Option Contract?

An option contract is the right (but not the obligation) to buy or sell a futures contract at some predetermined price within a specified time. There are two types of option contracts. An option to sell a futures contract (go "short") is called a **put** option while an option to buy a futures contract (go "long") is known as a **call** option. Options allow one to take advantage of changes in futures contracts without actually having a position in the futures market. The predetermined price of the contract is called the strike price.

It is important to know that for every option buyer there is an option seller. At any time before the option expires, the option buyer can exercise the option. For example, the buyer of a put option, as noted, has the right to convert the option position into a short futures position at the strike price. Correspondingly, the buyer of a call option can convert the option contract into a long futures position at the strike price. The option buyer pays a premium to the option seller to obtain the right to exercise the option. The option seller must post a margin and faces the possibility of margin calls if the market moves against his or her potential futures position. The option buyer can exercise the option or let it expire.

While live cattle futures contract are available for the months of February, April, June, August, October, and December, live cattle option contracts are available for every month. Strike prices are available in $2 per hundredweight intervals; however, nearby contracts may have $1 per hundredweight intervals. Option contracts terminate on the first Friday of the contract month.

Cattle producers typically use put options to establish a minimum expected price while maintaining the potential to take advantage of a price increase. To illustrate how an option contract works, assume that on November 7, 2012, the August 2013 live cattle futures contract was at $129 per hundredweight. A producer could sell (go short) this contract and establish this price plus or minus any expected basis difference. For example, if the historical basis on the August contract was -$1.00 per hundredweight, the expected price would be $128 per hundredweight. Another price risk opportunity would be to purchase an August put option contract. Table 55-2 shows the premiums for the various strike prices for the August put option contracts as of November 7, 2012. Table 55-3 shows the minimum expected price and minimum expected profit assuming a producer purchased a put options

Table 55-2. Strike Prices and Premiums for the August 2013 Futures Contract on November 7, 2012.

Strike Price, $/cwt	Premium, $/cwt
126	3.250
128	4.075
130	5.025
132	6.125
134	7.375

Table 55-3. Minimum Expected Price and Minimum Expected Profit Assuming the Purchase of an August Put Contract with a Strike Price of $126/cwt.

Item	$/cwt
August Live Cattle Put	$126.00
Premium	-3.25
Expected Basis	-1.00
Minimum Expected Price	$121.75
Breakeven	120.00
Minimum Expected Profit	$1.75

Table 55-4. Net Profit on the Cattle and the Put Contract with a Price Increase or Decrease.

Item	Price Decrease	Price Increase
	$/cwt	
August Put Contract	126.00	126.00
August Futures	119.00	133.00
Value of Put Contract	7.00	0.00
Premium	-3.25	-3.25
Profit on Put	3.75	-3.25
Cash Sales	118.00	132.00
Breakeven	120.00	120.00
Net Profit	1.75	8.75

contract with a $126 per hundredweight strike price and has a breakeven price of $120 per hundredweight. Table 55-4 shows the effect of an increase or decrease in the value of the futures contract on net profit.

This example shows how a put option can establish a minimum price while allowing the producer to reap the benefit of a price increase. One cost not included in this example is the broker's commission. There are many complex trading strategies built around options that are explained in greater detail in the references listed below.

References

CME Commodity Products. Strategies for CME Livestock Futures and Options. Website: *cmegroup.com.*

Feeder Cattle Basis Forecasts. *www.BeefBasis.com.*

Fincham, C., J. Mintert, and M.M. Waller. 1998. Introductions to Options. Website: *www.agmanager.info*. Kansas State University.

McElligott, J. and G.T. Tonsor. 2012. Fed Cattle Basis: An Updated Overview of Concepts and Applications. Website: *www.agmanager.info* Kansas State University.

Mintert, J., M. Waller, and R. Borchardt. 1999. Introduction to Futures Markets. Website: *www.agmanager.info* Kansas State University.

Mintert, J, K. Dhuyvetter, E.E. Davis, S. Bevers. 2002. Understanding and Using Feeder and Slaughter Cattle Basis. Website: *www.agmanager.info.* Kansas State University.

Murra, G. 1996. Futures Market – Basic. Website: *www.agmanager.info.* Kansas State University.

Petry, T. 2012. Futures and Options: Live Cattle and Feeder Cattle. Website: *www.ag.ndsu.edu/livestockeconomics.* North Dakota State University.

Sartwelle, J.D. and J. Mintert. 2002. Hedging Using Livestock Futures. Website: *www.agmanager.info.* Kansas State University.

The Options Guide – Options and Futures Trading Explained. Website: *www.theoptionsguide.com.*

Chapter 56

Interpreting a Feed Tag

Feed companies are required to supply a feed tag with each load of bagged or bulk feed. In addition, by-products sold as livestock feed must have a feed tag. The nutrients and other information that must appear on a feed tag are listed in guidelines published by the American Feed Control Officials. Furthermore, each state has guidelines for the information that must be included on feed tags distributed in that state. Figure 56-1 shows an example of the feed tag for a complete or supplemental beef feed.

Figure 56-1. A Sample Tag for a Complete or Supplemental Beef Feed.

Sample Feed Company
Beef Feed Range Cubes

Guaranteed Analysis

Crude Protein (Min) 20%
(This includes not more than 6 percent equivalent crude protein from non-protein nitrogen)

Crude Fat (Min) 2.0%

Crude Fiber (Max) 10.0%

Calcium (Min) 1.0%

Calcium (Max) 2.0%

Phosphorus (Min) 1.0%

Salt (Min) 2.0%

Salt (Max) 2.0%

Potassium (Min) 1.0%

Vitamin A (Min) .... 10,000 IU/lb.

Ingredients

Grain Products, Plant Protein Products, Molasses Products, Processed Grain By-Products, Urea, Vitamin A Supplement, Vitamin D3 Supplement, Vitamin E Supplement, Calcium Carbonate, Dicalcium Phosphate, Salt, Manganous Oxide, Ferrous Sulfate, Copper Oxide, Magnesium Oxide, Zinc Oxide, Cobalt Carbonate, Ethlenediamine Dihydroiodide (EDDI), and Potassium Chloride.

Feeding Directions:

Feed 3-4 pounds per head per day as a supplement on range or pasture. Provide plenty of fresh, clean water at all times. *Caution:* Use as directed. Do not feed additional salt.

Understanding the meaning of the information on a feed tag is useful when comparing different sources of a supplement or mineral. Furthermore, it is important to understand what valuable information is not listed on a feed tag.

Important Information on the Tag of a Protein Supplement or Complete Feed

Name of the Product

The name indicates the specific production situation for which the product was formulated. The suitability of the product for the production situation depends on the knowledge of the formulator, which may or may not be adequate.

Nutrient Guarantees

The concentration of some of the key nutrients must be listed on the tag. For some nutrients, only a minimum is required, while for others both a minimum and maximum must be listed. Maximums are required to prevent the use of low-cost ingredients as fillers. For example, calcium carbonate (limestone), because of its low cost, has been used as a filler to reduce the overall cost of the ingredients. Requiring a maximum level of calcium on the tag reduces the potential for this use of limestone. This section also shows the minimum and maximum salt inclusion level.

The nutrient guarantees on feed tags are on an "as fed" basis. It is often necessary to convert the nutrient levels on a feed tag to a dry matter basis before making comparisons or before ration formulation. Since the moisture content is not listed on the tag, an estimate of the moisture content must be made. For most protein supplements, a dry matter content of 89 percent should be close. For most mineral supplements, 98 percent is a good estimate of the dry matter content. A 20 percent range cube contains 22.5 percent crude protein ($20 \div 0.89 = 22.5$) on a dry matter basis.

Crude Protein

Total protein is based on the percentage of nitrogen times 6.25. It includes both natural protein (nitrogen contained in amino acids) and non-protein nitrogen.

Non-Protein Nitrogen

As shown on the sample tag, the protein equivalent from nonprotein nitrogen must be listed. For these cubes, 30 percent (6/20) of the protein is from nonprotein nitrogen. It is important to keep in mind that on feed tags, nonprotein nitrogen is only a measure of the amount of protein equivalent from urea — it does not include other sources of nitrogen that are not natural protein. In some cases, this type of nonprotein nitrogen can be significant, especially with liquid supplements.

Ingredients

A feed tag must contain a list of ingredients. However, it is legal to use group terms to describe ingredients rather than specifying the exact ingredient. For example, on the tag above, the term "Processed Grain By-Products" is a group term that could represent a broad range of ingredients with quite different nutrient composition. For instance, grain screenings or even grain dust could be in these cubes under one of these broad categories. Again, asking the supplier for a list of actual ingredients could be valuable in evaluating and comparing products.

Macro Minerals

While group terms can be used for some ingredients like grain by-products, specific sources of macro and trace minerals must be listed.

Trace Minerals

The ingredient section on the tag shows that trace minerals have been added to the supplement, but the inclusion levels are not listed. Trace mineral levels are not required on the tag of a protein supplement. In many cases, feed companies formulate the supplement to meet the total trace mineral requirements at the recommended feeding rate, while in other cases only small quantities of trace minerals are added so they can be listed on the tag as a selling point. Since including trace minerals in a protein supplement is a good way to supply them, the actual levels of trace minerals in the product should be requested from the supplier. A reputable manufacturer should be willing to supply this information.

Feeding Directions

The feeding directions section can give some indication of the level of fortification. In most, but not all cases, the recommended feeding rate will contain minimum levels of trace minerals and vitamin A based on NRC recommendations. The feeding directions section also will contain cautions for the use of the product, if appropriate.

Concentration of Drugs

While not shown on the example tag, the concentration of any drug included in the feed must be listed on the tag.

Figure 56-2. A Sample Tag for a Beef Mineral Supplement

Sample Feed Company
Beef Mineral for Beef Cattle
on Fescue Pasture

Guaranteed Analysis

Calcium (Min)	11.0%
Calcium (Max)	13.0%
Phosphorus (Min)	8%
Salt (Min)	18%
Salt (Max)	20%
Magnesium (Min)	1.0%
Potassium (Min)	2.5%
Copper (Min)	500 ppm
Selenium (Min)	10 ppm
Zinc (Min)	2,500 ppm
Vitamin A (Min)	120,000 IU/lb

Ingredients

Dicalcium Phosphate, Monocalcium Phosphate, Salt, Calcium Carbonate, Potassium Chloride, Ferrous Sulfate, Copper Sulfate, Magnesium Oxide, Zinc Sulfate, Ethylenediamine Dihydroiodide (EDDI), Cobalt Carbonate, Sodium Selenite, Vitamin A Supplement, Vitamin D3 Supplement, Vitamin E Supplement, Manganese Sulfate, Mineral Oil.

Feeding Directions:

This mineral should be fed in a rain-protected feeder. Provide fresh, clean water at all times. Do not feed additional salt. Feed to beef cattle on fescue pasture at the rate of 4 ounces per head per day.

Important Information on the Tag of a Mineral Supplement:

Name of Product

The name, in this case "Mineral for Beef Cattle on Fescue Pasture," (Figure 56-2) indicates the specific situation for which the mineral was formulated. However, because there are no published requirements for minerals for specific situations such as fescue pasture, this mineral may or may not fit the production situation described. Whether it is better than a generic mineral for the specific situation depends on the knowledge of the formulator. Unfortunately, in some feed companies, the nutritional training of formulators is not adequate enough to allow formulation of minerals for specific situations. However, the name does indicate the specific situation for which the mineral was formulated, which may be quite useful in some cases.

Guaranteed Analysis

This section shows the minimum and maximum levels guaranteed in the mineral. Again, the maximums are shown to prevent filling the product with low-cost ingredients with minimal feed value. The concentration of most, but not all, of the trace minerals must be shown on the tag for a mineral supplement. In this example, the levels for cobalt and iodine are not shown. There is some state-to-state variation with respect to the trace mineral levels that must be listed.

Ingredients

This section lists the specific mineral sources used in the product. Again, the sources should be screened for those that have low bioavailability. In addition, this section will show the inclusion of any organic (chelated) trace minerals. However, it will not show the specific level. In other words, an insignificant amount may be included as a selling point. An analogous situation exists with Vitamin D and E inclusion. The tag does not indicate the concentration, just that some was added to the mineral. The manufacturer should be willing to provide the levels of organic trace minerals and vitamins D and E in the mineral supplement. This section should be reviewed for any ingredients that are simply fillers.

Feeding Directions

In addition to the basic feeding instructions, this section usually shows the recommended feeding rate.

Interpreting Tags for By-Products

A feed tag must be provided for by-products as well as commercial feeds. These tags typically only show minimum guaranteed protein levels. In addition, if the by-product is a combination of by-products, the by-products included must be listed. There are several considerations in interpreting these tags on by-products:

The Nutrient Composition Shown on the Tag Is Often Well Below the Actual Composition

As pointed out previously, the composition of by-products is often variable because the inclusion level for different by-product streams is inconsistent.

To compensate for this variability, processors set the nutrient composition on their tags to the lowest level that the by-product will routinely meet. This means that the nutrient content of by-products often is much higher than shown on their tags. For example, a by-product with a tag showing 15 percent crude protein may actually contain 18 percent crude protein. The processor is showing 15 percent on the tag to ensure that all loads will meet specifications. This tendency for the actual composition to be higher than the tag level may or may not be of practical importance.

If the inclusion level in the ration is relatively low, this discrepancy between the tag and the actual composition will be of little importance. For example, if beef cows are being fed a couple of pounds of a by-product per day, the difference between 15 percent and 18 percent crude protein is insignificant. However, if the inclusion level is relatively high, the difference may be of practical importance. For example, in growing rations for replacement heifers or bulls, a by-product may constitute a fairly high percentage of the ration. Knowing the actual level of protein in the by-product allows more accurate ration formulation. In addition, if the actual level of protein is significantly higher than stated on the tag, the amount of protein required from other sources may be lower, reducing feed costs.

Importance of the Ingredient Section

Producers of by-products must list the ingredients included in the by-product. Unfortunately, group terms are often used which makes it difficult to determine exactly what was included. For example, the group term corn processing by-products includes a number of by-product streams. The percentage of ingredients is not required on the tag. It is almost impossible to know what was included in some by-products. However, the ingredient section may indicate that some ingredients were included that are not typical for the by-product. For example, a tag for corn distillers grains might show the inclusion of soybean processing by-products, which could have a significant impact on the nutritive value of the corn distillers grains. This type of inclusion increases the importance of a feed analysis.

Important Information Not Listed on a Feed Tag

There are several important pieces of information not included on a feed tag including the moisture and energy levels. Also, as noted previously, the actual inclusion rate for some minerals is not required.

Moisture

The fact that a moisture level is not required on a feed tag is not typically a problem for commercial feeds because dry feeds are usually 89 to 90 percent dry matter and minerals are 96 to 98 percent dry matter. By-products typically vary significantly in dry matter content. Consequently, an analysis of the moisture content of a representative sample of a by-product that constitutes a significant portion of the ration is recommended.

Energy

It is unfortunate that the energy level of a feed is not required on the tag because there can be significant differences in the energy level among similar products — at least based on the nutrients listed on the tag. For example, there are literally thousands of combinations of ingredients that can be used to produce a 20 percent crude protein range cube, and they differ markedly in energy content. This difference in energy content makes comparing sources of 20 percent range cubes difficult. In some cases, the energy level in the feed can be obtained from the supplier. Obtaining and using this information will make the comparisons of sources much more accurate.

References

Livestock Feed Labeling Guidelines. Association of American Feed Control Officials, Inc.

Simms, D.D. 2009. Feeding the Beef Cowherd for Maximum Profit. SMS Publishing. Amarillo, TX.

Chapter 57

Cattle Handling Facilities

A good cattle-working facility can save time and make it easy to perform routine management practices. Effective facilities take advantage of the behavior of cattle to make movement through the facility easy.

Cattle Behavior

Cattle have almost a full 360-degree field of vision with the exception of the area directly behind them. This makes them especially aware of movement on either side. While they have this wide range of vision, their depth perception is poor, and they have limited vertical vision. Because of this, they have difficulty focusing on objects or shadows on the ground, which accounts for their tendency to balk when they perceive a shadow. They prefer to move from dark to light areas rather than light to dark. And they move uphill better than they do downhill.

Cattle are herd animals with a tendency to follow the animal in front of them. A good working facility takes advantage of this herd instinct using a long holding alley before the chute. A curved alley is especially effective because the animals cannot see the activity at the front of the chute.

Cattle exhibit a comfort/flight zone that varies from animal to animal. Cattle that are handled gently with minimum noise exhibit a smaller flight zone than roughly handled cattle. The flight zone is larger when cattle are approached from the front than when approached from the rear. When a person enters the animal's flight zone, the animal will move away. Cattle also exhibit a point of balance that is essentially the shoulder. When a person enters the flight zone in front of the shoulder the animal backs up. Conversely, when a person enters the flight zone behind the shoulder, the animal moves forward.

Selecting a Site for the Facility

The handling facility should be accessible from all pastures and have road access if it is to be used for load out. The area should be well drained, and if there is a slope, the holding alley and working chute should be facing uphill. Access to electricity and water can be useful. A roof over the chute/squeeze allows cattle to be worked in inclement weather if necessary.

Components of a Good Working Facility

- **Holding Pen** – The size of this pen should be based on the size of the herd and number of cattle worked at one time. Table 57-1 shows the space requirements on a per head basis. Even for large herds, holding pens should probably hold no more than 50 to 60 head. It is easier to handle cattle in long narrow holding pens as compared to square pens of the same square footage.
- **Crowding Pen** – This pen should be solid sided and funnel down into a holding alley. A circular pen is widely used in the industry because, if properly designed, it avoids corners and dead ends. The radius should be at least 12 feet to allow the animals to turn around freely. It should be designed so the cattle can see the escape route (the holding alley) easily. The crowding gate should be solid sided.
- **Holding Alley** – This alley may be straight or curved with solid sides. A curved alley has the advantage of not allowing the cattle to see the activity at the working chute. The alley should hold at least three head (minimum length of 20 feet) and should have stops to prevent the cattle

Table 57-1 Recommended Sizes for Cattle Working Facilities.

Facility Component	Recommended Dimensions		
	Up to 600 lbs	600 to 1,200 lbs	More than 1,200 lbs
Holding Pen			
Space per had (sq. ft)	14	17	20
Pen Fence			
Height (in)	60	60	60
Post Spacing (ft)	8	8	8
Post Depth in ground (in)	30	30	30
Crowding Pen			
Space per head (sq. ft)	6	10	12
Post spacing	4-6	4-6	4-6
Solid wall height (in)	45	50	50-60
Working Chute			
Straight side (in)	18	22	28
Fully tapered – width at 32 in height (in)	18	22	28
Fully tapered – width at bottom (in)	15	16	18
Minimum length (ft)	20	20	20
Maximum curve angle (degrees)	15	15	15
Length for 16 foot outside radius (ft)	45	45	45
Solid wall height (in)	45	50	50-60
Overall height – top rail	55	60	60-72
Chute Fence			
Post Spacing (ft)	6	6	6
Post depth in ground (in)	36	36	36
Holding Chute/Squeeze			
Height (in)	45	50	50
Width			
Straight sides (in)	18	22	28
V-shaped sides, width at bottom (in)	6-8	8-12	14-16
Length – with head gate (ft)	5	5-8	5-8
Loading Chute			
Width (in)	26	26	26-30
Minimum length (ft)	12	12	12
Maximum rise (in/ft)	3.5	3.5	3.5
Radius of a curved chute (ft)	12-17	12-17	12-17
Spacing of 1X2-in hardwood cleats (in)	8	8	8

University of Kentucky Extension Bulletin AEN-82.

Figure 57-1 Diagram with the Components of a Good Cattle Working Facility.

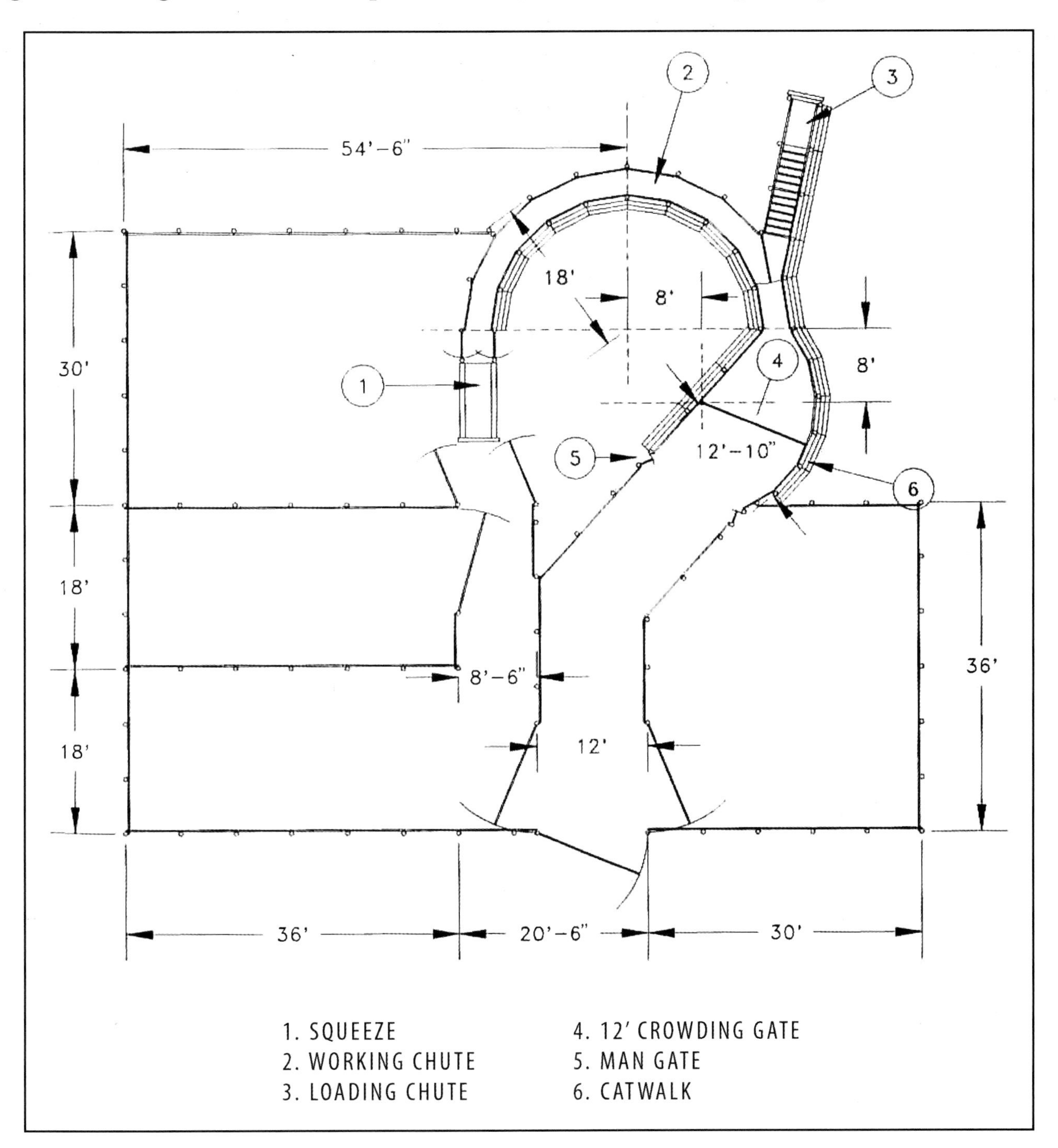

from backing. A catwalk on either the inside or outside of a circular holding alley will make it easier to move the cattle down the alley. Sloped sides help prevent balking. The recommended bottom width is 16 inches and the width at the top 28 inches. These widths may need to be a couple of inches wider for large-framed cattle.

- **Working Chute** – This chute should have a squeeze action that restrains the cattle while they are being worked on. It should be adjustable to allow it to be used for both cows and calves. Removable side panels are a nice feature because they allow easy access to the animal.

- **Head Gate** – There are numerous types of head gates that hold the animal between the head and shoulders.
- **Sorting Pens** – While this component may not be necessary for a small herd, having a couple of pens that cattle can be sorted into as they are released from the working chute can be valuable.

Other Possible Components

- **Cutting Gate in the Holding Alley** – A gate in the holding alley makes it easier to sort off cattle that do not need to be processed in the holding chute. Most of the time, it is easier to run the entire group through the crowding pen and sort them using this type of gate rather than trying to sort them in the holding pen.
- **Scales** – Scales can be part of the holding alley or part of the holding chute.
- **Service Gate behind the Holding Chute** – A walk-through gate behind the holding chute facilitates pregnancy testing, castration, and artificial insemination. This gate should have solid sides so the cattle in the holding alley cannot see the activity in front of them.
- **Loading Chute** – A loading chute off of the holding alley is a useful addition.

Figure 57-1 shows a diagram of a cattle working facility incorporating many of the features discussed above. This example may be too large for small operations, but the basic concepts can be applied to simpler designs. The following references show samples ranging from simple designs for small operations to designs for large feedlots.

References

Apple, K., R.L Huhnke, and S. Harp. Modern Corral Design. Oklahoma State University Cooperative Extension Service Publication E-938.

Beef Cattle Handling Facilities. 2004. Agriculture – Government of Saskatchewan Website.

Beef Housing and Equipment Handbook. 1987. Midwest Plans Service.

Bicudo, J., S. McNeill, and L. Turner, R. Burris, and J. Anderson. Cattle Handling Facilities: Planning, Components, and Layouts. University of Kentucky Extension Service Bulletin AEN-82.

Borg, R. 1993. Corrals for Handling Beef Cattle. Alberta Agriculture, Food and Rural Development Agdex 420/723-1.

Boyles, S., J. Fisher, and G. Fike. Cattle Handling and Working Facilities. The Ohio State University Extension Service Bulletin 906.

Grandin, T. Livestock Psychology and Handling-Facility Design. Great Plains Beef Cattle Handbook. GPE-5001.

Grandin, T. 1998. Cattle Handling Systems and Layout of Cattle Corrals and Races. Beef, Sept. Pg. 50-52.

Grandin, T. 1989. Behavioral Principles of Livestock Handling. Professional Animal Scientist. December. Pg 1-11.

Lane, C., R. Powell, B. White, and S. Glass. Handling Facilities for Beef Cattle. University of Tennessee Extension Bulletin SP690.

Chapter 58

Chemical Treatment of Roughages

Numerous universities and experiment stations have evaluated chemical treatment of roughages to improve their feed value. The most practical treatments include calcium oxide (CaO), calcium hydroxide ($Ca(OH)_2$) and anhydrous ammonia (NH_3). The calcium oxide and anhydrous ammonia combine with water to make calcium hydroxide ($Ca(OH)_2$) and ammonium hydroxide (NH_4OH), respectively. These strong bases act on the fiber breaking some of the linkages between lignin and hemicellulose. The result is an increase in digestibility of the fiber. Because of the increase in digestibility, intake of the treated roughage typically increases. The net result is a significant increase in the energy available to the animal from the roughage.

Several events have increased the interest in treating low-quality roughages to improve their nutritional value. First, the high price of and competition for corn has increased the search for alternative feedstuffs. Second, the frequent occurrence of droughts has increased the interest in using low-quality feedstuffs such as corn stover and wheat straw.

Consider treatment with calcium hydroxide ($Ca(OH)_2$) and anhydrous ammonia separately because the techniques used to treat the roughages are very different. In addition, the change in the roughage following treatment is somewhat different. For instance, with anhydrous ammonia, nitrogen is added to the roughage while it is not with calcium oxide or calcium hydroxide treatment.

Treating Roughages with Calcium Oxide or Calcium Hydroxide

The protocol for treating roughages with calcium oxide or calcium hydroxide consists of:

1. Grinding the roughage.
2. Dissolving the calcium oxide or calcium hydroxide in water to form a caustic solution.
3. Spraying the caustic solution on the roughage while adding water to achieve a 50 percent moisture level.
4. Storing the treated roughage anaerobically for at least 7 days. The roughage may be covered with plastic in a bunker or a stack or placed in an ag-bag to provide an anaerobic environment.

Unfortunately, there is considerable risk involved in this process. First, there is a tremendous amount of heat produced — the roughage may actually catch on fire. Consequently, adequate water must be available to put out a fire. Second, the caustic solution can cause damage if it gets on the skin or in the eyes. While the increase in the nutritional value of the roughage is increased markedly, some producers have made the decision to not treat with calcium oxide because of the human safety risk. In many cases producers are using calcium hydroxide rather than calcium oxide because it is safer.

Treatment Levels

The most common treatment level with calcium oxide has been 5 percent meaning that 100 pounds of calcium oxide is added per ton of dry roughage. Treating with calcium hydroxide requires 6.9 percent to get the same level of treatment obtained with 5 percent of calcium oxide.

Effect of Treatment with Calcium Oxide or Calcium Hydroxide on Digestibility

Table 58-1 shows the effect of treating with calcium oxide at the 5 percent level on in vitro dry

Table 58-1. Effects of Chemical Treatment of Roughages with Calcium Oxide on IVDMD.

Roughage	Digestibility, %		% Increase
	Control	5% CaO	
Corn Cobs	42.5[a]	54.8[b]	28.9
Wheat Straw	33.4[a]	50.2[b]	50.3
Corn Stalks	27.2[a]	40.0[b]	47.1

[ab]Within a row, values with different superscripts differ (P<.05). Nebraska Beef Report, 2011.

matter digestibility (IVDMD) in a Nebraska research trial. The treatment made a significant difference in the digestibility of all three low-quality roughages, which would result in much higher energy levels and probably an increase in intake.

Use of Calcium Oxide or Calcium Hydroxide Treated Corn Stalks in Finishing Diets

Researchers in both Iowa and Nebraska have evaluated replacing the roughage and some of the corn on finishing cattle performance. In these trials, they fed the treated corn stalks at 20 percent of the diet dry matter replacing all of the other roughage and approximately 10 percent of the corn. Table 58-2 shows the performance and carcass characteristics of the cattle in the Nebraska research trial. The cattle receiving the 20 percent treated corn stover performed similarly to the control cattle (fed 10 percent roughage)and had similar carcass characteristics. Moreover, as shown, they returned the same profit at $3.00 corn and much higher profit at $4.50 and $6.00 corn, respectively.

The research in Iowa also showed significant increases in profit per head for the cattle fed the 20 percent treated corn stover diet compared to the control.

Use of Calcium Oxide or Calcium Hydroxide Treated Corn Stalks in Growing Diets

Unfortunately, the effectiveness of calcium oxide treated roughages has not been adequately evaluated in typical growing programs. Based on the improved digestibility following treatment, calcium oxide treated residues should be effective in growing cattle rations.

Use of Calcium Oxide or Calcium Hydroxide Treated Corn Stalks in Beef Cow and Replacement Heifer Diets

Again, sufficient research has not been conducted to evaluate the usefulness of calcium oxide treated roughages in beef cow and replacement heifer diets. As with growing cattle, calcium oxide treated roughages should be effective in these diets.

Table 58-2. Performance and Carcass Characteristics of Finishing Cattle in a Trial Conducted at the University of Nebraska.

Item	Control	Corn Stover	
		Treated	Untreated
Initial BW, lbs	785	791	780
Final BW, lbs	1,313[a]	1,325[a]	1,267[b]
ADG, lbs	3.78[a]	3.83[a]	3.49[b]
DMI, lbs	25.8	26.1	25.1
Feed/Gain, lbs	6.83[a]	6.82[a]	7.18[b]
Profit-$3.00[c]	0.00	-0.05	-13.12
Profit-$4.50[c]	0.00	13.68	-6.70
Profit-$6.00[c]	0.00	27.33	-0.16

[ab]Within a row, values with different superscripts differ (P<.05). [c]At each price of corn, the profit for the control cattle was set to 0.00 for comparison purposes. Nebraska Beef Report 2012.

Treating Roughages with Anhydrous Ammonia

Anhydrous ammonia has been used to treat low-quality roughages and silages for many years. Ammoniation of dry, low quality roughages is an excellent way of increasing their feed value. It is also a great way to stretch feed supplies, especially when the production of high-quality roughage is limited by drought. Ammoniation of silages can also be practical in some situations.

Ammoniation of Dry Roughages

The Effect of Anhydrous Ammonia on Dry Roughages

Anhydrous ammonia attaches to the moisture in the forage forming a base, ammonium hydroxide, which then acts on the roughage, especially the fibrous fractions. The major effects of treatment with anhydrous ammonia include:

1. **Increased Crude Protein Content** – The protein content in the roughage is increased by approximately 4 percentage units using a standard treatment rate of 60 pounds of anhydrous ammonia per ton of forage. This means that with a typical wheat straw at 4 percent crude protein, treatment with anhydrous ammonia will essentially double the crude protein content. It should be noted that the additional protein is all non-protein nitrogen, which limits its usefulness.
2. **Increased Intake** – Cows will typically consume an additional 15 to 25 percent of an ammoniated feed compared to the same roughage not ammoniated. This increased intake is a function of both greater palatability and faster passage rate through the digestive system resulting from higher digestibility.
3. **Increased Digestibility** – As a general rule, ammoniation increases digestibility by 8 to 15 percent units.

Summary – While the increase in crude protein is of some value, the greatest benefit of ammoniation is the increase in the total energy available to the animal as a result of the increase in both intake and digestibility.

Performance of Beef Cows Fed Ammoniated Wheat Straw

Table 58-3 shows the results of an Oklahoma State University trial comparing untreated straw to ammoniated straw when fed to beef cows. It demonstrates the increase in digestibility, intake, and overall increase in energy intake as shown by the increase in weight gain.

Performance of Growing Cattle Fed Ammoniated Roughages

Ammoniated low-quality roughages, especially ammoniated wheat straw, have been evaluated in numerous research trials with growing cattle. As a general statement, performance has been reduced when the percentage of the ammoniated roughage has exceeded 25 percent of the ration. At higher levels, the ammoniated roughage reduces the energy level in the diet reducing performance significantly.

Table 58-3. Effect of Ammoniation on Mean Body Weight Gains and Condition Score Changes, and Straw Dry Matter Digestibility and Intake by Beef Cows.

Item	Untreated Straw	Ammoniated Straw
Straw CP, % DM Basis	4.2	8.7
Straw DM digestibility, %	47.0	56.6
Straw DM Intake		
Lb/day	12.3	15.1
% of Body Weight	1.50	1.81
ADG, lb	0.09	0.40
Condition score change	+.05	+.12

Oklahoma State University

Figure 58-1. Illustration Showing the Typical Technique Used to Treat Dry Roughages with Anhydrous Ammonia

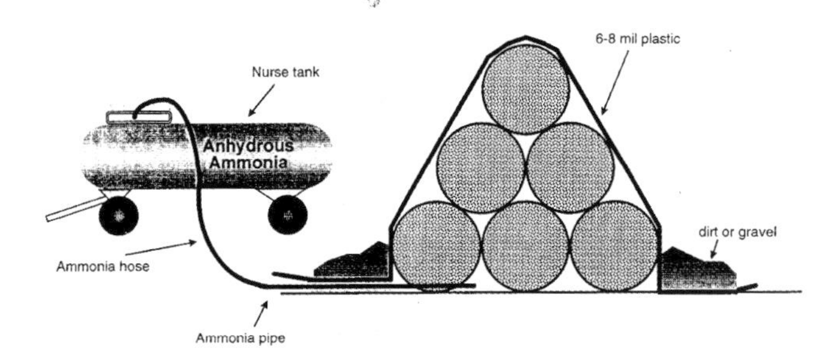

Treatment Technique, Storage, and Feeding

1. **Treatment Rate** – While several levels of treatment have been evaluated, the most common rate has been 3 percent, which is 60 pounds of anhydrous ammonia per ton of roughage. With a typical 5,000-gallon nurse tank, each 1 percent on the fill gauge is 50 pounds of anhydrous.

2. **Treatment Technique** – The material to be treated should be stacked on level ground since during treatment the liquid anhydrous ammonia flows to the lowest point and has a tendency to seep under the plastic creating a hazard. The stack is then covered with black plastic (usually 6 mil) using dirt to seal the plastic against the ground. Enough dirt should be used to make a good seal and hold the plastic in place for at least a few days. A single sheet of plastic 40 feet by 100 feet will cover a stack approximately 13 bales long.

   After the stack has been thoroughly sealed with dirt, a pipe from the anhydrous tank is inserted under the plastic and the area of insertion sealed with dirt (Figure 58-1). The anhydrous ammonia is slowly released under the plastic. The plastic balloons as a result of the anhydrous going into the gaseous state and from the heat generated by the reaction of the anhydrous ammonia with the water in the roughage. After a day or so, the pressure inside the plastic lessens. If the plastic becomes too loose, additional dirt can be added to tighten it and reduce the chance of the wind tearing the plastic.

   Almost any type of forage bale can be treated, but penetration in dense stacks or loafs may be limited. Penetration in large round or small square bales is usually good. A trench silo is also an effective treatment structure allowing large quantities of roughage to be treated using a single sheet of plastic.

3. **Duration of Treatment** – As with most chemical reactions, temperature has a major effect on the speed of the reaction. In warm weather, it only takes a few days to treat a stack of bales. In cooler weather, it may take a couple of weeks or a month to adequately treat the material. The roughage exhibits a dark brown color following treatment, and the surface of the upper bales usually is wet due to moisture migration and condensation.

4. **Importance of Uniform Moisture in the Roughage** – Since the anhydrous ammonia attaches to the available moisture, it is important that the moisture in the roughage is evenly dispersed to avoid uneven treatment. For example, the outsides of bales that have been rained on are treated to a much greater extent than the inside of the bale. This reduces the overall treatment of the roughage.

5. **Storage Following Ammoniation** – The chemical reaction that increases the digestibility of the roughages by ruminal microbes also makes the roughage more susceptible to attack by molds and other organisms if stored outside uncovered. In other words, the "storability" of the roughage is reduced, and the cost incurred in ammoniating roughages warrants taking extra care to protect them from weathering.

6. **Feeding** – Ammoniated forages can be fed in similar fashion to typical roughages such as in bale feeders or on the ground. However, ammoniated roughages are notorious for being difficult to grind, and an old hay grinder or one with dull blades may not be able to grind ammoniated roughages effectively.

Toxicity Problems with Ammoniated Roughages

There have been numerous reports of toxicity problems when feeding ammoniated roughages. These problems have occurred when medium-quality roughages were ammoniated and fed. While the exact mechanism for this toxicity is not understood, numerous compounds are formed during ammoniation when the roughage contains soluble sugars such as those found in medium-quality forages. Some of these compounds, when consumed, cause "crazy cow syndrome" or "bovine bonkers" where the animals exhibit extreme hyperexcitability. Symptoms include circling and convulsions. Affected cattle often run into fences, gates, posts, and other objects injuring themselves. Furthermore, it has been demonstrated that these compounds can be passed through the milk affecting calves while the cows appear normal.

This toxicity has been observed with a wide range of roughages with the only common characteristic being their higher quality. Thus, only poor quality roughages, i.e., straws and stovers, should be ammoniated. Aftermath feeds harvested after a grain crop has been removed should be safe. Very low-quality hays might be ammoniated without a problem, but the cows should be observed closely when the material is first fed. Ammoniated hays should not be fed to lactating cows. Diluting potential problem roughages with untreated roughages has been practiced with some success.

Feeding and Supplementing Ammoniated Dry Roughages

Characteristics of Ammoniated Forages

1. **Energy Content Based on Feed Analysis** – Ammoniated forages have special characteristics that must be considered, especially with respect to supplementation. For example, compare the analyses of wheat straw before and after ammoniation. A typical feed analysis for wheat straw might indicate 4.5 percent crude protein and an NEm of 43 Mcal per hundredweight. Following ammoniation, the analysis might be 9.0 percent crude protein and an NEm of 43 Mcal per hundredweight. Obviously, the same energy value for both does not make sense considering ammoniation increases the energy content of dry roughages as discussed previously. This illustrates a problem with current feed analysis techniques where energy values are based on acid detergent fiber (ADF) levels. Ammoniation does not change the percentage of acid detergent fiber, it simply makes the fiber more digestible, which results in more energy. Since the level of acid detergent fiber is the same in both the treated and untreated wheat straw, the laboratory analysis will show the same estimates of energy. A more appropriate NEm for the ammoniated wheat straw would be 50 Mcal per hundredweight. A good rule-of-thumb is to increase the NEm of ammoniated roughages by 20 percent over the analysis or book value for the same roughage not ammoniated.
2. **Value of Added Protein** – Another important point is the type of protein added during ammoniation. It is clearly nonprotein nitrogen, which as mentioned several times, is poorly used in a low-quality forage diet. Thus, while ammoniated wheat straw at 9 percent crude protein should have enough crude protein for a cow in late gestation, research has shown that these cows respond favorably to additional natural protein. A good rule-of-thumb is to only give credit for one-half of the added protein in ration formulation. For example, in the previous example, 4.5 percent units of protein were added during ammoniation. Using this rule-of-thumb, only 2.25 percent should be added resulting in an effective crude protein content of 6.75 percent.
3. **Mineral and Vitamin Content** – Since roughages suitable for ammoniation should be low-quality, they tend to have low mineral and vitamin levels. Consequently, a good mineral and vitamin supplementation program should accompany the feeding of ammoniated roughages.

Supplementation of Ammoniated Roughages

The special characteristics of ammoniated roughages make it difficult to formulate a supplementation program. Is the first limiting nutrient energy or protein? A study at Kansas State University (Table 58-3) compared grain to a higher-protein supplement fed to cows consuming ammoniated wheat straw in late gestation. The feeding period was essentially the last 60 days before calving.

The control cows received only ammoniated wheat straw. Milo was used as the grain and soybean meal was blended with milo to create the 20 percent crude protein supplement. All groups received a mineral/vitamin supplement to meet their requirements. The trial was conducted in drylot, which means that the energy required for activity was minimal. The ammoniated wheat straw came close to maintaining body condition on the control cows under these conditions. It should be noted, however, that weather stress during this trial was minimal.

All of the supplementation treatments increased cow weight gain and body condition score. In addition, even though the ammoniated wheat straw had a level of crude protein that should have been adequate, the cows responded to additional natural protein by gaining more weight. The 6 pounds of grain treatment and the 3 pounds of 20 percent crude protein treatment contained essentially the same amount of crude protein. However, the 6 pounds of grain treatment contained about twice the energy content.

A digestion trial showed that the 6 pounds of grain treatment reduced intake and digestibility of the ammoniated wheat straw compared to the other treatments. Thus, the additional energy intake from the grain was offset by a reduction in the energy from the AWS — a result similar to that observed when 5 to 6 pounds of grain is fed to cows grazing low-quality roughages.

The reproductive performance of cows on all treatments was similar in the subsequent breeding season and calf-weaning weights were similar. Thus, supplementation of the ammoniated wheat straw would not have been justified from a profitability standpoint. The cows in this study averaged a condition score 5.3 at the start of the study, and the average condition score of all treatments was at least a 5 following the 60-day treatment period. If the cows had entered the study in lower body condition or the weather had been more stressful, the weight and body condition changes might have influenced subsequent reproduction.

Ammoniation of Silages

Anhydrous ammonia also has been used to treat silages, but the rate of treatment and expected benefits differ markedly from those expected from treating dry roughages. For example, a typical rate of treatment for silages is 7 to 12 pounds per ton rather than the 60 pounds per ton applied to dry roughages. Moreover, the primary benefit of ammoniating silages is the increase in protein level with little effect on fiber digestibility or energy level in the roughage. Ammoniating silage does reduce spoilage and increase bunk life, however.

Toxicity problems have not been reported with ammoniated silages, probably because the treatment rate is much lower than that used with dry roughages.

Applying Anhydrous Ammonia to Silage

The most common method of treatment is to run the anhydrous ammonia through a Cold Flo® converter and apply the resulting liquid directly to the silage during cutting. Approximately 70 to 80 percent of the ammonia attaches to the wet silage if the application rate is no higher than 12 pounds per ton. Another method of application is to use an aqueous ammonia solution and apply it to the silage. Application of anhydrous reduces the rate of silage production per hour, which should be considered in making the decision to ammoniate or not.

Effect of Treating Silage with Anhydrous Ammonia

As noted, the main benefits of treating silage are to increase the protein content and reduce spoilage as

Table 58-3. Influence of Supplementation on Weight Change and Body Condition Score in Beef Cows Consuming Ammoniated Wheat Straw.

Item	Control	Supplement		
	No Suppl.	3 lbs Grain	6 lbs Grain	3 lbs 20% CP
Body Weight Change, lbs	47.4	74.7	100.8	96.8
Body Condition Score Change	-0.21	-0.11	0.19	0.16

Kansas State University

Table 58-4. Effects of Anhydrous Ammonia on Corn Silage – A Summary of Several Trials Conducted at Michigan State University.

Item	Lbs of Ammonia Per Ton		
	0	4	10
Ammonia Applied, %	0	0.2	0.5
Crude Protein, %[a]	7.8	10.3	13.0
Spoilage, %	14	12	7

[a]*DM Basis*

Table 58-5. Comparative Cost of Crude Protein for Several Protein Sources.

Protein Source	Price Per Ton, $	Crude Protein, %	Lbs CP Per Ton	Cost Per Lb of CP
Alfalfa Hay	230	19	380	$0.61
Soybean Meal	480	49	480	$0.49
Urea	400	281	5,620	$0.07
Anhydrous (75% Recovery)	1,067	512	10,240	$0.10

shown in Table 58-4. As shown, even relatively small amounts of anhydrous can raise the protein content significantly.

Feeding Value of the Nitrogen Added Via Anhydrous Ammonia

The additional protein in treated silages is obviously not "natural" protein and thus performance has been shown to be slightly lower than when soybean meal or another natural protein source is fed. However, performance has only been reduced slightly compared to natural protein and has been shown to be superior to the use of urea as a protein source. Thus, anhydrous ammonia may be an economical source of protein.

Economics of Treating Silage with Anhydrous Ammonia

There are several factors that influence the economics of treating silage with anhydrous ammonia including the effect on the amount of silage put up per hour and the cost of the equipment required to apply the anhydrous. However, the main consideration is the cost of the added protein compared to other sources. Table 58-5 shows the cost per pound of crude protein for several common protein sources. Recently the cost of anhydrous has increased markedly, meaning that the cost of crude protein from anhydrous must be compared to other protein sources.

Effect of Treating Silages with Anhydrous Ammonia on Storage Losses

It should be noted that research conducted at Kansas State University indicated that dry matter losses are higher in treated silage than in untreated silage. Conversely, the treated silage exhibited lower losses during feeding because of less spoilage. However, the loss of dry matter during storage was much greater than the reduction in spoilage losses during feeding.

Situations Where Ammoniation of Silages Should Be Considered

Since silages typically fed to beef cattle, like corn and forage sorghums, normally have 6 to 8 percent crude protein, minimal protein supplementation is required for nonlactating beef cows. On the other hand, lactating beef cows and growing cattle have higher protein requirements than commonly found in corn or forage sorghum silages. Thus, ammoniated silages have their greatest value in diets for these classes of cattle.

References

Beck, T.J., D.D. Simms, R.C. Cochran, R.T. Brandt, Jr., E.S. Vanzant, and G.L. Kuhl. 1992. Supplementation of Ammoniated Wheat Straw: Performance and Forage Utilization Characteristics in Beef Cattle Receiving Energy and Protein Supplements. JAS 70:349-357.

Bolsen, B. and H. Ilg. 1981. Ensila Plus and Cold-flo Additives for Corn Silage. KSU Cattlemen's Day Report.

Bolsen, B. and H. Ilg. 1981. Sila-bac, Cold-flo, and Sodium Hydroxide for Forage Sorghum Silage. KSU Cattlemen's Day Report.

Goering, H.K. and D.R. Walco. 1981. Ammonia Addition to Whole-Corn Plant at Ensiling. Proceedings off the Maryland Nutrition Conference for Feed Manufacturers.

Horn, G.W., C.L. Streeter, K.S. Lusby, D.W. Pace, and J. Zorrilla-Rios. 1983. "Stack Method" of Ammoniation of Wheat Straw and Effects on Performance of Cows and Growing Steers. Proceedings O-K Cattle Conference.

Kuhl, G.L. 1981. Ammoniation of Dry Forages for Beef Cattle. 7th Annual O-K Beef Cattle Conference Proceedings.

Kuhl, G., C. Carlson, G. Williamson, and B. Jurgensen. 1981. Effectiveness of Cold-Flo Anhydrous Ammonia with Forage Sorghum Silage. South Dakota State University Cattle Feeders Day Report.

Russell, J., D. Loy, J. Anderson, and M.J. Cecava. 2011. Potential of Chemically Treated Corn Stover and Modified Distillers Grains as a Partial Replacement for Corn Grain in Feedlot Diets. Iowa State University Animal Industry Report.

Shreck, A.L., C.D. Buckner, G. E. Erickson, T.J. Klopfenstein, and M.J. Cecava. 2011. Digestibility of Crop Residues after Chemical Treatment and Anaerobic Storage. Nebraska Beef Report.

Shreck, A.L., B.L. Nuttelman, W. A. Griffin, G. E. Erickson, T.J. Klopfenstein, and M.J. Cecava. 2012. Chemical Treatment of Low-quality Forages to Replace Corn in Cattle Finishing Diets. Nebraska Beef Report.

Troxel, T.R. Hay Ammoniation. Great Plains Beef Cattle Handbook Factsheet GPE-2150.

Weiss, W.P., H.R. Conrad, C.M. Martin, R.F. Cross, and W.L. Shocky. 1986. Etiology of Ammoniated Hay Toxicosis. JAS 63:525-532.

Chapter 59

Industry Issues

Animal Rights

While the primary focus of the major animal rights organization, The Humane Society of the United States (HSUS), has been on the poultry and swine industries, there can be little doubt that their stated goals will eventually make the beef industry a higher priority target. To understand the animal rights movement, it must be distinguished from the animal welfare movement. Animal welfare organizations support the humane treatment and use of animals. They recognize animal ownership and the responsibilities that it entails. They believe that animal cruelty is wrong. Conversely, animal rights advocates believe that animals have rights equal to humans. Thus, animals should not be used for food, clothing, research, or as pets. The animal rights movement is well-funded and understands how to use political pressure to promote their agenda.

The Humane Society of the United States was originally an animal welfare organization supporting local animal shelters. In recent years it has moved closer to embracing the animal rights philosophy. It has assets of over $160 million dollars to promote its activities. People who think HSUS supports local humane societies in caring for abandoned pets donated much of this money, when in fact little of HSUS's budget is given to local shelters. Many supporters of HSUS are vegetarians and have a "vegan" agenda. The bulk of HSUS's budget is spent on lobbying to pass legislation aimed at changing "factory farming."

Beef Safety

There are two major concerns regarding the safety of beef:

1. Contamination with *E. Coli* O157:H7
2. Bovine Spongiform Encephalopathy (BSE) commonly called "Mad Cow" disease.

E. coli O157

E. coli O157:H7 is one of hundreds of strains of *E.coli* that live in the intestines of healthy animals and humans, most of which are harmless. However, *E. coli* O157:H7 can cause serious illness in humans. According to the Centers for Disease Control and Prevention, the number of cases of foodborne illness causes by *E. coli* O157:H7 has declined markedly in the past 10 years. This reduction has resulted primarily from cleaner slaughter methods, increased microbial testing, and education of food service personnel and consumers about the risk of consuming undercooked, ground beef.

Bovine Spongiform Encephalopathy (BSE)

BSE is a neurological disease thought to be caused by a misfolded protein called a prion. It is not passed from animal to animal or from animals to humans by contact. It is passed through consumption of infected tissue. The prion is found primarily in the brain and spinal tissue. Preventative measures have greatly reduced the potential for the passage of the prion. For example, in 1997 a ban went into effect prohibiting the inclusion of ruminant meat and bone meal in beef feeds. In 2003 the first case of "mad cow" disease was reported in the United States in a cow in Washington that had been imported from Canada. Because of numerous human deaths in Britain from eating contaminated beef and sheep, the identification of "mad cow" disease in the United States depressed the cattle market for several months. Since the initial case was reported, a few other cases have been reported, but it appears that the preventative measures taken to avoid the disease have been effective.

Health Concerns

Over the years, there have been claims that eating beef causes heart disease and some types of cancer. Consider examining these claims more closely.

Beef Consumption and Heart Disease

Major risk factors related to coronary heart disease are smoking, genetics, high blood cholesterol, hypertension, lack of physical activity, and obesity. The primary basis for claiming that beef consumption increases the risk of developing coronary heart disease is the presence of saturated fat in beef. As a general rule, saturated fats tend to cause increases in blood cholesterol levels while unsaturated fats (the predominate fats in plants) tend to lower cholesterol levels. However, while this is generally true, some saturated fats have a neutral effect on blood cholesterol levels. For example, one-third of the saturated fat in beef is stearic acid, which has been shown to have a neutral effect on cholesterol levels. Furthermore, roughly one-half of the fat in beef is unsaturated. A meta-analysis of 21 studies looking at the relationship between saturated fat intake and cardiovascular disease conducted at Children's Hospital Oakland Research Institute, Oakland, CA showed that there was no relationship between saturated fat intake and cardiovascular disease. In addition, a meta-analysis conducted at Harvard School of Public Health indicated no relationship between eating red meat and cardiovascular disease. However, there did appear to be an increase in cardiovascular disease with consumption of high levels of processed meats like hot dogs and bacon.

Currently, Americans eat an average of less than 3 ounces of beef per day. At this level of consumption, it is unlikely that beef is having a major influence on blood cholesterol levels and coronary heart disease. The other risk factors like genetics, lack of activity, and obesity are certainly having a greater influence on high blood cholesterol and coronary heart disease than beef.

Beef Consumption and Cancer

Some studies have shown a trend for excessive consumption of red meat to increase the risk of cancer, especially colon cancer. Epidemiological studies looking for a cause-and-effect relationship are difficult to interpret because of the many factors that must be considered and the interactions between these factors. In other words, trying to determine that a single dietary item causes cancer is difficult when considering all the dietary items and lifestyle factors that may influence the risk of cancer. While there may be a link between beef consumption and colon cancer, the increased risk is extremely low.

Environmental Issues

Livestock's Contribution to Greenhouse Gas Emissions

In 2006, the Food and Agriculture Organization of the United Nations (FAO) published *Livestock's Long Shadow – Environmental Issues and Options.* In this publication, livestock were blamed for producing 18 percent of global greenhouse gas (GHG) emissions, which was more than the transportation industry. Reviews of this study have shown it to be flawed in several ways; however, this number is still being quoted by those critical of the livestock industry. The EPA estimates that livestock production accounts for 2.8 percent of the greenhouse gas emissions in the United States – much less than the transportation industry. Furthermore, greenhouse gas emissions by the livestock industry have been declining because fewer animals (especially beef and dairy) are in the U.S. herds than in the past. In other words, improvements in productivity have decreased greenhouse gas emissions by the livestock industry in the United States.

Grazing on Public Lands

Historically, cattle producers have been allowed to graze on public lands, especially in the western United States. Use of these lands for grazing is coming under pressure as other users of these lands view grazing as detrimental to the environment. As an example, the Forest Service has for years operated "under the principles of the Multiple-Use, Sustained-Yield Act of 1960, that set out the process for development and revision of the land management plans and guidelines and standards." However, in March 2012, the Forest Service finalized a new forest service rule that generally focuses on and elevates ecosystem services, sustainability, preservation, and even "spiritual values" over multiple-use. It is clear that use of public lands for grazing will be under pressure in coming years.

Beef Quality Issues

While a good steak is at the top of many people's idea of a great meal, a tough or tasteless steak is a bad experience, especially considering the cost of beef. Unsatisfactory eating experiences with beef occur way too often. Over the past 20 years there has been an effort to identify the issues and address them. Central to this effort has been a series of Beef Quality Audits with the first one conducted in 1991 and the most recent in 2011. Each audit begins with a series of questionnaires sent to all segments of the beef industry. In addition to surveying the attitudes of segments of the industry, data are collected on a large number of carcasses to characterize the carcasses being produced.

Table 59-1 shows the results of the survey of purveyors, retailers, and restaurateurs for all five NBQAs. These data show that the industry has made some progress, but some problems still exist. For example, in 1991

Table 59-1. Greatest Quality Challenges for Purveyors, Retailers, and Restaurateurs as Reported in the National Beef Quality Audits of 1991, 1995, 2000, and 2005.

Rank	NBQA-1991	NBQA-1995	NBQA-2000	NBQA-2005	NBQA-2011
1	External Fat	Overall uniformity	Overall uniformity	Traceability	Food Safety
2	Seam fat	Overall palatability	Carcass weights	Overall uniformity	Eating Satisfaction
3	Overall palatability	Marbling	Tenderness	Instrument grading	How and where cattle were raised
4	Tenderness	Tenderness	Marbling	Market signals	Lean, Fat, Bone
5	Overall cutability	External and Seam Fat	Reduced quality due to Implanting	Segmentation	Weight and size
6	Marbling	Cut weights	External Fat	Carcass weights	Cattle genetics

Adapted from the Executive Summary of the 2011 National Beef Quality Audit.

Table 59-2. Ranking of Quality Categories by Sector (Relative Importance of Specified Quality Categories).

Quality Issue	Government & Allied Industry	Feeders	Packers	Foodservice/ Distributors/ Further Processors	Retailers	Overall
Food safety	25%	11%	35%	42%	39%	28%
Eating satisfaction	24%	9%	20%	24%	29%	20%
Cattle genetics	14%	15%	7%	1%	3%	9%
Weight and size	10%	19%	7%	7%	5%	11%
How and Where the cattle were raised	9%	22%	12%	10%	10%	13%
Visual characteristics	9%	9%	6%	7%	10%	8%
Lean, fat, and bone	9%	15%	13%	10%	5%	11%

Adapted from the Executive Summary of the 2011 National Beef Quality Audit.

Table 59-3. Lost Opportunities Per Head Due to Nonconformance with Ideal Targets for Quality (NBQA 2011).

Quality Issue	Value
Quality Grade	($25.25)
Yield Grade	($ 5.77)
Carcass Weight	($ 6.75)
Hide/Branding	($ 0.74)
Offal	($ 5.15)
Total	($43.66)

Adapted from the Executive Summary of the 2011 National Beef Quality Audit.

external fat was the number one concern whereas in 2011 it was not even on the list of quality concerns. In addition, injection sites and bruise damage have dropped out of the top ten challenges. However, carcass weights were a major concern in the 2005 audit, and they have increased in recent years. These heavy carcass weights lead to ribeyes that are too large, often forcing supermarkets to cut steaks too thin to keep the portion size down. Thin steaks are harder to cook properly causing tough, dry steaks in too many cases.

Table 59-2 shows the ranking of quality issues by sector. As might be expected, sectors differed in how they ranked quality issues. In many cases these differences result in mixed signals between segments and the overall industry.

Table 59-3 shows the amount lost per head because of nonconformance with ideal targets as identified in the National Beef Quality Audit 2011. As shown by far the most significant loss results from cattle with low-quality grade.

References

2011 National Beef Quality Audit. Website: *BQA.org*.

Critical Analysis of Livestock's Long Shadow. Website: *www.ExploreBeef.org*

Steinfeld, H, P. Gerber, T. Wassenaar, V. Castel, M. Rosales, and C. deHaan. 2006. Livestock's Long Shadow – Environmental Issues and Options. Food and Agriculture Organization (FAO).

Appendix Tables

Contents

Appendix Table 1. Nutrient Requirements of Mature Beef Cows to Maintain Moderate Body Condition

Months Since Calving	Milk (lbs./day)	Diet Nutrient Density						Daily Nutrients per Animal					
		DMI (lbs./day)	TDN (% DM)	NEm (Mcal/lb.)	CP (% DM)	Ca (% DM)	P (% DM)	TDN (lbs.)	NEm (Mcal)	CP (lbs.)	Ca (lb.)	P (lb.)	Vit. A (1,000s IU)
900-lb. Mature Weight													
10-lb. Peak Milk													
1	8.3	20.2	56.2	0.55	8.9	0.24	0.17	11.4	11.1	1.79	0.049	0.033	36
2	10.0	20.6	57.0	0.56	9.3	0.25	0.17	11.7	11.7	1.92	0.053	0.035	36
3	9.0	21.4	54.8	0.53	8.6	0.23	0.16	11.7	11.3	1.84	0.051	0.035	38
4	7.2	21.0	53.8	0.51	8.1	0.22	0.15	11.3	10.8	1.70	0.046	0.031	37
5	5.4	20.5	52.8	0.50	7.7	0.20	0.14	10.8	10.2	1.57	0.042	0.029	36
6	3.9	20.2	52.0	0.49	7.2	0.18	0.13	10.5	9.8	1.45	0.037	0.026	36
20-lb. Peak Milk													
1	16.7	22.6	60.2	0.61	10.8	0.30	0.20	13.6	13.8	2.44	0.068	0.044	40
2	20.0	23.5	61.6	0.64	11.5	0.33	0.21	14.5	14.9	2.70	0.077	0.051	42
3	18.0	23.8	59.2	0.60	10.7	0.30	0.20	14.1	14.3	2.54	0.073	0.046	42
4	14.4	22.9	57.6	0.57	9.9	0.28	0.18	13.2	13.1	2.26	0.064	0.042	41
5	10.8	21.9	55.8	0.55	9.1	0.25	0.17	12.2	12.0	1.99	0.055	0.037	39
6	7.8	21.1	54.3	0.52	8.3	0.22	0.15	11.5	11.0	1.76	0.046	0.033	37
30-lb. Peak Milk													
1	25.0	24.9	63.6	0.67	12.4	0.36	0.23	15.8	16.5	3.09	0.088	0.057	44
2	30.0	26.3	65.4	0.69	13.2	0.39	0.24	17.2	18.2	3.48	0.101	0.064	47
3	27.0	26.3	62.9	0.65	12.3	0.36	0.23	16.5	17.2	3.24	0.095	0.060	47
4	21.6	24.8	60.8	0.62	11.4	0.33	0.21	15.1	15.5	2.83	0.082	0.053	44
5	16.2	23.4	58.5	0.59	10.3	0.29	0.19	13.7	13.7	2.41	0.068	0.044	41
6	11.7	22.2	56.4	0.56	9.3	0.25	0.17	12.5	12.3	2.06	0.057	0.037	39

Appendix Table 1 modified from University of Arkansas Extension Bulletin MP391. Appreciation is expressed to Dr. Shane Gadberry for permissioin to use this table.

Appendix Table 1 continues on page 307.

Appendix Table 1. Nutrient Requirements of Mature Beef Cows to Maintain Moderate Body Condition (Continued)

Months Since Calving	Milk (lbs./day)	Diet Nutrient Density						Daily Nutrients per Animal					
		DMI (lbs./day)	TDN (% DM)	NEm (Mcal/lb.)	CP (% DM)	Ca (% DM)	P (% DM)	TDN (lbs.)	NEm (Mcal)	CP (lbs.)	Ca (lb.)	P (lb.)	Vit. A (1,000s IU)
900-lb. Mature Weight													
10, 20, 30-lb. Peak Milk													
7	0.0	18.0	46.8	0.40	6.5	0.15	0.12	8.4	7.2	1.17	0.028	0.022	23
8	0.0	18.3	47.2	0.41	6.6	0.15	0.12	8.6	7.5	1.20	0.028	0.022	23
9	0.0	18.7	47.8	0.42	6.7	0.15	0.12	8.9	7.9	1.25	0.028	0.022	24
10	0.0	19.4	48.8	0.43	6.9	0.24	0.15	9.5	8.3	1.34	0.046	0.029	25
11	0.0	19.4	51.8	0.48	7.6	0.24	0.15	10.0	9.3	1.47	0.046	0.029	25
12	0.0	19.7	55.5	0.54	8.6	0.24	0.15	10.9	10.6	1.70	0.046	0.029	25
1,000-lb. Mature Wt.													
10-lb. Peak Milk													
1	8.3	21.6	55.8	0.55	8.7	0.24	0.17	12.1	11.8	1.88	0.051	0.035	38
2	10.0	22.1	56.6	0.56	9.1	0.25	0.17	12.5	12.3	2.01	0.055	0.037	39
3	9.0	23.0	54.3	0.52	8.4	0.23	0.16	12.5	12.0	1.93	0.053	0.037	41
4	7.2	22.5	53.4	0.51	8.0	0.22	0.15	12.0	11.5	1.80	0.049	0.033	40
5	5.4	22.1	52.5	0.49	7.5	0.20	0.14	11.6	10.9	1.66	0.044	0.031	39
6	3.9	21.7	51.8	0.48	7.1	0.19	0.14	11.2	10.5	1.55	0.040	0.029	38
20-lb. Peak Milk													
1	16.7	24.0	59.6	0.60	10.5	0.30	0.20	14.3	14.5	2.53	0.073	0.049	42
2	20.0	25.0	60.9	0.62	11.2	0.32	0.21	15.2	15.6	2.79	0.079	0.053	44
3	18.0	25.4	58.6	0.59	10.4	0.30	0.19	14.9	15.0	2.64	0.075	0.049	45
4	14.4	24.4	57.0	0.56	9.7	0.27	0.18	13.9	13.8	2.36	0.066	0.044	43
5	10.8	23.5	55.4	0.54	8.9	0.24	0.17	13.0	12.7	2.08	0.057	0.040	42
6	7.8	22.7	54.0	0.52	8.2	0.22	0.15	12.3	11.7	1.85	0.051	0.035	40

Appendix Table 1 continues on page 308.

Appendix Table 1. Nutrient Requirements of Mature Beef Cows to Maintain Moderate Body Condition (Continued)

Months Since Calving	Milk (lbs./day)	Diet Nutrient Density: DMI (lbs./day)	TDN (% DM)	NEm (Mcal/lb.)	CP (% DM)	Ca (% DM)	P (% DM)	Daily Nutrients per Animal: TDN (lbs.)	NEm (Mcal)	CP (lbs.)	Ca (lb.)	P (lb.)	Vit. A (1,000s IU)
1,000-lb. Mature Wt.													
10-lb. Peak Milk													
1	25.0	26.4	62.8	0.65	12.1	0.35	0.22	16.6	17.2	3.18	0.093	0.060	47
2	30.0	27.8	64.5	0.68	12.9	0.38	0.24	17.9	18.9	3.57	0.104	0.066	49
3	27.0	27.8	62.1	0.64	12.0	0.35	0.22	17.3	17.9	3.34	0.097	0.062	49
4	21.6	26.4	60.1	0.61	11.1	0.32	0.21	15.9	16.1	2.92	0.084	0.055	47
5	16.2	24.9	57.9	0.58	10.0	0.28	0.19	14.4	14.4	2.50	0.071	0.046	44
6	11.7	23.7	55.9	0.55	9.1	0.25	0.17	13.2	13.0	2.16	0.060	0.040	42
10, 20, 30-lb. Peak Milk													
7	0.0	19.5	46.8	0.40	6.5	0.16	0.12	9.1	7.8	1.26	0.031	0.024	25
8	0.0	19.8	47.2	0.41	6.6	0.16	0.12	9.3	8.1	1.30	0.031	0.024	25
9	0.0	20.3	47.9	0.42	6.7	0.16	0.12	9.7	8.5	1.35	0.031	0.024	26
10	0.0	21.1	48.9	0.44	6.9	0.24	0.15	10.3	9.3	1.45	0.051	0.032	27
11	0.0	21.0	52.1	0.49	7.7	0.24	0.15	10.9	10.3	1.61	0.051	0.032	27
12	0.0	21.4	55.9	0.55	8.7	0.24	0.15	12.0	11.8	1.86	0.051	0.032	27
1,100-lb. Mature Wt.													
10-lb. Peak Milk													
1	8.3	23.1	55.6	0.54	8.5	0.24	0.17	12.8	12.5	1.97	0.055	0.037	41
2	10.0	23.5	56.3	0.55	8.9	0.25	0.17	13.2	13.0	2.10	0.060	0.040	42
3	9.0	24.5	54.1	0.52	8.2	0.23	0.16	13.3	12.7	2.02	0.055	0.040	43
4	7.2	24.1	53.2	0.50	7.8	0.21	0.15	12.8	12.1	1.89	0.051	0.037	43
5	5.4	23.6	52.3	0.49	7.4	0.20	0.14	12.3	11.6	1.75	0.046	0.033	42
6	3.9	23.3	51.6	0.48	7.0	0.19	0.14	12.0	11.2	1.64	0.044	0.031	41

Appendix Table 1 continues on page 309.

Appendix Table 1. Nutrient Requirements of Mature Beef Cows to Maintain Moderate Body Condition (Continued)

Months Since Calving	Milk (lbs./day)	Diet Nutrient Density						Daily Nutrients per Animal					
		DMI (lbs./day)	TDN (% DM)	NEm (Mcal/lb.)	CP (% DM)	Ca (% DM)	P (% DM)	TDN (lbs.)	NEm (Mcal)	CP (lbs.)	Ca (lb.)	P (lb.)	Vit. A (1,000s IU)
1,100-lb. Mature Wt. 20-lb. Peak Milk													
1	16.7	25.4	59.1	0.60	10.3	0.29	0.20	15.0	15.2	2.62	0.075	0.051	45
2	20.0	26.4	60.4	0.62	10.9	0.31	0.21	15.9	16.3	2.88	0.084	0.055	47
3	18.0	26.9	58.1	0.58	10.2	0.29	0.19	15.6	15.6	2.73	0.077	0.051	48
4	14.4	26.0	56.6	0.56	9.4	0.27	0.18	14.7	14.5	2.45	0.068	0.046	46
5	10.8	25.0	55.0	0.53	8.7	0.24	0.17	13.8	13.3	2.17	0.060	0.042	44
6	7.8	24.2	53.7	0.51	8.1	0.22	0.15	13.0	12.4	1.95	0.053	0.037	43
1,100-lb. Mature Wt. 30-lb. Peak Milk													
1	25.0	27.8	62.2	0.64	11.8	0.34	0.22	17.3	17.9	3.27	0.095	0.062	49
2	30.0	29.2	63.9	0.67	12.5	0.37	0.24	18.7	19.5	3.66	0.108	0.068	52
3	27.0	29.4	61.5	0.63	11.7	0.34	0.22	18.1	18.6	3.43	0.099	0.064	52
4	21.6	27.9	59.5	0.60	10.8	0.31	0.20	16.6	16.8	3.01	0.086	0.057	49
5	16.2	26.4	57.4	0.57	9.8	0.28	0.19	15.2	15.1	2.59	0.073	0.049	47
6	11.7	25.3	55.6	0.54	8.9	0.25	0.17	14.1	13.7	2.25	0.062	0.042	45
10-, 20-, 30-lb. Peak Milk													
7	0.0	20.9	46.8	0.40	6.5	0.16	0.12	9.8	8.4	1.36	0.033	0.026	27
8	0.0	21.2	47.2	0.41	6.6	0.16	0.12	10.0	8.7	1.40	0.033	0.026	27
9	0.0	21.8	47.9	0.42	6.7	0.16	0.12	10.4	9.2	1.45	0.033	0.026	28
10	0.0	22.6	48.9	0.44	7.0	0.25	0.16	11.1	9.9	1.56	0.057	0.036	29
11	0.0	22.5	52.1	0.49	7.7	0.25	0.16	11.7	11.0	1.73	0.057	0.036	29
12	0.0	23.0	56.0	0.55	8.7	0.25	0.16	12.9	12.7	2.00	0.057	0.036	29

Appendix Table 1 continues on page 310.

Appendix Table 1. Nutrient Requirements of Mature Beef Cows to Maintain Moderate Body Condition (Continued)

Months Since Calving	Milk (lbs./day)	Diet Nutrient Density: DMI (lbs./day)	TDN (% DM)	NEm (Mcal/lb.)	CP (% DM)	Ca (% DM)	P (% DM)	Daily Nutrients per Animal: TDN (lbs.)	NEm (Mcal)	CP (lbs.)	Ca (lb.)	P (lb.)	Vit. A (1,000s IU)
1200-lb. Mature Wt.													
10-lb. Peak Milk													
1	8.3	24.4	55.3	0.54	8.4	0.24	0.17	13.5	13.1	2.06	0.057	0.040	43
2	10.0	24.9	56.0	0.55	8.8	0.25	0.17	13.9	13.7	2.19	0.620	0.042	44
3	9.0	26.0	53.7	0.51	8.1	0.23	0.16	14.0	13.4	2.12	0.060	0.042	46
4	7.2	25.6	52.9	0.50	7.7	0.21	0.15	13.5	12.8	1.98	0.055	0.040	45
5	5.4	25.1	52.1	0.49	7.3	0.20	0.14	13.1	12.2	1.84	0.051	0.035	44
6	3.9	24.8	51.5	0.48	7.0	0.19	0.14	12.8	11.8	1.73	0.046	0.035	44
20-lb. Peak Milk													
1	16.7	26.8	58.7	0.59	10.1	0.29	0.19	15.7	15.8	2.71	0.077	0.053	47
2	20.0	27.8	59.9	0.61	10.7	0.31	0.21	16.7	16.9	2.97	0.086	0.057	49
3	18.0	28.4	57.6	0.57	9.9	0.29	0.19	16.4	16.3	2.82	0.082	0.055	50
4	14.4	27.4	56.2	0.55	9.3	0.26	0.18	15.4	15.1	2.54	0.073	0.049	48
5	10.8	26.5	54.7	0.53	8.5	0.24	0.17	14.5	14.0	2.26	0.064	0.044	47
6	7.8	25.7	53.4	0.51	7.9	0.22	0.15	13.7	13.1	2.04	0.055	0.040	45
30-lb. Peak Milk													
1	25.0	29.2	61.6	0.64	11.5	0.34	0.22	18.0	18.6	3.36	0.099	0.064	52
2	30.0	30.6	63.2	0.66	12.3	0.36	0.23	19.3	20.2	3.75	0.110	0.071	54
3	27.0	30.8	60.8	0.62	11.4	0.34	0.22	18.7	19.2	3.52	0.104	0.066	54
4	21.6	29.4	59.0	0.59	10.6	0.31	0.20	17.3	17.5	3.10	0.090	0.060	52
5	16.2	27.9	57.0	0.56	9.6	0.27	0.18	15.9	15.8	2.68	0.077	0.051	49
6	11.7	26.7	55.2	0.54	8.8	0.25	0.17	14.7	14.4	2.34	0.066	0.046	47

Appendix Table 1 continues on page 311.

Appendix Table 1. Nutrient Requirements of Mature Beef Cows to Maintain Moderate Body Condition (Continued)

Months Since Calving	Milk (lbs./day)	Diet Nutrient Density						Daily Nutrients per Animal					
		DMI (lbs./day)	TDN (% DM)	NEm (Mcal/lb.)	CP (% DM)	Ca (% DM)	P (% DM)	TDN (lbs.)	NEm (Mcal)	CP (lbs.)	Ca (lb.)	P (lb.)	Vit. A (1,000s IU)
1200-lb. Mature Wt.													
10, 20, 30-lb. Peak Milk													
7	0.0	22.4	46.9	0.40	6.5	0.16	0.13	10.5	9.0	1.45	0.037	0.029	28
8	0.0	22.8	47.3	0.41	6.5	0.16	0.13	10.8	9.3	1.49	0.037	0.029	29
9	0.0	23.3	47.9	0.42	6.7	0.16	0.13	11.2	9.8	1.56	0.037	0.029	30
10	0.0	24.3	49.0	0.44	6.9	0.25	0.16	11.9	10.7	1.67	0.062	0.040	31
11	0.0	24.1	52.3	0.49	7.7	0.25	0.16	12.6	11.8	1.86	0.062	0.040	31
12	0.0	24.6	56.2	0.55	8.8	0.25	0.16	13.8	13.5	2.16	0.062	0.040	31
1,300-lb. Mature Wt.													
10-lb. Peak Milk													
1	8.3	25.8	55.1	0.53	8.3	0.23	0.17	14.2	13.8	2.15	0.060	0.042	46
2	10.0	26.3	55.8	0.55	8.7	0.25	0.17	14.7	14.3	2.28	0.064	0.046	47
3	9.0	27.5	53.5	0.51	8.0	0.23	0.16	14.7	14.0	2.20	0.062	0.044	49
4	7.2	27.0	52.7	0.50	7.6	0.21	0.15	14.2	13.4	2.06	0.057	0.042	48
5	5.4	26.6	52.0	0.49	7.3	0.20	0.15	13.8	12.9	1.93	0.053	0.040	47
6	3.9	26.3	51.4	0.48	6.9	0.19	0.14	13.5	12.5	1.82	0.051	0.037	47
20-lb. Peak Milk													
1	16.7	28.2	58.4	0.59	9.9	0.29	0.19	16.5	16.5	2.80	0.082	0.055	50
2	20.0	29.1	59.5	0.60	10.5	0.31	0.21	17.3	17.6	3.06	0.088	0.060	51
3	18.0	29.9	57.2	0.57	9.7	0.28	0.19	17.1	16.9	2.91	0.084	0.057	53
4	14.4	28.9	55.9	0.55	9.1	0.26	0.18	16.2	15.8	2.63	0.075	0.051	51
5	10.8	28.0	54.4	0.52	8.4	0.24	0.17	15.2	14.6	2.35	0.066	0.046	50
6	7.8	27.2	53.3	0.51	7.8	0.22	0.15	14.5	13.7	2.13	0.060	0.042	48

Appendix Table 1 continues on page 312.

Appendix Table 1. Nutrient Requirements of Mature Beef Cows to Maintain Moderate Body Condition (Continued)

Months Since Calving	Milk (lbs./day)	Diet Nutrient Density: DMI (lbs./day)	TDN (% DM)	NEm (Mcal/lb.)	CP (% DM)	Ca (% DM)	P (% DM)	Daily Nutrients per Animal: TDN (lbs.)	NEm (Mcal)	CP (lbs.)	Ca (lb.)	P (lb.)	Vit. A (1,000s IU)
1,300-lb. Mature Wt.													
30-lb. Peak Milk													
1	25.0	30.6	61.2	0.63	11.3	0.33	0.22	18.7	19.2	3.45	0.101	0.066	54
2	30.0	32.0	62.7	0.65	12.0	0.36	0.23	20.1	20.8	3.84	0.115	0.073	57
3	27.0	32.3	60.3	0.62	11.2	0.33	0.21	19.5	19.9	3.61	0.106	0.068	57
4	21.6	30.8	58.5	0.59	10.4	0.30	0.20	18.1	18.1	3.19	0.093	0.062	55
5	16.2	29.4	56.6	0.56	9.4	0.27	0.18	16.6	16.4	2.77	0.079	0.055	52
6	11.7	28.2	55.0	0.53	8.6	0.24	0.17	15.5	15.0	2.43	0.068	0.049	50
10-, 20-, 30-lb. Peak Milk													
7	0.0	23.8	46.9	0.40	6.5	0.17	0.13	11.2	9.5	1.54	0.040	0.031	30
8	0.0	24.2	47.3	0.41	6.5	0.17	0.13	11.4	9.9	1.58	0.040	0.031	31
9	0.0	24.8	48.0	0.42	6.7	0.17	0.13	11.9	10.4	1.66	0.040	0.031	31
10	0.0	25.8	49.0	0.44	6.9	0.26	0.16	12.6	11.4	1.78	0.066	0.042	32
11	0.0	25.6	52.5	0.49	7.8	0.26	0.16	13.4	12.5	1.99	0.066	0.042	33
12	0.0	26.1	56.5	0.56	8.9	0.26	0.16	14.7	14.6	2.31	0.066	0.042	33
1,400-lb. Mature Wt.													
10-lb. Peak Milk													
1	8.3	27.1	54.9	0.53	8.2	0.23	0.17	14.9	14.4	2.23	0.064	0.046	48
2	10.0	27.6	55.5	0.54	8.6	0.25	0.17	15.3	15.0	2.36	0.068	0.049	49
3	9.0	28.9	53.3	0.51	7.9	0.23	0.16	15.4	14.6	2.29	0.066	0.046	51
4	7.2	28.5	52.5	0.49	7.6	0.21	0.15	15.0	14.1	2.15	0.062	0.044	50
5	5.4	28.0	51.8	0.48	7.2	0.20	0.15	14.5	13.5	2.02	0.057	0.042	50
6	3.9	27.7	51.2	0.47	6.9	0.19	0.14	14.2	13.1	1.91	0.053	0.040	49

Appendix Table 1 continues on page 313.

Appendix Table 1. (Continued) Nutrient Requirements of Mature Beef Cows to Maintain Moderate Body Condition

Months Since Calving	Milk (lbs./day)	Diet Nutrient Density						Daily Nutrients per Animal					
		DMI (lbs./day)	TDN (% DM)	NEm (Mcal/lb.)	CP (% DM)	Ca (% DM)	P (% DM)	TDN (lbs.)	NEm (Mcal)	CP (lbs.)	Ca (lb.)	P (lb.)	Vit. A (1,000s IU)
1,400-lb. Mature Wt.													
20-lb. Peak Milk													
1	16.7	29.5	58.0	0.58	9.8	0.28	0.19	17.1	17.1	2.88	0.084	0.057	52
2	20.0	30.5	59.1	0.60	10.3	0.30	0.20	18.0	18.2	3.14	0.093	0.062	54
3	18.0	31.3	56.8	0.56	9.6	0.28	0.19	17.8	17.6	2.99	0.088	0.060	55
4	14.4	30.3	55.5	0.54	8.9	0.26	0.18	16.8	16.4	2.70	0.079	0.053	54
5	10.8	29.4	54.1	0.52	8.3	0.24	0.17	15.9	15.3	2.44	0.071	0.049	52
6	7.8	28.6	53.0	0.50	7.7	0.22	0.16	15.2	14.4	2.21	0.062	0.044	51
30-lb. Peak Milk													
1	25.0	31.9	60.7	0.62	11.1	0.33	0.22	19.4	19.8	3.53	0.104	0.068	56
2	30.0	33.3	62.2	0.64	11.8	0.35	0.23	20.7	21.5	3.93	0.117	0.075	59
3	27.0	33.7	59.8	0.61	11.0	0.32	0.21	20.2	20.5	3.69	0.110	0.073	60
4	21.6	32.3	58.1	0.58	10.2	0.30	0.20	18.8	18.8	3.27	0.097	0.064	57
5	16.2	30.8	56.2	0.55	9.3	0.27	0.18	17.3	17.0	2.86	0.084	0.057	54
6	11.7	29.6	54.7	0.53	8.5	0.24	0.17	16.2	15.7	2.52	0.073	0.051	52
10-, 20-, 30-lb. Peak Milk													
7	0.0	25.2	46.9	0.40	6.5	0.17	0.13	11.8	10.1	1.63	0.044	0.033	32
8	0.0	25.6	47.3	0.41	6.5	0.17	0.13	12.1	10.5	1.67	0.044	0.033	33
9	0.0	26.2	48.0	0.42	6.7	0.17	0.13	12.6	11.0	1.75	0.044	0.033	33
10	0.0	27.3	49.1	0.44	6.9	0.26	0.16	13.4	12.0	1.89	0.071	0.044	35
11	0.0	27.0	52.6	0.50	7.8	0.26	0.16	14.2	13.5	2.11	0.071	0.044	34
12	0.0	27.6	56.6	0.56	8.9	0.26	0.16	15.6	15.5	2.45	0.071	0.044	35

Appendix Table 2. Nutrient Requirements to Increase Body Condition Score of Mature Cows from 4 to 5 During the Last 90 Days

Mature Weight (at BCS 5)	DMI (lbs./day)	Diet Nutrient Density					Daily Nutrients per Animal					
		TDN (% DM)	NEm (Mcal/lb.)	CP (% DM)	Ca (% DM)	P (% DM)	TDN (lb.)	NEm (Mcal)	CP (lb.)	Ca (lb.)	P (lb.)	Vitamin A (1,000s IU)
1,000	20.5	60	0.59	7.7	0.36	0.20	12.3	12.1	1.57	0.074	0.040	26
1,100	22.0	60	0.58	7.5	0.35	0.20	13.2	12.8	1.65	0.078	0.043	28
1,200	23.5	59	0.58	7.4	0.34	0.19	13.9	13.6	1.74	0.081	0.045	30

Appendix Table 2 modified from University of Arkansas Extension Bulletin MP391.
Appreciation is expressed to Dr. Shane Gadberry for permissioin to use this table.

Appendix Table 3. Nutrient Requirements to Increase Body Condition Score of Open Mature Cows

				Diet Nutrient Density					Daily Nutrients per Animal					
Mature Weight (at BCS 5)	Current BCS	Days to Gain 1 BCS	DMI (lb./day)	TDN (% DM)	NEm (Mcal/lb.)	CP (% DM)	Ca (%DM)	P (% DM)	TDN (lb.)	NEm (Mcal)	CP (lb.)	Ca (lb.)	P (lb.)	Vit. A (1000's IU)
1,000	3	30	18.8	64	0.65	6.1	0.28	0.16	12.0	12.2	1.14	0.053	0.031	24
		60	17.7	57	0.55	6.4	0.23	0.15	10.1	9.8	1.14	0.041	0.026	28
	4	30	20.5	66	0.67	5.9	0.27	0.16	13.5	13.8	1.21	0.056	0.033	26
		60	19.0	58	0.56	6.4	0.23	0.15	11.0	10.7	1.21	0.044	0.028	24
1,100	3	30	20.3	65	0.66	6.0	0.29	0.17	13.2	13.4	1.22	0.058	0.034	26
		60	19.0	58	0.56	6.4	0.24	0.15	11.0	10.6	1.22	0.045	0.029	24
	4	30	22.2	67	0.69	5.9	0.28	0.16	14.9	15.3	1.30	0.062	0.036	28
		60	20.4	58	0.57	6.4	0.24	0.15	11.8	11.6	1.30	0.048	0.030	26
1,200	3	30	21.0	65	0.69	6.2	0.30	0.18	13.7	14.5	1.30	0.063	0.037	27
		60	20.3	58	0.56	6.4	0.24	0.15	11.8	11.3	1.30	0.048	0.031	26
	4	30	23.5	67	0.68	5.9	0.28	0.16	15.7	16.0	1.38	0.065	0.038	30
		60	21.8	58	0.56	6.3	0.23	0.15	12.6	12.3	1.38	0.051	0.033	28

Appendix Table 3 modified from University of Arkansas Extension Bulletin MP391.
Appreciation is expressed to Dr. Shane Gadberry for permissioin to use this table.

Appendix Table 4. Nutrient Requirements of Two-Year-Old Heifers (20 lbs. peak milk as a mature cow)

				Diet Nutrient Density					Daily Nutrients per Animal					
Months Since Calving	Body Wt. (lbs.)	Milk (lbs./day)	DMI (lbs./day)	TDN (% DM)	NEm (Mcal/lb.)	CP (% DM)	Ca (% DM)	P (% DM)	TDN (lb.)	NEm (Mcal)	CP (lbs.)	Ca (lb.)	P (lb.)	Vit. A (1000's IU)
900-lb. Mature Weight														
1	729	12.3	19.1	61.4	0.63	10.8	0.31	0.20	11.7	12.0	2.07	0.059	0.038	34
2	738	14.8	19.9	62.6	0.65	11.4	0.33	0.21	12.5	12.9	2.27	0.066	0.042	35
3	748	13.3	20.4	60.3	0.62	10.6	0.30	0.19	12.3	12.6	2.16	0.061	0.039	36
4	758	10.7	19.8	58.8	0.59	9.9	0.28	0.18	11.6	11.7	1.96	0.055	0.036	35
5	769	8.0	19.3	57.4	0.57	9.2	0.26	0.17	11.1	11.0	1.77	0.050	0.033	34
6	781	5.8	18.8	56.2	0.55	8.6	0.23	0.15	10.6	10.3	1.61	0.043	0.028	33
7	794	0.0	17.4	48.5	0.43	6.8	0.17	0.12	8.4	7.5	1.18	0.030	0.021	22
8	810	0.0	17.5	49.2	0.44	7.0	0.17	0.12	8.6	7.7	1.22	0.030	0.021	22
9	830	0.0	17.7	50.5	0.46	7.2	0.17	0.12	8.9	8.1	1.28	0.030	0.021	22
10	854	0.0	17.9	52.4	0.49	7.7	0.27	0.16	9.4	8.8	1.38	0.048	0.029	23
11	885	0.0	18.3	55.2	0.54	8.3	0.27	0.16	10.1	9.9	1.52	0.049	0.029	23
12	925	0.0	18.9	58.7	0.59	9.3	0.26	0.16	11.1	11.2	1.75	0.049	0.030	24
1,000-lb. Mature Weight														
1	810	12.3	20.4	61.0	0.63	10.6	0.31	0.19	12.4	12.9	2.16	0.063	0.039	36
2	820	14.8	21.2	62.1	0.64	11.1	0.32	0.20	13.2	13.6	2.36	0.068	0.042	38
3	831	13.3	21.8	59.8	0.61	10.4	0.30	0.19	13.0	13.3	2.26	0.065	0.041	39
4	843	10.7	21.2	58.5	0.59	9.7	0.28	0.18	12.4	12.5	2.06	0.059	0.038	38
5	854	8.0	20.7	57.1	0.57	9.0	0.25	0.17	11.8	11.8	1.87	0.052	0.035	37
6	868	5.8	20.3	56.0	0.55	8.4	0.23	0.16	11.4	11.2	1.71	0.047	0.032	36
7	883	0.0	18.8	48.6	0.43	6.9	0.18	0.13	9.1	8.1	1.29	0.034	0.024	24
8	900	0.0	18.9	49.4	0.44	7.0	0.18	0.13	9.3	8.3	1.33	0.034	0.025	24
9	922	0.0	19.1	50.7	0.46	7.3	0.18	0.13	9.7	8.8	1.39	0.034	0.025	24
10	949	0.0	19.4	52.7	0.50	7.7	0.28	0.17	10.2	9.7	1.50	0.054	0.033	25
11	984	0.0	19.9	55.5	0.54	8.3	0.28	0.16	11.0	10.7	1.66	0.056	0.033	25
12	1,028	0.0	20.6	59.1	0.60	9.3	0.27	0.16	12.2	12.4	1.92	0.056	0.033	26

Appendix Table 4 modified from University of Arkansas Extension Bulletin MP391.
Appreciation is expressed to Dr. Shane Gadberry for permissioin to use this table.

Appendix Table 4 continues on page 317.

Appendix Table 4. Nutrient Requirements of Two-Year-Old Heifers (20 lbs. peak milk as a mature cow) (Continued)

				Diet Nutrient Density					Daily Nutrients per Animal					
Months Since Calving	Body Wt. (lbs.)	Milk (lbs./day)	DMI (lbs./day)	TDN (% DM)	NEm (Mcal/lb.)	CP (% DM)	Ca (% DM)	P (% DM)	TDN (lb.)	NEm (Mcal)	CP (lbs.)	Ca (lb.)	P (lb.)	Vit. A (1000's IU)
1,100-lb Mature Weight														
1	891	12.3	21.6	60.7	0.62	10.4	0.30	0.19	13.1	13.4	2.25	0.065	0.041	38
2	902	14.8	22.5	61.7	0.64	10.9	0.32	0.20	13.9	14.4	2.45	0.072	0.045	40
3	915	13.3	23.2	59.5	0.60	10.1	0.30	0.19	13.8	13.9	2.35	0.070	0.044	41
4	927	10.7	22.6	58.2	0.58	9.6	0.27	0.18	13.2	13.1	2.16	0.061	0.041	40
5	940	8.0	22.1	56.9	0.56	8.9	0.25	0.17	12.6	12.4	1.96	0.055	0.038	39
6	954	5.8	21.7	55.9	0.55	8.3	0.23	0.16	12.1	11.9	1.81	0.050	0.035	38
7	971	0.0	20.2	48.8	0.43	6.9	0.18	0.13	9.9	8.7	1.39	0.036	0.026	26
8	990	0.0	20.4	49.5	0.45	7.0	0.18	0.13	10.1	9.2	1.43	0.037	0.027	26
9	1,014	0.0	20.6	50.8	0.47	7.3	0.18	0.13	10.5	9.7	1.50	0.037	0.027	26
10	1,044	0.0	20.9	52.8	0.50	7.7	0.29	0.17	11.0	10.5	1.61	0.060	0.036	27
11	1,082	0.0	21.4	55.7	0.54	8.4	0.28	0.17	11.9	11.6	1.79	0.060	0.036	27
12	1,130	0.0	22.1	59.4	0.60	9.4	0.27	0.16	13.1	13.3	2.07	0.060	0.036	28
1,200-lb. Mature Weight														
1	972	12.3	22.9	60.4	0.62	10.2	0.30	0.19	13.8	14.2	2.34	0.069	0.044	41
2	984	14.8	23.8	61.4	0.63	10.7	0.32	0.20	14.6	15.0	2.55	0.076	0.048	42
3	998	13.3	24.5	59.2	0.60	10.0	0.29	0.19	14.5	14.7	2.44	0.071	0.047	43
4	1,011	10.7	24.0	58.0	0.58	9.4	0.27	0.18	13.9	13.9	2.25	0.065	0.043	42
5	1,025	8.0	23.4	56.8	0.56	8.8	0.25	0.17	13.3	13.1	2.05	0.059	0.040	41
6	1,041	5.8	23.0	55.8	0.55	8.3	0.23	0.16	12.8	12.7	1.90	0.053	0.037	41
7	1,059	0.0	21.5	48.9	0.44	6.9	0.19	0.13	10.5	9.5	1.48	0.041	0.028	27
8	1,080	0.0	21.7	49.7	0.45	7.1	0.19	0.13	10.8	9.8	1.53	0.041	0.028	28
9	1,106	0.0	22.0	51.0	0.47	7.3	0.19	0.13	11.2	10.3	1.61	0.042	0.029	28
10	1,139	0.0	22.3	53.1	0.50	7.8	0.29	0.17	11.8	11.2	1.73	0.065	0.038	28
11	1,180	0.0	22.8	55.9	0.55	8.5	0.29	0.17	12.7	12.5	1.93	0.066	0.039	29
12	1,233	0.0	23.7	59.7	0.61	9.4	0.28	0.17	14.1	14.5	2.23	0.066	0.040	30

Appendix Table 4 continues on page 318.

Appendix Table 4. Nutrient Requirements of Two-Year-Old Heifers (20 lbs. peak milk as a mature cow) (Continued)

				Diet Nutrient Density					Daily Nutrients per Animal					
Months Since Calving	Body Wt. (lbs.)	Milk (lbs./day)	DMI (lbs./day)	TDN (% DM)	NEm (Mcal/lb.)	CP (% DM)	Ca (% DM)	P (% DM)	TDN (lb.)	NEm (Mcal)	CP (lbs.)	Ca (lb.)	P (lb.)	Vit. A (1000's IU)
1,300-lb. Mature Weight														
1	1,053	12.3	24.1	60.2	0.61	10.1	0.30	0.19	14.5	14.7	2.43	0.072	0.046	43
2	1,066	14.8	25.0	61.1	0.63	10.5	0.31	0.20	15.3	15.8	2.63	0.078	0.050	44
3	1,081	13.3	25.8	58.9	0.59	9.8	0.29	0.19	15.2	15.2	2.53	0.075	0.049	46
4	1,095	10.7	25.3	57.8	0.58	9.2	0.27	0.18	14.6	14.7	2.34	0.068	0.046	45
5	1,111	8.0	24.8	56.6	0.56	8.7	0.25	0.17	14.0	13.9	2.15	0.062	0.042	44
6	1,128	5.8	24.4	55.7	0.54	8.2	0.23	0.16	13.6	13.2	1.99	0.056	0.039	43
7	1,147	0.0	22.9	49.0	0.44	6.9	0.19	0.13	11.2	10.1	1.58	0.044	0.030	29
8	1,170	0.0	23.1	49.8	0.45	7.1	0.19	0.13	11.5	10.4	1.63	0.044	0.030	29
9	1,199	0.0	23.3	51.2	0.47	7.3	0.19	0.13	11.9	11.0	1.71	0.044	0.030	30
10	1,234	0.0	23.7	53.3	0.51	7.8	0.30	0.18	12.6	12.1	1.85	0.071	0.043	30
11	1,279	0.0	24.3	56.2	0.55	8.5	0.29	0.17	13.7	13.4	2.06	0.071	0.043	31
12	1,336	0.0	25.2	60.0	0.61	9.5	0.28	0.17	15.1	15.4	2.40	0.071	0.043	32
1,400-lb. Mature Weight														
1	1,134	12.3	25.3	60.0	0.61	10.0	0.30	0.19	15.2	15.4	2.52	0.076	0.048	45
2	1,148	14.8	26.2	60.9	0.62	10.4	0.31	0.20	16.0	16.2	2.72	0.081	0.052	46
3	1,164	13.3	27.1	58.7	0.59	9.7	0.29	0.19	15.9	16.0	2.62	0.079	0.051	48
4	1,179	10.7	26.6	57.6	0.57	9.1	0.27	0.16	15.3	16.2	2.43	0.072	0.048	47
5	1,196	8.0	26.1	56.5	0.56	8.5	0.25	0.17	14.7	14.6	2.23	0.065	0.044	46
6	1,214	5.8	25.7	55.7	0.54	8.1	0.23	0.16	14.3	13.9	2.08	0.059	0.041	46
7	1,235	0.0	24.2	49.1	0.44	6.9	0.19	0.14	11.9	10.6	1.67	0.046	0.034	31
8	1,260	0.0	24.4	49.9	0.45	7.0	0.19	0.14	12.2	11.0	1.72	0.046	0.034	31
9	1,291	0.0	24.7	51.3	0.47	7.3	0.19	0.14	12.7	11.6	1.81	0.047	0.035	31
10	1,329	0.0	25.1	53.4	0.51	7.8	0.30	0.18	13.4	12.8	1.96	0.075	0.045	32
11	1,377	0.0	25.7	56.4	0.56	8.5	0.30	0.18	14.5	14.4	2.19	0.077	0.046	33
12	1,438	0.0	26.7	60.2	0.61	9.5	0.29	0.17	16.1	16.3	2.54	0.077	0.046	34

Appendix Table 5. Nutrient Requirements of Pregnant Replacement Heifers

			Diet Nutrient Density					Daily Nutrients per Animal					
Months Since Conception	Body Wt. (lbs.)	DMI (lbs./day)	TDN (% DM)	NEm (Mcal/lb)	CP (% DM)	Ca (% DM)	P (% DM)	TDN (lb.)	NEm (Mcal)	CP (lbs.)	Ca (lb.)	P (lb.)	Vit. A (1000's IU)
900-lb. Mature Weight													
1	563	15.5	50.2	0.46	7.2	0.22	0.16	7.8	7.1	1.11	0.034	0.025	20
2	585	15.9	50.3	0.46	7.1	0.21	0.16	8.0	7.3	1.13	0.034	0.025	20
3	609	16.3	50.4	0.46	7.1	0.21	0.16	8.2	7.5	1.16	0.034	0.026	21
4	636	16.8	50.7	0.46	7.2	0.21	0.16	8.5	7.7	1.21	0.035	0.027	21
5	665	17.3	51.3	0.47	7.3	0.20	0.16	8.9	8.1	1.27	0.035	0.028	22
6	700	17.8	52.3	0.49	7.6	0.20	0.16	9.3	8.7	1.36	0.036	0.028	23
7	741	18.4	54.1	0.51	8.1	0.31	0.23	10.0	9.4	1.49	0.057	0.042	23
8	792	19.1	57.1	0.56	8.9	0.31	0.22	10.9	10.7	1.70	0.059	0.042	24
9	854	19.7	62.0	0.63	10.3	0.30	0.22	12.2	12.4	2.02	0.059	0.043	25
1,000-lb. Mature Weight													
1	625	16.7	50.1	0.46	7.2	0.22	0.17	8.4	7.7	1.20	0.037	0.028	21
2	650	17.2	50.2	0.46	7.2	0.22	0.17	8.6	7.9	1.23	0.038	0.029	22
3	676	17.7	50.4	0.46	7.2	0.22	0.17	8.9	8.1	1.27	0.038	0.030	22
4	704	18.2	50.7	0.46	7.2	0.21	0.17	9.2	8.4	1.31	0.038	0.031	23
5	735	18.7	51.3	0.47	7.3	0.21	0.17	9.6	8.8	1.37	0.039	0.032	24
6	772	19.4	52.3	0.49	7.6	0.20	0.16	10.1	9.5	1.47	0.039	0.032	25
7	815	20.0	54.0	0.52	8.0	0.32	0.23	10.8	10.4	1.60	0.064	0.046	25
8	866	20.7	56.8	0.56	8.7	0.31	0.23	11.8	11.6	1.81	0.064	0.048	26
9	929	21.3	61.3	0.63	10.0	0.31	0.22	13.1	13.4	2.13	0.066	0.048	27

Appendix Table 5 modified from University of Arkansas Extension Bulletin MP391. Appreciation is expressed to Dr. Shane Gadberry for permissioin to use this table.

Appendix Table 5 continues on Page 320.

Appendix Table 5. Nutrient Requirements of Pregnant Replacement Heifers (Continued)

			Diet Nutrient Density					Daily Nutrients per Animal					
Months Since Conception	Body Wt. (lbs.)	DMI (lbs./day)	TDN (% DM)	NEm (Mcal/lb)	CP (% DM)	Ca (% DM)	P (% DM)	TDN (lb.)	NEm (Mcal)	CP (lbs.)	Ca (lb.)	P (lb.)	Vit. A (1000's IU)
1,100-lb. Mature Weight													
1	687	18.0	50.3	0.46	7.2	0.23	0.18	9.1	8.3	1.30	0.041	0.032	23
2	714	18.5	50.4	0.46	7.2	0.22	0.17	9.3	8.5	1.33	0.041	0.032	23
3	742	19.0	50.5	0.46	7.2	0.22	0.17	9.6	8.7	1.36	0.042	0.032	24
4	773	19.5	50.8	0.47	7.2	0.22	0.17	9.9	9.2	1.41	0.043	0.033	25
5	806	20.1	51.3	0.48	7.3	0.21	0.17	10.3	9.6	1.47	0.043	0.034	26
6	845	20.8	52.3	0.49	7.5	0.21	0.17	10.9	10.2	1.57	0.044	0.035	26
7	890	21.5	53.9	0.52	7.9	0.32	0.23	11.6	11.2	1.70	0.069	0.049	27
8	944	22.3	56.5	0.56	8.6	0.31	0.22	12.6	12.5	1.92	0.069	0.049	28
9	1,009	22.9	60.6	0.62	9.8	0.30	0.22	13.9	14.2	2.24	0.069	0.050	29
1,200-lb. Mature Weight													
1	750	19.3	50.5	0.46	7.2	0.23	0.18	9.7	8.9	1.39	0.044	0.035	25
2	779	19.8	50.5	0.46	7.2	0.23	0.18	10.0	9.1	1.42	0.046	0.036	25
3	809	20.3	50.7	0.46	7.2	0.22	0.18	10.3	9.3	1.46	0.046	0.036	26
4	842	20.9	50.9	0.47	7.2	0.22	0.17	10.6	9.8	1.51	0.046	0.036	27
5	878	21.5	51.4	0.48	7.3	0.22	0.17	11.1	10.3	1.57	0.047	0.037	27
6	918	22.2	52.3	0.49	7.5	0.21	0.17	11.6	10.9	1.67	0.047	0.038	28
7	966	23.0	53.8	0.51	7.9	0.31	0.23	12.4	11.7	1.81	0.071	0.053	29
8	1,022	23.7	56.2	0.55	8.5	0.31	0.22	13.3	13.0	2.02	0.073	0.053	30
9	1,089	24.4	59.9	0.61	9.6	0.30	0.22	14.6	14.9	2.35	0.073	0.054	31

Appendix Table 5 continues on Page 321.

Table 5. Nutrient Requirements of Pregnant Replacement Heifers (Continued)

			Diet Nutrient Density					Daily Nutrients per Animal					
Months Since Conception	Body Wt. (lbs.)	DMI (lbs./day)	TDN (% DM)	NEm (Mcal/lb)	CP (% DM)	Ca (% DM)	P (% DM)	TDN (lb.)	NEm (Mcal)	CP (lbs.)	Ca (lb.)	P (lb.)	Vit. A (1000's IU)
1,300-lb. Mature Weight													
1	812	20.5	50.6	0.46	7.2	0.24	0.18	10.4	9.4	1.48	0.049	0.037	26
2	843	21.0	50.7	0.46	7.2	0.23	0.18	10.6	9.7	1.51	0.049	0.038	27
3	876	21.6	50.8	0.47	7.2	0.23	0.18	11.0	10.2	1.56	0.050	0.039	27
4	911	22.2	51.0	0.47	7.2	0.22	0.18	11.3	10.4	1.61	0.050	0.040	28
5	949	22.9	51.5	0.48	7.3	0.22	0.18	11.8	11.0	1.67	0.050	0.041	29
6	992	23.6	52.4	0.49	7.5	0.22	0.17	12.4	11.6	1.77	0.052	0.041	30
7	1,041	24.4	53.7	0.51	7.9	0.31	0.23	13.1	12.4	1.92	0.076	0.056	31
8	1,099	25.2	56.0	0.55	8.5	0.30	0.22	14.1	13.9	2.13	0.076	0.056	32
9	1,169	25.9	59.5	0.60	9.5	0.30	0.22	15.4	15.5	2.45	0.078	0.057	33
1,400-lb. Mature Weight													
1	874	21.7	50.7	0.47	7.3	0.24	0.18	11.0	10.2	1.57	0.052	0.039	28
2	907	22.3	50.8	0.47	7.2	0.24	0.18	11.3	10.5	1.61	0.054	0.040	28
3	942	22.9	50.9	0.47	7.2	0.23	0.18	11.7	10.8	1.65	0.054	0.041	29
4	979	23.5	51.2	0.47	7.2	0.23	0.18	12.0	11.0	1.70	0.054	0.042	30
5	1,020	24.2	51.6	0.48	7.3	0.22	0.18	12.5	11.6	1.77	0.054	0.044	31
6	1,065	24.9	52.4	0.49	7.5	0.22	0.18	13.0	12.2	1.86	0.055	0.045	32
7	1,117	25.8	53.7	0.51	7.8	0.31	0.23	13.9	13.2	2.01	0.080	0.059	33
8	1,177	26.6	55.8	0.55	8.4	0.30	0.22	14.8	14.6	2.23	0.080	0.059	34
9	1,249	27.4	59.0	0.60	9.3	0.30	0.22	16.2	16.4	2.56	0.082	0.060	35

Appendix Table 6. Nutrient Requirements of Growing Steer and Heifer Calves

			Diet Nutrient Density						Daily Nutrients per Animal						
Body Wt. (lbs)	ADG (lbs)	DMI (lbs/day)	TDN (% DM)	NEm (Mcal/lb)	NEg (Mcal/lb)	CP (% DM)	Ca (% DM)	P (% DM)	TDN (lbs)	NEm (Mcal)	NEg (Mcal)	CP (lbs)	Ca (lbs)	P (lbs)	Vit. A (1000's IU)
1,100 lbs at Finishing															
300	0.5	7.9	54	0.50	0.24	9.2	0.30	0.16	4.3	3.07	0.42	0.73	0.024	0.013	8
	1.0	8.4	59	0.57	0.31	11.4	0.46	0.23	5.0	3.07	0.90	0.95	0.039	0.019	8
	1.5	8.6	64	0.64	0.37	13.6	0.62	0.29	5.5	3.07	1.40	1.17	0.053	0.025	9
	2.0	8.6	69	0.72	0.44	16.2	0.79	0.36	5.9	3.07	1.92	1.39	0.068	0.031	9
	2.5	8.5	75	0.81	0.52	18.9	0.96	0.40	6.4	3.07	2.46	1.61	0.082	0.034	9
	3.0	8.2	83	0.92	0.62	22.2	1.17	0.51	6.8	3.07	3.00	1.83	0.096	0.042	8
400	0.5	9.8	54	0.50	0.24	8.7	0.27	0.15	5.3	3.81	0.52	0.85	0.026	0.015	10
	1.0	10.4	59	0.57	0.31	10.4	0.39	0.20	6.1	3.81	1.12	1.08	0.040	0.021	10
	1.5	10.7	64	0.64	0.37	12.1	0.50	0.24	6.8	3.81	1.74	1.30	0.053	0.026	11
	2.0	10.7	69	0.72	0.44	14.1	0.62	0.29	7.4	3.81	2.39	1.51	0.066	0.031	11
	2.5	10.6	75	0.81	0.52	16.3	0.75	0.34	8.0	3.81	3.05	1.72	0.079	0.036	11
	3.0	10.2	83	0.92	0.62	19.0	0.90	0.41	8.5	3.81	3.72	1.94	0.092	0.042	10
500	0.5	11.6	54	0.50	0.24	8.4	0.25	0.15	6.3	4.50	0.62	0.97	0.029	0.017	12
	1.0	12.2	59	0.57	0.31	9.8	0.34	0.18	7.2	4.50	1.32	1.19	0.041	0.022	12
	1.5	12.6	64	0.64	0.37	11.2	0.42	0.22	8.1	4.50	2.06	1.41	0.054	0.027	13
	2.0	12.7	69	0.72	0.44	12.8	0.52	0.25	8.8	4.50	2.82	1.63	0.066	0.032	13
	2.5	12.5	75	0.81	0.52	14.7	0.62	0.30	9.4	4.50	3.60	1.84	0.077	0.037	13
	3.0	12.1	83	0.92	0.62	16.9	0.74	0.35	10.0	4.50	4.40	2.05	0.089	0.042	12
600	0.5	13.2	54	0.50	0.24	8.2	0.23	0.14	7.1	5.16	0.71	1.08	0.031	0.019	13
	1.0	14.0	59	0.57	0.31	9.4	0.30	0.17	8.3	5.16	1.51	1.31	0.043	0.024	14
	1.5	14.4	64	0.64	0.37	10.6	0.38	0.20	9.2	5.16	2.36	1.53	0.054	0.028	14
	2.0	14.6	69	0.72	0.44	11.9	0.44	0.22	10.1	5.16	3.23	1.74	0.065	0.033	15
	2.5	14.4	75	0.81	0.52	13.6	0.52	0.26	10.8	5.16	4.13	1.95	0.075	0.037	14
	3.0	13.8	83	0.92	0.62	15.7	0.62	0.30	11.5	5.16	5.04	2.17	0.086	0.041	14

Appendix Table 6 modified from University of Arkansas Extension Bulletin MP391.
Appreciation is expressed to Dr. Shane Gadberry for permission to use this table.

Appendix Table 6 continues on page 323

Appendix Table 6. Nutrient Requirements of Growing Steer and Heifer Calves (Continued)

			Diet Nutrient Density						Daily Nutrients per Animal						
Body Wt. (lbs)	ADG (lbs)	DMI (lbs/day)	TDN (% DM)	NEm (Mcal/lb)	NEg (Mcal/lb)	CP (% DM)	Ca (% DM)	P (% DM)	TDN (lbs)	NEm (Mcal)	NEg (Mcal)	CP (lbs)	Ca (lbs)	P (lbs)	Vit. A (1000's IU)
1,100 lbs at Finishing															
700	0.5	14.9	54	0.50	0.24	8.0	0.22	0.14	8.0	5.79	0.79	1.19	0.033	0.021	15
	1.0	15.8	59	0.57	0.31	9.0	0.28	0.16	9.3	5.79	1.70	1.42	0.044	0.026	16
	1.5	16.2	64	0.64	0.37	10.1	0.33	0.19	10.4	5.79	2.65	1.64	0.054	0.030	16
	2.0	16.3	69	0.72	0.44	11.4	0.39	0.21	11.2	5.79	3.63	1.85	0.064	0.034	16
	2.5	16.1	75	0.81	0.52	12.8	0.46	0.24	12.1	5.79	4.64	2.06	0.074	0.038	16
	3.0	15.5	83	0.92	0.62	14.6	0.54	0.27	12.9	5.79	5.66	2.27	0.084	0.042	16
1,200 lbs at Finishing															
300	0.5	7.8	54	0.49	0.24	9.4	0.31	0.17	4.2	3.07	0.39	0.73	0.025	0.013	8
	1.0	8.3	58	0.56	0.30	11.5	0.48	0.23	4.8	3.07	0.84	0.95	0.040	0.019	8
	1.5	8.6	63	0.63	0.36	13.7	0.63	0.29	5.4	3.07	1.31	1.17	0.054	0.025	9
	2.0	8.6	68	0.70	0.42	16.2	0.80	0.36	5.8	3.07	1.80	1.40	0.069	0.031	9
	2.5	8.6	73	0.78	0.50	18.7	0.96	0.43	6.3	3.07	2.30	1.61	0.083	0.037	9
	3.0	8.3	80	0.88	0.58	22.0	1.18	0.52	6.6	3.07	2.81	1.83	0.098	0.043	8
400	0.5	9.7	54	0.49	0.24	8.8	0.28	0.16	5.2	3.81	0.49	0.85	0.027	0.015	10
	1.0	10.3	58	0.56	0.30	10.4	0.39	0.20	6.0	3.81	1.04	1.07	0.041	0.021	10
	1.5	10.6	63	0.63	0.36	12.2	0.51	0.25	6.7	3.81	1.63	1.30	0.054	0.026	11
	2.0	10.7	68	0.70	0.42	14.1	0.63	0.30	7.3	3.81	2.23	1.51	0.068	0.032	11
	2.5	10.7	73	0.78	0.50	16.1	0.76	0.35	7.8	3.81	2.85	1.72	0.081	0.037	11
	3.0	10.4	80	0.88	0.58	18.7	0.90	0.41	8.3	3.81	3.49	1.94	0.094	0.043	10
500	0.5	11.5	54	0.49	0.24	8.4	0.25	0.15	6.2	4.50	0.58	0.97	0.029	0.017	12
	1.0	12.2	58	0.56	0.30	9.8	0.34	0.18	7.1	4.50	1.23	1.19	0.042	0.022	12
	1.5	12.6	63	0.63	0.36	11.2	0.43	0.22	7.9	4.50	1.93	1.41	0.055	0.028	13
	2.0	12.6	68	0.70	0.42	12.9	0.53	0.26	8.6	4.50	2.64	1.63	0.067	0.033	13
	2.5	12.6	73	0.78	0.50	14.6	0.63	0.30	9.2	4.50	3.37	1.84	0.079	0.038	13
	3.0	12.2	80	0.88	0.58	16.8	0.75	0.35	9.8	4.50	4.12	2.05	0.092	0.043	12

Appendix Table 6 continues on page 324

Appendix Table 6. Nutrient Requirements of Growing Steer and Heifer Calves (Continued)

			Diet Nutrient Density						Daily Nutrients per Animal						
Body Wt. (lbs)	ADG (lbs)	DMI (lbs/day)	TDN (% DM)	NEm (Mcal/lb)	NEg (Mcal/lb)	CP (% DM)	Ca (% DM)	P (% DM)	TDN (lbs)	NEm (Mcal)	NEg (Mcal)	CP (lbs)	Ca (lbs)	P (lbs)	Vit. A (1000's IU)
1,200 lbs at Finishing															
600	0.5	13.2	54	0.49	0.24	8.2	0.24	0.15	7.1	5.16	0.66	1.08	0.031	0.019	13
	1.0	14.0	58	0.56	0.30	9.3	0.31	0.17	8.1	5.16	1.42	1.31	0.043	0.024	14
	1.5	14.4	63	0.63	0.36	10.6	0.38	0.20	9.1	5.16	2.21	1.52	0.055	0.029	14
	2.0	14.4	68	0.70	0.42	12.1	0.46	0.23	9.8	5.16	3.03	1.74	0.067	0.034	14
	2.5	14.4	73	0.78	0.50	13.5	0.54	0.26	10.5	5.16	3.87	1.95	0.078	0.038	14
	3.0	14.0	80	0.88	0.58	15.4	0.64	0.31	11.2	5.16	4.73	2.16	0.089	0.043	14
700	0.5	14.8	54	0.49	0.24	8.0	0.23	0.14	8.0	5.79	0.74	1.18	0.034	0.021	15
	1.0	15.7	58	0.56	0.30	9.0	0.29	0.17	9.1	5.79	1.59	1.42	0.045	0.026	16
	1.5	16.2	63	0.63	0.36	10.1	0.34	0.19	10.2	5.79	2.48	1.64	0.056	0.030	16
	2.0	16.3	68	0.70	0.42	11.3	0.41	0.21	11.1	5.79	3.40	1.85	0.067	0.035	16
	2.5	16.2	73	0.78	0.50	12.7	0.47	0.24	11.8	5.79	4.34	2.05	0.077	0.039	16
	3.0	15.8	80	0.88	0.58	14.4	0.55	0.27	12.6	5.79	5.30	2.27	0.087	0.043	16
1,300 lbs at Finishing															
300	0.5	7.9	50	0.50	0.20	9.1	0.30	0.16	4.0	3.0	0.4	0.72	0.024	0.013	8
	1.0	8.4	54	0.57	0.26	11.2	0.48	0.23	4.5	3.0	0.8	0.94	0.040	0.019	9
	1.5	8.6	58	0.64	0.32	13.6	0.65	0.30	5.0	3.0	1.2	1.17	0.056	0.026	9
	2.0	8.6	63	0.72	0.38	16.0	0.81	0.37	5.4	3.0	1.7	1.38	0.070	0.031	9
	2.5	8.6	68	0.80	0.44	18.5	0.98	0.44	5.8	3.0	2.1	1.59	0.085	0.038	9
	3.0	8.3	74	0.90	0.53	21.6	1.19	0.52	6.1	3.0	2.6	1.80	0.100	0.044	9
400	0.5	9.8	50	0.50	0.20	8.5	0.27	0.16	4.9	3.8	0.4	0.83	0.027	0.015	10
	1.0	10.4	54	0.57	0.26	10.2	0.40	0.21	5.6	3.8	1.0	1.07	0.042	0.021	11
	1.5	10.7	58	0.64	0.32	12.0	0.52	0.25	6.2	3.8	1.5	1.28	0.056	0.027	11
	2.0	10.7	63	0.72	0.38	14.0	0.65	0.31	6.7	3.8	2.1	1.50	0.070	0.033	11
	2.5	10.6	68	0.80	0.45	16.0	0.78	0.36	7.2	3.8	2.6	1.70	0.083	0.038	11
	3.0	10.2	75	0.91	0.53	18.7	0.95	0.43	7.7	3.8	3.2	1.91	0.097	0.044	11
500	0.5	11.6	51	0.50	0.20	8.3	0.26	0.15	5.9	4.5	0.6	0.96	0.030	0.018	12
	1.0	12.3	55	0.57	0.26	9.6	0.35	0.19	6.8	4.5	1.2	1.17	0.043	0.023	13
	1.5	12.6	59	0.64	0.32	11.1	0.45	0.22	7.4	4.5	1.8	1.39	0.056	0.028	13
	2.0	12.7	63	0.72	0.38	12.6	0.54	0.26	8.0	4.5	2.4	1.60	0.069	0.033	13
	2.5	12.5	68	0.81	0.45	14.5	0.66	0.31	8.5	4.5	3.1	1.81	0.082	0.039	13
	3.0	12.1	75	0.91	0.53	16.7	0.78	0.36	9.1	4.5	3.8	2.01	0.094	0.044	13

Appendix Table 6 continues on page 325

Appendix Table 6. Nutrient Requirements of Growing Steer and Heifer Calves (Continued)

			Diet Nutrient Density						Daily Nutrients per Animal						
Body Wt. (lbs)	ADG (lbs)	DMI (lbs/day)	TDN (% DM)	NEm (Mcal/lb)	NEg (Mcal/lb)	CP (% DM)	Ca (% DM)	P (% DM)	TDN (lbs)	NEm (Mcal)	NEg (Mcal)	CP (lbs)	Ca (lbs)	P (lbs)	Vit. A (1000's IU)
600	0.5	13.3	51	0.50	0.20	8.0	0.24	0.15	6.8	5.1	0.6	1.07	0.032	0.020	14
	1.0	14.0	54	0.57	0.26	9.1	0.31	0.17	7.6	5.1	1.3	1.27	0.044	0.024	14
	1.5	14.5	59	0.64	0.32	10.3	0.39	0.20	8.6	5.1	2.1	1.50	0.057	0.030	15
	2.0	14.6	63	0.72	0.38	11.7	0.47	0.24	9.2	5.1	2.8	1.70	0.069	0.034	15
	2.5	14.4	68	0.80	0.44	13.3	0.56	0.27	9.8	5.1	3.6	1.91	0.081	0.039	15
	3.0	13.9	75	0.90	0.53	15.3	0.66	0.32	10.4	5.1	4.3	2.13	0.092	0.043	14
700	0.5	14.9	51	0.50	0.20	7.8	0.23	0.15	7.6	5.7	0.7	1.17	0.034	0.022	15
	1.0	15.8	55	0.57	0.26	8.8	0.29	0.17	8.7	5.7	1.5	1.40	0.047	0.027	16
	1.5	16.2	58	0.64	0.32	9.9	0.35	0.19	9.4	5.7	2.3	1.60	0.057	0.031	17
	2.0	16.3	63	0.72	0.38	11.2	0.42	0.22	10.3	5.7	3.2	1.82	0.069	0.036	17
	2.5	16.2	68	0.80	0.44	12.4	0.49	0.24	11.0	5.7	4.0	2.01	0.079	0.040	17
	3.0	15.6	75	0.90	0.52	14.2	0.57	0.28	11.7	5.7	4.9	2.22	0.090	0.044	16
1,400 lbs at Finishing															
300	0.5	7.8	50	0.49	0.20	9.3	0.33	0.17	3.9	3.0	0.3	0.72	0.025	0.014	8
	1.0	8.3	53	0.56	0.25	11.3	0.49	0.24	4.4	3.0	0.7	0.94	0.040	0.020	9
	1.5	8.6	58	0.63	0.31	13.6	0.66	0.30	5.0	3.0	1.2	1.17	0.056	0.026	9
	2.0	8.7	62	0.70	0.36	15.8	0.81	0.37	5.4	3.0	1.6	1.37	0.071	0.032	9
	2.5	8.6	67	0.78	0.43	18.6	1.01	0.45	5.8	3.0	2.0	1.60	0.087	0.038	9
	3.0	8.4	73	0.88	0.51	21.5	1.2	0.52	6.1	3.0	2.5	1.81	0.101	0.044	9
400	0.5	9.8	50	0.50	0.20	8.6	0.29	0.16	4.9	3.8	0.4	0.85	0.028	0.016	10
	1.0	10.3	54	0.56	0.25	10.3	0.41	0.21	5.6	3.8	0.9	1.06	0.042	0.021	11
	1.5	10.6	58	0.63	0.31	12.1	0.53	0.26	6.1	3.8	1.4	1.28	0.057	0.027	11
	2.0	10.7	62	0.70	0.36	13.9	0.66	0.31	6.6	3.8	1.9	1.49	0.070	0.033	11
	2.5	10.6	67	0.79	0.43	16.1	0.8	0.37	7.1	3.8	2.5	1.71	0.084	0.038	11
	3.0	10.3	73	0.88	0.51	18.6	0.96	0.43	7.5	3.8	3.1	1.92	0.099	0.044	11
500	0.5	11.5	50	0.50	0.20	8.3	0.26	0.15	5.8	4.5	0.5	0.95	0.030	0.017	12
	1.0	12.2	54	0.56	0.25	9.6	0.36	0.19	6.6	4.5	1.1	1.17	0.043	0.023	13
	1.5	12.6	58	0.63	0.30	11.0	0.45	0.23	7.3	4.5	1.7	1.38	0.057	0.028	13
	2.0	12.7	62	0.70	0.30	12.6	0.55	0.27	7.8	4.5	2.3	1.60	0.070	0.034	13
	2.5	12.6	67	0.78	0.43	14.3	0.66	0.31	8.4	4.5	2.9	1.80	0.083	0.039	13
	3.0	12.3	73	0.87	0.50	16.4	0.78	0.36	9.0	4.5	3.6	2.02	0.096	0.045	13

Appendix Table 6 continues on page 326

Appendix Table 6. Nutrient Requirements of Growing Steer and Heifer Calves (Continued)

			Diet Nutrient Density						Daily Nutrients per Animal						
Body Wt. (lbs)	ADG (lbs)	DMI (lbs/day)	TDN (% DM)	NEm (Mcal/lb)	NEg (Mcal/lb)	CP (% DM)	Ca (% DM)	P (% DM)	TDN (lbs)	NEm (Mcal)	NEg (Mcal)	CP (lbs)	Ca (lbs)	P (lbs)	Vit. A (1000's IU)
600	0.5	13.2	50	0.50	0.20	8.0	0.24	0.15	6.6	5.1	0.6	1.06	0.032	0.020	14
	1.0	14.0	54	0.56	0.25	9.1	0.32	0.18	7.6	5.1	1.2	1.28	0.045	0.025	14
	1.5	14.4	58	0.63	0.31	10.4	0.4	0.21	8.4	5.1	1.9	1.50	0.058	0.030	15
	2.0	14.5	62	0.70	0.37	11.8	0.49	0.24	9.0	5.1	2.6	1.71	0.070	0.035	15
	2.5	14.4	67	0.78	0.43	13.3	0.57	0.28	9.6	5.1	3.4	1.91	0.082	0.040	15
	3.0	14.1	72	0.87	0.50	15.0	0.67	0.32	10.2	5.1	4.1	2.12	0.095	0.045	15
700	0.5	14.8	50	0.50	0.20	7.9	0.23	0.15	7.4	5.7	0.6	1.16	0.035	0.022	15
	1.0	15.7	54	0.56	0.25	8.8	0.3	0.17	8.5	5.7	1.4	1.38	0.046	0.027	16
	1.5	16.2	58	0.63	0.30	9.9	0.36	0.19	9.4	5.7	2.2	1.60	0.059	0.031	17
	2.0	16.3	63	0.70	0.36	11.1	0.43	0.22	10.3	5.7	3.0	1.81	0.070	0.036	17
	2.5	16.2	67	0.78	0.43	12.5	0.51	0.25	10.9	5.7	3.8	2.02	0.082	0.041	17
	3.0	15.8	73	0.87	0.50	14.1	0.59	0.29	11.5	5.7	4.6	2.23	0.093	0.045	16

Appendix Table 7. Nutrient Requirements of Bull Calves (less than 12 months of age, 2,000-lb. mature weight)

			Diet Nutrient Density						Daily Nutrients per Animal						
Body Wt. (lbs.)	ADG (lbs.)	DMI (lbs./day)	TDN (% DM)	NEm (Mcal/lb.)	NEg (Mcal/lb.)	CP (% DM)	Ca (% DM)	P (% DM)	TDN (lbs.)	NEm (Mcal)	NEg (Mcal)	CP (lb.)	Ca (lb.)	P (lb.)	Vit. A (1000's IU)
300	0.5	8.0	55	0.51	0.25	9.1	0.31	0.16	4.4	3.53	0.39	0.73	0.025	0.013	8
	1.0	8.3	58	0.56	0.30	11.4	0.48	0.23	4.8	3.53	0.84	0.95	0.040	0.019	8
	1.5	8.5	61	0.60	0.34	13.8	0.64	0.29	5.2	3.53	1.31	1.17	0.054	0.025	9
	2.0	8.6	65	0.65	0.38	16.3	0.80	0.36	5.6	3.53	1.80	1.40	0.069	0.031	9
	2.5	8.7	68	0.71	0.43	18.5	0.95	0.43	5.9	3.53	2.30	1.61	0.083	0.037	9
	3.0	8.6	72	0.76	0.48	21.3	1.14	0.50	6.2	3.53	2.81	1.83	0.098	0.043	9
400	0.5	9.9	55	0.51	0.25	8.6	0.27	0.15	5.4	4.38	0.49	0.85	0.027	0.015	10
	1.0	10.3	58	0.56	0.30	10.5	0.40	0.20	6.0	4.38	1.04	1.08	0.041	0.021	10
	1.5	10.5	61	0.60	0.34	12.4	0.51	0.25	6.4	4.38	1.63	1.30	0.054	0.026	11
	2.0	10.7	65	0.65	0.38	14.1	0.64	0.30	7.0	4.38	2.23	1.51	0.068	0.032	11
	2.5	10.7	68	0.71	0.43	16.2	0.76	0.35	7.3	4.38	2.85	1.73	0.081	0.037	11
	3.0	10.7	72	0.76	0.48	18.1	0.88	0.40	7.7	4.38	3.49	1.94	0.094	0.043	11
500	0.5	11.7	55	0.51	0.25	8.3	0.25	0.15	6.4	5.18	0.58	0.97	0.029	0.017	12
	1.0	12.2	58	0.56	0.30	9.8	0.34	0.18	7.1	5.18	1.23	1.19	0.042	0.022	12
	1.5	12.5	61	0.60	0.34	11.3	0.44	0.22	7.6	5.18	1.93	1.41	0.055	0.028	13
	2.0	12.6	65	0.65	0.38	12.9	0.53	0.26	8.2	5.18	2.64	1.63	0.067	0.033	13
	2.5	12.7	68	0.71	0.43	14.5	0.63	0.29	8.6	5.18	3.37	1.84	0.080	0.037	13
	3.0	12.6	72	0.76	0.48	16.3	0.73	0.34	9.1	5.18	4.12	2.05	0.092	0.043	13
600	0.5	13.4	55	0.51	0.25	8.1	0.24	0.14	7.4	5.93	0.66	1.08	0.032	0.019	13
	1.0	13.9	58	0.56	0.30	9.4	0.32	0.17	8.1	5.93	1.42	1.31	0.044	0.024	14
	1.5	14.3	61	0.60	0.34	10.7	0.38	0.20	8.7	5.93	2.21	1.53	0.055	0.029	14
	2.0	14.5	65	0.65	0.38	12.0	0.46	0.23	9.4	5.93	3.03	1.74	0.067	0.034	15
	2.5	14.5	68	0.71	0.43	13.4	0.54	0.26	9.9	5.93	3.87	1.95	0.078	0.038	15
	3.0	14.5	72	0.76	0.48	14.9	0.61	0.30	10.4	5.93	4.73	2.16	0.089	0.043	15

Appendix Table 7 modified from University of Arkansas Extension Bulletin MP391.
Appreciation is expressed to Dr. Shane Gadberry for permissioin to use this table.

Appendix Table 7 continues on page 328.

Appendix Table 7. Nutrient Requirements of Bull Calves (less than 12 months of age, 2,000-lb. mature weight) (Continued)

			Diet Nutrient Density						Daily Nutrients per Animal						
Body Wt. (lbs.)	ADG (lbs.)	DMI (lbs./day)	TDN (% DM)	NEm (Mcal/lb.)	NEg (Mcal/lb.)	CP (% DM)	Ca (% DM)	P (% DM)	TDN (lbs.)	NEm (Mcal)	NEg (Mcal)	CP (lbs.)	Ca (lb.)	P (lb.)	Vit. A (1000's IU)
700	0.5	15.1	55	0.51	0.25	7.9	0.23	0.14	8.3	6.66	0.74	1.19	0.034	0.021	15
	1.0	15.6	58	0.56	0.30	9.1	0.29	0.17	9.0	6.66	1.59	1.42	0.045	0.026	16
	1.5	16.0	61	0.60	0.34	10.3	0.35	0.19	9.8	6.66	2.48	1.64	0.056	0.030	16
	2.0	16.3	65	0.65	0.38	11.4	0.40	0.21	10.6	6.66	3.40	1.86	0.066	0.035	16
	2.5	16.3	68	0.71	0.43	12.7	0.47	0.24	11.1	6.66	4.34	2.07	0.077	0.039	16
	3.0	16.3	72	0.76	0.48	13.9	0.53	0.26	11.7	6.66	5.30	2.27	0.087	0.043	16
800	0.5	16.7	55	0.51	0.25	7.7	0.22	0.14	9.2	7.36	0.82	1.28	0.036	0.023	17
	1.0	17.3	58	0.56	0.30	8.7	0.27	0.16	10.0	7.36	1.76	1.51	0.047	0.028	17
	1.5	17.7	61	0.60	0.34	9.7	0.32	0.18	10.8	7.36	2.74	1.72	0.057	0.032	18
	2.0	18.0	65	0.65	0.38	10.7	0.37	0.19	11.7	7.36	3.76	1.93	0.066	0.035	18
	2.5	18.1	68	0.71	0.43	11.8	0.42	0.22	12.3	7.36	4.80	2.13	0.076	0.039	18
	3.0	18.0	72	0.76	0.48	12.9	0.47	0.24	13.0	7.36	5.86	2.33	0.085	0.043	18
900	0.5	18.2	55	0.51	0.25	7.5	0.21	0.14	10.0	8.04	0.90	1.37	0.039	0.026	18
	1.0	18.9	58	0.56	0.30	8.3	0.25	0.15	11.0	8.04	1.92	1.57	0.048	0.029	19
	1.5	19.4	61	0.60	0.34	9.1	0.29	0.17	11.8	8.04	2.99	1.77	0.057	0.033	19
	2.0	19.6	65	0.65	0.38	9.9	0.34	0.19	12.7	8.04	4.11	1.95	0.066	0.037	20
	2.5	19.7	68	0.71	0.43	10.9	0.38	0.20	13.4	8.04	5.24	2.14	0.075	0.040	20
	3.0	19.6	72	0.76	0.48	11.9	0.42	0.22	14.1	8.04	6.40	2.33	0.083	0.043	20

Appendix Table 8. Nutrient Requirements of Growing Yearlings - Bulls[a], Steers[b], and Heifers[b]

			Diet Nutrient Density						Daily Nutrients per Animal						
Body Wt. (lbs.)	ADG (lbs.)	DMI (lb./day)	TDN (% DM)	NEm (Mcal/lb.)	NEg (Mcal/lb.)	CP (% DM)	Ca (% DM)	P (% DM)	TDN (lb.)	NEm (Mcal)	NEg (Mcal)	CP (lb.)	Ca (lb.)	P (lb.)	Vit. A (1000's IU)
1,000 lbs. at Finishing															
550	0.64	15.2	50	0.45	0.20	7.1	0.21	0.13	7.6	4.83	0.93	1.08	0.032	0.020	15
	1.77	16.1	60	0.61	0.35	9.8	0.36	0.19	9.7	4.83	2.85	1.58	0.058	0.031	16
	2.68	15.7	70	0.76	0.48	12.4	0.49	0.24	11.0	4.83	4.48	1.95	0.077	0.038	16
	3.34	14.8	80	0.90	0.61	14.9	0.61	0.29	11.8	4.83	5.71	2.21	0.090	0.043	15
	3.75	13.7	90	1.04	0.72	17.3	0.73	0.34	12.3	4.83	6.48	2.37	0.100	0.047	14
600	0.64	16.2	50	0.45	0.20	7.0	0.21	0.13	8.1	5.16	0.99	1.13	0.034	0.021	16
	1.77	17.2	60	0.61	0.35	9.5	0.34	0.18	10.3	5.16	3.04	1.63	0.058	0.031	17
	2.68	16.8	70	0.76	0.48	11.9	0.45	0.23	11.8	5.16	4.79	2.00	0.076	0.039	17
	3.34	15.8	80	0.90	0.61	14.3	0.56	0.27	12.6	5.16	6.10	2.26	0.088	0.043	16
	3.75	14.6	90	1.04	0.72	16.5	0.66	0.32	13.1	5.16	6.92	2.41	0.096	0.047	15
650	0.64	17.3	50	0.45	0.20	6.9	0.20	0.12	8.7	5.48	1.06	1.19	0.035	0.021	17
	1.77	18.2	60	0.61	0.35	9.2	0.32	0.17	10.9	5.48	3.22	1.67	0.058	0.031	18
	2.68	17.8	70	0.76	0.48	11.5	0.42	0.21	12.5	5.48	5.08	2.05	0.075	0.037	18
	3.34	16.8	80	0.90	0.61	13.7	0.52	0.26	13.4	5.48	6.47	2.30	0.087	0.044	17
	3.75	15.5	90	1.04	0.72	15.9	0.61	0.30	14.0	5.48	7.35	2.46	0.095	0.047	16
700	0.64	18.2	50	0.45	0.20	6.8	0.19	0.12	9.1	5.79	1.12	1.24	0.035	0.022	18
	1.77	19.3	60	0.61	0.35	8.8	0.30	0.16	11.6	5.79	3.41	1.70	0.058	0.031	19
	2.68	18.8	70	0.76	0.48	10.9	0.39	0.20	13.2	5.79	5.37	2.05	0.073	0.038	19
	3.34	17.8	80	0.90	0.61	13.0	0.48	0.24	14.2	5.79	6.84	2.31	0.085	0.043	18
	3.75	16.4	90	1.04	0.72	15.0	0.56	0.28	14.8	5.79	7.77	2.46	0.092	0.046	16

[a]Use expected mature weight of bull times 0.60 to choose weight to use for growing yearlings bulls.
[b]1,000 to 1,400 lbs. at finishing (28 percent body fat) or maturity (replacement heifers).

Appendix Table 8 modified from University of Arkansas Extension Bulletin MP391.
Appreciation is expressed to Dr. Shane Gadberry for permissioin to use this table.

Appendix Table 8 continues on page 330.

Appendix Table 8. Nutrient Requirements of Growing Yearlings - Bulls[a], Steers[b], and Heifers[b] (Continued)

			Diet Nutrient Density						Daily Nutrients per Animal						
Body Wt. (lbs.)	ADG (lbs.)	DMI (lb./day)	TDN (% DM)	NEm (Mcal/lb.)	NEg (Mcal/lb.)	CP (% DM)	Ca (% DM)	P (% DM)	TDN (lb.)	NEm (Mcal)	NEg (Mcal)	CP (lb.)	Ca (lb.)	P (lb.)	Vit. A (1000's IU)
1,000 lbs. at Finishing															
750	0.64	19.2	50	0.45	0.20	6.7	0.19	0.12	9.6	6.10	1.18	1.29	0.036	0.023	19
	1.77	20.3	60	0.61	0.35	8.5	0.28	0.16	12.2	6.10	3.59	1.73	0.057	0.032	20
	2.68	19.8	70	0.76	0.48	10.3	0.37	0.19	13.9	6.10	5.66	2.04	0.073	0.038	20
	3.34	18.7	80	0.90	0.61	12.2	0.45	0.23	15.0	6.10	7.21	2.28	0.084	0.043	19
	3.75	17.3	90	1.04	0.72	14.0	0.52	0.26	15.6	6.10	8.18	2.42	0.090	0.045	17
800	0.64	20.2	50	0.45	0.20	6.5	0.19	0.12	10.1	6.40	1.23	1.31	0.038	0.024	20
	1.77	21.3	60	0.61	0.35	8.1	0.27	0.15	12.8	6.40	3.77	1.73	0.058	0.032	21
	2.68	20.8	70	0.76	0.48	9.8	0.34	0.18	14.6	6.40	5.94	2.04	0.071	0.037	21
	3.34	19.6	80	0.90	0.61	11.5	0.42	0.22	15.7	6.40	7.56	2.25	0.082	0.043	20
	3.75	18.1	90	1.04	0.72	13.2	0.48	0.25	16.3	6.40	8.59	2.39	0.087	0.045	18
1,100 lbs. at Finishing															
605	0.68	16.3	50	0.45	0.20	7.2	0.22	0.13	8.2	5.19	1.00	1.17	0.036	0.021	16
	1.88	17.3	60	0.61	0.35	10.0	0.36	0.19	10.4	5.19	3.04	1.73	0.062	0.033	17
	2.86	16.9	70	0.76	0.48	12.7	0.49	0.24	11.8	5.19	4.82	2.15	0.083	0.041	17
	3.56	15.9	80	0.90	0.61	15.3	0.61	0.29	12.7	5.19	6.12	2.43	0.097	0.046	16
	4.00	14.7	90	1.04	0.72	17.8	0.72	0.34	13.2	5.19	6.96	2.62	0.106	0.050	15
660	0.68	17.5	50	0.45	0.20	7.1	0.21	0.13	8.8	5.54	1.06	1.24	0.037	0.023	18
	1.88	18.4	60	0.61	0.35	9.7	0.34	0.18	11.0	5.54	3.24	1.78	0.063	0.033	18
	2.86	18.0	70	0.76	0.48	12.3	0.45	0.23	12.6	5.54	5.14	2.21	0.081	0.041	18
	3.56	17.0	80	0.90	0.61	14.7	0.56	0.27	13.6	5.54	6.54	2.50	0.095	0.046	17
	4.00	15.7	90	1.04	0.72	17.1	0.66	0.32	14.1	5.54	7.43	2.68	0.104	0.050	16

[a]Use expected mature weight of bull times 0.60 to choose weight to use for growing yearlings bulls.

[b]1,000 to 1,400 lbs. at finishing (28 percent body fat) or maturity (replacement heifers).

Appendix Table 8 continues on page 331.

Appendix Table 8. Nutrient Requirements of Growing Yearlings - Bulls[a], Steers[b], and Heifers[b] (Continued)

			Diet Nutrient Density						Daily Nutrients per Animal						
Body Wt. (lbs.)	ADG (lbs.)	DMI (lb./day)	TDN (% DM)	NEm (Mcal/lb.)	NEg (Mcal/lb.)	CP (% DM)	Ca (% DM)	P (% DM)	TDN (lb.)	NEm (Mcal)	NEg (Mcal)	CP (lb.)	Ca (lb.)	P (lb.)	Vit. A (1000's IU)
1,100 lbs. at Finishing															
715	0.68	18.5	50	0.45	0.20	6.9	0.20	0.13	9.3	5.89	1.13	1.28	0.037	0.024	19
	1.88	19.6	60	0.61	0.35	9.2	0.32	0.17	11.8	5.89	3.45	1.80	0.063	0.033	20
	2.86	19.1	70	0.76	0.48	11.5	0.42	0.21	13.4	5.89	5.46	2.20	0.080	0.040	19
	3.56	18.1	80	0.90	0.61	13.7	0.52	0.26	14.5	5.89	6.94	2.48	0.094	0.047	18
	4.00	16.7	90	1.04	0.72	15.9	0.61	0.30	15.0	5.89	7.89	2.66	0.102	0.050	17
770	0.68	19.6	50	0.45	0.20	6.8	0.20	0.12	9.8	6.22	1.19	1.33	0.039	0.024	20
	1.88	20.7	60	0.61	0.35	8.8	0.30	0.16	12.4	6.22	3.64	1.82	0.062	0.033	21
	2.86	20.2	70	0.76	0.48	10.9	0.39	0.20	14.1	6.22	5.77	2.20	0.079	0.040	20
	3.56	19.1	80	0.90	0.61	12.9	0.48	0.24	15.3	6.22	7.34	2.46	0.092	0.046	19
	4.00	17.6	90	1.04	0.72	14.8	0.56	0.28	15.8	6.22	8.34	2.60	0.099	0.049	18
825	0.68	20.6	50	0.45	0.20	6.6	0.19	0.12	10.3	6.55	1.26	1.36	0.039	0.025	21
	1.88	21.8	60	0.61	0.35	8.4	0.28	0.16	13.1	6.55	3.84	1.83	0.061	0.035	22
	2.86	21.3	70	0.76	0.48	10.3	0.37	0.19	14.9	6.55	6.08	2.19	0.079	0.040	21
	3.56	20.1	80	0.90	0.61	12.1	0.44	0.23	16.1	6.55	7.73	2.43	0.088	0.046	20
	4.00	18.6	90	1.04	0.72	13.9	0.52	0.26	16.7	6.55	8.78	2.59	0.097	0.048	19
880	0.68	21.7	50	0.45	0.20	6.5	0.19	0.12	10.9	6.88	1.32	1.41	0.041	0.026	22
	1.88	22.9	60	0.61	0.35	8.1	0.27	0.15	13.7	6.88	4.03	1.85	0.062	0.034	23
	2.86	22.4	70	0.76	0.48	9.8	0.34	0.18	15.7	6.88	6.38	2.20	0.076	0.040	22
	3.56	21.1	80	0.90	0.61	11.4	0.42	0.22	16.9	6.88	8.11	2.41	0.089	0.046	21
	4.00	19.5	90	1.04	0.72	13.1	0.48	0.25	17.6	6.88	9.22	2.55	0.094	0.049	20

[a]Use expected mature weight of bull times 0.60 to choose weight to use for growing yearlings bulls.

[b]1,000 to 1,400 lbs. at finishing (28 percent body fat) or maturity (replacement heifers).

Appendix Table 8 continues on page 332.

Appendix Table 8. Nutrient Requirements of Growing Yearlings - Bulls[a], Steers[b], and Heifers[b] (Continued)

			Diet Nutrient Density						Daily Nutrients per Animal						
Body Wt. (lbs.)	ADG (lbs.)	DMI (lb./day)	TDN (% DM)	NEm (Mcal/lb.)	NEg (Mcal/lb.)	CP (% DM)	Ca (% DM)	P (% DM)	TDN (lb.)	NEm (Mcal)	NEg (Mcal)	CP (lb.)	Ca (lb.)	P (lb.)	Vit. A (1000's IU)
1,200 lbs. at Finishing															
660	0.72	17.5	50	0.45	0.20	7.3	0.22	0.13	8.8	5.54	1.06	1.28	0.039	0.023	18
	2.00	18.4	60	0.61	0.35	10.2	0.36	0.19	11.0	5.54	3.25	1.88	0.066	0.035	18
	3.04	18.0	70	0.76	0.48	13.0	0.49	0.24	12.6	5.54	5.15	2.34	0.088	0.043	18
	3.78	17.0	80	0.90	0.61	15.8	0.61	0.29	13.6	5.54	6.54	2.69	0.104	0.049	17
	4.25	15.7	90	1.04	0.72	18.4	0.72	0.34	14.1	5.54	7.44	2.89	0.113	0.053	16
720	0.72	18.6	50	0.45	0.20	7.1	0.21	0.13	9.3	5.92	1.13	1.32	0.039	0.024	19
	2.00	19.7	60	0.61	0.35	9.7	0.34	0.18	11.8	5.92	3.47	1.91	0.067	0.035	20
	3.04	19.2	70	0.76	0.48	12.2	0.45	0.23	13.4	5.92	5.50	2.34	0.086	0.044	19
	3.78	18.2	80	0.90	0.61	14.6	0.56	0.27	14.6	5.92	6.98	2.66	0.102	0.049	18
	4.25	16.8	90	1.04	0.72	17.0	0.66	0.32	15.1	5.92	7.94	2.86	0.111	0.054	17
780	0.72	19.8	50	0.45	0.20	6.9	0.20	0.13	9.9	6.28	1.20	1.37	0.040	0.026	20
	2.00	20.9	60	0.61	0.35	9.2	0.32	0.17	12.5	6.28	3.69	1.92	0.067	0.036	21
	3.04	20.4	70	0.76	0.48	11.4	0.42	0.21	14.3	6.28	5.84	2.33	0.086	0.043	20
	3.78	19.3	80	0.90	0.61	13.6	0.52	0.26	15.4	6.28	7.41	2.62	0.100	0.050	19
	4.25	17.8	90	1.04	0.72	15.8	0.61	0.30	16.0	6.28	8.43	2.81	0.109	0.053	18
840	0.72	20.9	50	0.45	0.20	6.8	0.20	0.13	10.5	6.64	1.27	1.42	0.042	0.027	21
	2.00	22.1	60	0.61	0.35	8.8	0.30	0.16	13.3	6.64	3.90	1.94	0.071	0.035	22
	3.04	21.6	70	0.76	0.48	10.8	0.39	0.20	15.1	6.64	6.17	2.33	0.091	0.043	22
	3.78	20.4	80	0.90	0.61	12.8	0.48	0.24	16.3	6.64	7.84	2.61	0.106	0.049	20
	4.25	18.8	90	1.04	0.72	14.7	0.56	0.28	16.9	6.64	8.91	2.76	0.115	0.053	19

[a]Use expected mature weight of bull times 0.60 to choose weight to use for growing yearlings bulls.

[b]1,000 to 1,400 lbs. at finishing (28 percent body fat) or maturity (replacement heifers).

Appendix Table 8 continues on page 333.

Appendix Table 8. Nutrient Requirements of Growing Yearlings - Bulls[a], Steers[b], and Heifers[b] (Continued)

			Diet Nutrient Density						Daily Nutrients per Animal						
Body Wt. (lbs.)	ADG (lbs.)	DMI (lb./day)	TDN (% DM)	NEm (Mcal/lb.)	NEg (Mcal/lb.)	CP (% DM)	Ca (% DM)	P (% DM)	TDN (lb.)	NEm (Mcal)	NEg (Mcal)	CP (lb.)	Ca (lb.)	P (lb.)	Vit. A (1000's IU)
1,200 lbs. at Finishing															
900	0.72	22.0	50	0.45	0.20	6.6	0.19	0.12	11.0	6.99	1.34	1.45	0.042	0.026	22
	2.00	23.3	60	0.61	0.35	8.4	0.28	0.16	14.0	6.99	4.11	1.96	0.065	0.037	23
	3.04	22.7	70	0.76	0.48	10.2	0.37	0.19	15.9	6.99	6.50	2.32	0.084	0.043	23
	3.78	21.5	80	0.90	0.61	12.0	0.44	0.23	17.2	6.99	8.25	2.58	0.095	0.049	22
	4.25	19.8	90	1.04	0.72	13.8	0.52	0.26	17.8	6.99	9.39	2.73	0.103	0.051	20
960	0.72	23.1	50	0.45	0.20	6.5	0.19	0.12	11.6	7.34	1.40	1.50	0.044	0.028	23
	2.00	24.4	60	0.61	0.35	8.1	0.27	0.15	14.6	7.34	4.31	1.98	0.066	0.037	24
	3.04	23.9	70	0.76	0.48	9.7	0.34	0.19	16.7	7.34	6.82	2.32	0.081	0.045	24
	3.78	22.5	80	0.90	0.61	11.3	0.41	0.22	18.0	7.34	8.66	2.54	0.092	0.050	23
	4.25	20.8	90	1.04	0.72	13.0	0.48	0.25	18.7	7.34	9.85	2.70	0.100	0.052	21
1,300 lbs. at Finishing															
715	0.76	18.5	50	0.45	0.20	7.3	0.22	0.13	9.3	5.89	1.13	1.35	0.041	0.024	19
	2.11	19.6	60	0.61	0.35	10.2	0.36	0.19	11.8	5.89	3.45	2.00	0.071	0.037	20
	3.21	19.1	70	0.76	0.48	13.0	0.49	0.24	13.4	5.89	5.47	2.48	0.094	0.046	19
	3.99	18.1	80	0.90	0.61	15.7	0.61	0.29	14.5	5.89	6.94	2.84	0.110	0.052	18
	4.48	16.7	90	1.04	0.72	18.3	0.72	0.34	15.0	5.89	7.88	3.06	0.120	0.057	17
780	0.76	19.8	50	0.45	0.20	7.1	0.21	0.13	9.9	6.28	1.20	1.41	0.042	0.026	20
	2.11	20.9	60	0.61	0.35	9.6	0.34	0.18	12.5	6.28	3.68	2.01	0.071	0.038	21
	3.21	20.4	70	0.76	0.48	12.1	0.45	0.23	14.3	6.28	5.84	2.47	0.092	0.047	20
	3.99	19.3	80	0.90	0.61	14.5	0.56	0.27	15.4	6.28	7.41	2.80	0.108	0.052	19
	4.48	17.8	90	1.04	0.72	16.9	0.66	0.32	16.0	6.28	8.41	3.01	0.117	0.057	18

[a]Use expected mature weight of bull times 0.60 to choose weight to use for growing yearlings bulls.
[b]1,000 to 1,400 lbs. at finishing (28 percent body fat) or maturity (replacement heifers).

Appendix Table 8 continues on page 334.

Appendix Table 8. Nutrient Requirements of Growing Yearlings - Bulls[a], Steers[b], and Heifers[b] (Continued)

			Diet Nutrient Density						Daily Nutrients per Animal						
Body Wt. (lbs.)	ADG (lbs.)	DMI (lb./day)	TDN (% DM)	NEm (Mcal/lb.)	NEg (Mcal/lb.)	CP (% DM)	Ca (% DM)	P (% DM)	TDN (lb.)	NEm (Mcal)	NEg (Mcal)	CP (lb.)	Ca (lb.)	P (lb.)	Vit. A (1000's IU)
1,300 lbs. at Finishing															
845	0.76	21.0	50	0.45	0.20	6.9	0.21	0.13	10.5	6.67	1.28	1.45	0.044	0.027	21
	2.11	22.2	60	0.61	0.35	9.1	0.32	0.17	13.3	6.67	3.91	2.02	0.071	0.038	22
	3.21	21.7	70	0.76	0.48	11.4	0.42	0.22	15.2	6.67	6.20	2.47	0.091	0.048	22
	3.99	20.5	80	0.90	0.61	13.6	0.51	0.26	16.4	6.67	7.87	2.79	0.105	0.053	21
	4.48	18.9	90	1.04	0.72	15.7	0.60	0.30	17.0	6.67	8.93	2.97	0.113	0.057	19
910	0.76	22.2	50	0.45	0.20	6.7	0.20	0.13	11.1	7.05	1.35	1.49	0.044	0.029	22
	2.11	23.5	60	0.61	0.35	8.7	0.30	0.17	14.1	7.05	4.13	2.04	0.071	0.040	24
	3.21	22.9	70	0.76	0.48	10.7	0.39	0.20	16.0	7.05	6.55	2.45	0.089	0.046	23
	3.99	21.6	80	0.90	0.61	12.7	0.48	0.24	17.3	7.05	8.32	2.74	0.104	0.052	22
	4.48	20.0	90	1.04	0.72	14.6	0.56	0.28	18.0	7.05	9.44	2.92	0.112	0.056	20
975	0.76	23.4	50	0.45	0.20	6.6	0.20	0.13	11.7	7.43	1.42	1.54	0.047	0.030	23
	2.11	24.7	60	0.61	0.35	8.3	0.28	0.16	14.8	7.43	4.35	2.05	0.069	0.040	25
	3.21	24.1	70	0.76	0.48	10.2	0.37	0.19	16.9	7.43	6.90	2.46	0.089	0.046	24
	3.99	22.8	80	0.90	0.61	11.9	0.44	0.23	18.2	7.43	8.76	2.71	0.100	0.052	23
	4.48	21.0	90	1.04	0.72	13.7	0.52	0.26	18.9	7.43	9.94	2.88	0.109	0.055	21
1,040	0.76	24.5	50	0.45	0.20	6.5	0.19	0.13	12.3	7.80	1.49	1.59	0.047	0.032	25
	2.11	25.9	60	0.61	0.35	8.0	0.27	0.15	15.5	7.80	4.57	2.07	0.070	0.039	26
	3.21	25.3	70	0.76	0.48	9.6	0.34	0.19	17.7	7.80	7.24	2.43	0.086	0.048	25
	3.99	23.9	80	0.90	0.61	11.3	0.41	0.22	19.1	7.80	9.19	2.70	0.098	0.053	24
	4.48	22.1	90	1.04	0.72	12.9	0.48	0.25	19.9	7.80	10.44	2.85	0.106	0.055	22

[a]Use expected mature weight of bull times 0.60 to choose weight to use for growing yearlings bulls.

[b]1,000 to 1,400 lbs. at finishing (28 percent body fat) or maturity (replacement heifers).

Appendix Table 8 continues on page 335.

Appendix Table 8. Nutrient Requirements of Growing Yearlings - Bulls[a], Steers[b], and Heifers[b] (Continued)

			Diet Nutrient Density						Daily Nutrients per Animal						
Body Wt. (lbs.)	ADG (lbs.)	DMI (lb./day)	TDN (% DM)	NEm (Mcal/lb.)	NEg (Mcal/lb.)	CP (% DM)	Ca (% DM)	P (% DM)	TDN (lb.)	NEm (Mcal)	NEg (Mcal)	CP (lb.)	Ca (lb.)	P (lb.)	Vit. A (1000's IU)
1,400 lbs. at Finishing															
770	0.80	19.6	50	0.45	0.20	7.3	0.22	0.13	9.8	6.22	1.19	1.43	0.043	0.025	20
	2.22	20.7	60	0.61	0.35	10.1	0.36	0.19	12.4	6.22	3.65	2.09	0.075	0.039	21
	3.38	20.2	70	0.76	0.48	12.9	0.49	0.24	14.1	6.22	5.79	2.61	0.099	0.048	20
	4.20	19.1	80	0.90	0.61	15.6	0.61	0.29	15.3	6.22	7.34	2.98	0.117	0.055	19
	4.72	17.6	90	1.04	0.72	18.1	0.72	0.34	15.8	6.22	8.34	3.19	0.127	0.060	18
840	0.80	20.9	50	0.45	0.20	7.1	0.21	0.13	10.5	6.64	1.27	1.48	0.044	0.027	21
	2.22	22.1	60	0.61	0.35	9.6	0.34	0.18	13.3	6.64	3.89	2.12	0.075	0.040	22
	3.38	21.6	70	0.76	0.48	12.1	0.45	0.23	15.1	6.64	6.18	2.61	0.097	0.050	22
	4.20	20.4	80	0.90	0.61	14.5	0.56	0.27	16.3	6.64	7.84	2.96	0.114	0.055	20
	4.72	18.8	90	1.04	0.72	16.8	0.65	0.32	16.9	6.64	8.91	3.16	0.122	0.060	19
910	0.80	22.2	50	0.45	0.20	6.9	0.21	0.13	11.1	7.05	1.35	1.53	0.047	0.029	22
	2.22	23.5	60	0.61	0.35	9.1	0.32	0.17	14.1	7.05	4.13	2.14	0.075	0.040	24
	3.38	22.9	70	0.76	0.48	11.3	0.42	0.22	16.0	7.05	6.56	2.59	0.096	0.050	23
	4.20	21.6	80	0.90	0.61	13.5	0.51	0.26	17.3	7.05	8.32	2.92	0.110	0.056	22
	4.72	20.0	90	1.04	0.72	15.6	0.60	0.30	18.0	7.05	9.46	3.12	0.120	0.060	20
980	0.80	23.5	50	0.45	0.20	6.7	0.20	0.13	11.8	7.46	1.43	1.57	0.047	0.031	24
	2.22	24.8	60	0.61	0.35	8.7	0.30	0.17	14.9	7.46	4.37	2.16	0.074	0.042	25
	3.38	24.2	70	0.76	0.48	10.7	0.39	0.20	16.9	7.46	6.93	2.59	0.094	0.048	24
	4.20	22.9	80	0.90	0.61	12.6	0.47	0.24	18.3	7.46	8.80	2.89	0.108	0.055	23
	4.72	21.1	90	1.04	0.72	14.5	0.56	0.28	19.0	7.46	10.00	3.06	0.118	0.059	21

[a]Use expected mature weight of bull times 0.60 to choose weight to use for growing yearlings bulls.
[b]1,000 to 1,400 lbs. at finishing (28 percent body fat) or maturity (replacement heifers).

Appendix Table 8 continues on page 336.

Appendix Table 8. Nutrient Requirements of Growing Yearlings - Bulls[a], Steers[b], and Heifers[b] (Continued)

			Diet Nutrient Density						Daily Nutrients per Animal						
Body Wt. (lbs.)	ADG (lbs.)	DMI (lb./day)	TDN (% DM)	NEm (Mcal/lb.)	NEg (Mcal/lb.)	CP (% DM)	Ca (% DM)	P (% DM)	TDN (lb.)	NEm (Mcal)	NEg (Mcal)	CP (lb.)	Ca (lb.)	P (lb.)	Vit. A (1000's IU)
1,400 lbs. at Finishing															
1,050	0.80	24.7	50	0.45	0.20	6.6	0.20	0.13	12.4	7.85	1.50	1.63	0.049	0.032	25
	2.22	26.1	60	0.61	0.35	8.3	0.28	0.16	15.7	7.85	4.60	2.17	0.073	0.042	26
	3.38	25.5	70	0.76	0.48	10.1	0.37	0.20	17.9	7.85	7.30	2.58	0.094	0.051	26
	4.20	24.1	80	0.90	0.61	11.9	0.44	0.23	19.3	7.85	9.26	2.87	0.106	0.055	24
	4.72	22.2	90	1.04	0.72	13.6	0.51	0.26	20.0	7.85	10.53	3.02	0.113	0.058	22
1,120	0.80	25.9	50	0.45	0.20	6.5	0.19	0.13	13.0	8.24	1.58	1.68	0.049	0.034	26
	2.22	27.4	60	0.61	0.35	8.0	0.27	0.16	16.4	8.24	4.83	2.19	0.074	0.044	27
	3.38	26.8	70	0.76	0.48	9.6	0.34	0.19	16.8	8.24	7.66	2.57	0.091	0.051	27
	4.20	25.3	80	0.90	0.61	11.2	0.41	0.22	20.2	8.24	9.72	2.83	0.104	0.056	25
	4.72	23.3	90	1.04	0.72	12.8	0.48	0.25	21.0	8.24	11.05	2.98	0.112	0.058	23

[a]Use expected mature weight of bull times 0.60 to choose weight to use for growing yearlings bulls.
[b]1,000 to 1,400 lbs. at finishing (28 percent body fat) or maturity (replacement heifers).

Appendix Table 9. Nutrient Requirements of Yearling and Breeding Bulls[a,b]

			Diet Nutrient Density						Daily Nutrients per Animal					
Body Wt. (lbs.)	ADG (lbs.)	DMI (lbs./day)	TDN (% DM)	NEm (Mcal/lb.)	NEg (Mcal/lb.)	CP (% DM)	Ca (% DM)	P (% DM)	TDN (lb.)	NEm (Mcal)	NEg (Mcal)	CP (lb.)	Ca (lb.)	P (lb.)
1,700-lb. Mature Weight														
900	0.44	22.0	50	0.45	0.20	6.0	0.16	0.11	11.0	8.04	0.88	1.32	0.036	0.025
	1.55	23.3	60	0.61	0.35	7.3	0.23	0.14	14.0	8.04	3.51	1.71	0.054	0.032
	2.46	22.7	70	0.76	0.48	8.8	0.30	0.16	15.9	8.04	5.82	1.99	0.068	0.037
	3.12	21.5	80	0.90	0.61	10.2	0.36	0.19	17.2	8.04	7.55	2.19	0.077	0.041
1,000	0.44	23.8	50	0.45	0.20	5.9	0.16	0.11	11.9	8.71	0.95	1.40	0.039	0.027
	1.55	25.2	60	0.61	0.35	6.9	0.22	0.13	15.1	8.71	3.79	1.74	0.055	0.033
	2.46	24.6	70	0.76	0.48	8.1	0.27	0.15	17.2	8.71	6.30	1.99	0.067	0.038
	3.12	23.2	80	0.90	0.61	9.3	0.32	0.18	18.6	8.71	8.17	2.16	0.074	0.041
1,100	0.44	25.6	50	0.45	0.20	5.8	0.16	0.11	12.8	9.35	1.02	1.48	0.041	0.029
	1.55	27.0	60	0.61	0.35	6.6	0.20	0.13	16.2	9.35	4.08	1.78	0.055	0.034
	2.46	26.4	70	0.76	0.48	7.5	0.25	0.14	18.5	9.35	6.76	1.99	0.065	0.038
	3.12	24.9	80	0.90	0.61	8.6	0.29	0.16	19.9	9.35	8.78	2.13	0.072	0.041
1,200	0.44	27.3	50	0.45	0.20	5.7	0.16	0.11	13.7	9.98	1.09	1.55	0.044	0.031
	1.55	28.9	60	0.61	0.35	6.3	0.19	0.12	17.3	9.98	4.35	1.81	0.056	0.036
	2.46	28.2	70	0.76	0.48	7.1	0.23	0.14	19.7	9.98	7.22	1.99	0.064	0.039
	3.12	26.6	80	0.90	0.61	7.9	0.26	0.15	21.3	9.98	9.37	2.10	0.069	0.041
1,300	0.44	29.0	50	0.45	0.20	5.6	0.16	0.11	14.5	10.60	1.16	1.63	0.046	0.033
	1.55	30.7	60	0.61	0.35	6.0	0.19	0.12	18.4	10.60	4.62	1.85	0.057	0.037

[a]For bulls that are at least 12 months of age and weigh more than 50 percent of their mature weight.

[b]Vitamin A requirements per pound of dry feed are 1,000 IUs for growing bulls and 1,770 IUs for breeding bulls.

Appendix Table 1 modified from University of Arkansas Extension Bulletin MP391.
Appreciation is expressed to Dr. Shane Gadberry for permissioin to use this table.

Appendix Table 9 continues on page 338.

Appendix Table 9. Nutrient Requirements of Yearling and Breeding Bulls[a,] (Continued)

			Diet Nutrient Density						Daily Nutrients per Animal					
Body Wt. (lbs.)	ADG (lbs.)	DMI (lbs./day)	TDN (% DM)	NEm (Mcal/lb.)	NEg (Mcal/lb.)	CP (% DM)	Ca (% DM)	P (% DM)	TDN (lb.)	NEm (Mcal)	NEg (Mcal)	CP (lb.)	Ca (lb.)	P (lb.)
1,700-lb. Mature Weight														
1,400	0.44	30.7	50	0.45	0.20	5.5	0.16	0.11	15.4	11.20	1.23	1.70	0.049	0.035
	1.55	32.4	60	0.61	0.35	5.8	0.18	0.12	19.4	11.20	4.88	1.88	0.057	0.03
1,500	0.44	32.3	50	0.45	0.20	5.5	0.16	0.11	16.2	11.80	1.29	1.77	0.051	0.037
	1.55	34.1	60	0.61	0.35	5.6	0.17	0.12	20.5	11.80	5.14	1.92	0.058	0.040
1,600	0.44	33.9	50	0.45	0.20	5.4	0.16	0.12	17.0	12.38	1.36	1.84	0.054	0.039
	1.55	35.8	60	0.61	0.35	5.4	0.16	0.11	21.5	12.38	5.40	1.95	0.059	0.041
1,700	0.00	32.9	46	0.39	0.00	5.6	0.16	0.12	15.1	12.96	0.00	1.83	0.052	0.040
	0.44	35.5	50	0.45	0.20	5.4	0.16	0.12	17.8	12.96	1.42	1.91	0.056	0.041
2,000-lb. Mature Weight														
1,000	0.49	23.8	50	0.45	0.20	6.1	0.17	0.12	11.9	8.71	0.95	1.44	0.041	0.028
	1.73	25.2	60	0.61	0.35	7.5	0.25	0.14	15.1	8.71	3.79	1.89	0.062	0.036
	2.75	24.6	70	0.76	0.48	9.1	0.32	0.17	17.2	8.71	6.30	2.23	0.078	0.043
	3.49	23.2	80	0.90	0.61	10.5	0.38	0.20	18.6	8.71	8.18	2.46	0.088	0.047
1,100	0.49	25.6	50	0.45	0.20	5.9	0.17	0.12	12.8	9.35	1.02	1.52	0.043	0.030
	1.73	27.0	60	0.61	0.35	7.1	0.23	0.14	16.2	9.35	4.07	1.92	0.062	0.037
	2.75	26.4	70	0.76	0.48	8.4	0.29	0.16	18.5	9.35	6.77	2.22	0.076	0.043
	3.49	24.9	80	0.90	0.61	9.8	0.35	0.19	19.9	9.35	8.79	2.43	0.086	0.047
1,200	0.49	27.3	50	0.45	0.20	5.8	0.17	0.12	13.7	9.98	1.09	1.59	0.046	0.032
	1.73	28.9	60	0.61	0.35	6.8	0.22	0.13	17.3	9.98	4.34	1.96	0.063	0.039
	2.75	28.2	70	0.76	0.48	7.9	0.27	0.16	19.7	9.98	7.22	2.22	0.075	0.044
	3.49	26.6	80	0.90	0.61	9.0	0.32	0.18	21.3	9.98	9.38	2.40	0.084	0.047

[a]For bulls that are at least 12 months of age and weigh more than 50 percent of their mature weight.

[b]Vitamin A requirements per pound of dry feed are 1,000 IUs for growing bulls and 1,770 IUs for breeding bulls.

Appendix Table 9 continues on page 339.

Appendix Table 9. Nutrient Requirements of Yearling and Breeding Bulls[a,b] (Continued)

			Diet Nutrient Density						Daily Nutrients per Animal					
Body Wt. (lbs.)	ADG (lbs.)	DMI (lbs./day)	TDN (% DM)	NEm (Mcal/lb.)	NEg (Mcal/lb.)	CP (% DM)	Ca (% DM)	P (% DM)	TDN (lb.)	NEm (Mcal)	NEg (Mcal)	CP (lb.)	Ca (lb.)	P (lb.)
2,000-lb. Mature Weight														
1,300	0.49	29.0	50	0.45	0.20	5.8	0.17	0.12	14.5	10.60	1.16	1.67	0.048	0.034
	1.73	30.7	60	0.61	0.35	6.5	0.21	0.13	18.4	10.60	4.61	2.00	0.063	0.040
	2.75	30.0	70	0.76	0.48	7.4	0.25	0.15	21.0	10.60	7.67	2.22	0.074	0.044
	3.49	28.3	80	0.90	0.61	8.4	0.29	0.17	22.6	10.60	9.96	2.38	0.081	0.047
1,400	0.49	30.7	50	0.45	0.20	5.7	0.16	0.12	15.4	11.20	1.22	1.74	0.051	0.036
	1.73	32.4	60	0.61	0.35	6.3	0.20	0.13	19.4	11.20	4.88	2.03	0.064	0.041
1,500	0.49	32.3	50	0.45	0.20	5.6	0.16	0.12	16.2	11.80	1.29	1.81	0.053	0.038
	1.73	34.1	60	0.61	0.35	6.0	0.19	0.13	20.5	11.80	5.14	2.06	0.065	0.043
1,600	0.49	33.9	50	0.45	0.20	5.5	0.17	0.12	17.0	12.38	1.35	1.88	0.056	0.040
	1.73	35.8	60	0.61	0.35	5.8	0.18	0.12	21.5	12.38	5.39	2.09	0.066	0.044
1,700	0.49	35.5	50	0.45	0.20	5.5	0.16	0.12	17.8	12.96	1.41	1.95	0.058	0.042
	1.73	37.5	60	0.61	0.35	5.7	0.18	0.12	22.5	12.96	5.64	2.13	0.066	0.046
1,800	0.49	37.0	50	0.45	0.20	5.5	0.16	0.12	18.5	13.53	1.48	2.02	0.061	0.044
	1.73	39.1	60	0.61	0.35	5.5	0.17	0.12	23.5	13.53	5.89	2.16	0.067	0.047
1,900	0.49	38.6	50	0.45	0.20	5.4	0.16	0.12	19.3	14.09	1.54	2.09	0.063	0.047
	1.73	40.8	60	0.61	0.35	5.4	0.17	0.12	24.5	14.09	6.13	2.19	0.068	0.049
2,000	0.00	37.2	46	0.39	0.00	5.6	0.17	0.13	17.1	14.64	0.00	2.07	0.062	0.047
	0.49	40.1	50	0.45	0.20	5.2	0.16	0.12	20.1	14.64	1.60	2.15	0.065	0.049

[a]For bulls that are at least 12 months of age and weigh more than 50 percent of their mature weight.

[b]Vitamin A requirements per pound of dry feed are 1,000 IUs for growing bulls and 1,770 IUs for breeding bulls.

Appendix Table 9 continues on page 340.

Appendix Table 9. Nutrient Requirements of Yearling and Breeding Bulls[a,b] (Continued)

			Diet Nutrient Density						Daily Nutrients per Animal					
Body Wt. (lbs.)	ADG (lbs.)	DMI (lbs./day)	TDN (% DM)	NEm (Mcal/lb.)	NEg (Mcal/lb.)	CP (% DM)	Ca (% DM)	P (% DM)	TDN (lb.)	NEm (Mcal)	NEg (Mcal)	CP (lb.)	Ca (lb.)	P (lb.)
2,300-lb. Mature Weight														
1,200	0.54	27.3	50	0.45	0.20	6.0	0.18	0.12	13.7	9.98	1.09	1.63	0.048	0.032
	1.91	28.9	60	0.61	0.35	7.3	0.24	0.14	17.3	9.98	4.36	2.10	0.070	0.041
	3.03	28.2	70	0.76	0.48	8.7	0.30	0.17	19.7	9.98	7.23	2.45	0.086	0.048
	3.84	26.6	80	0.90	0.61	10.1	0.36	0.20	21.3	9.98	9.38	2.68	0.097	0.052
1,300	0.54	29.0	50	0.45	0.20	5.9	0.17	0.12	14.5	10.60	1.16	1.71	0.050	0.035
	1.91	30.7	60	0.61	0.35	7.0	0.23	0.14	18.4	10.60	4.63	2.14	0.070	0.043
	3.03	30.0	70	0.76	0.48	8.2	0.28	0.16	21.0	10.60	7.68	2.45	0.085	0.049
	3.84	28.3	80	0.90	0.61	9.4	0.34	0.19	22.6	10.60	9.96	2.66	0.095	0.053
1,400	0.54	30.7	50	0.45	0.20	5.8	0.17	0.12	15.4	11.20	1.22	1.78	0.052	0.037
	1.91	32.4	60	0.61	0.35	6.7	0.22	0.14	19.4	11.20	4.90	2.17	0.071	0.044
	3.03	31.7	70	0.76	0.48	7.7	0.26	0.15	22.2	11.20	8.12	2.45	0.084	0.049
	3.84	29.9	80	0.90	0.61	8.8	0.31	0.18	23.9	11.20	10.53	2.64	0.092	0.053
1,500	0.54	32.3	50	0.45	0.20	5.7	0.17	0.12	16.2	11.80	1.29	1.85	0.055	0.039
	1.91	34.1	60	0.61	0.35	6.5	0.21	0.13	20.5	11.80	5.16	2.20	0.072	0.045
1,600	0.54	33.9	50	0.45	0.20	5.7	0.17	0.12	17.0	12.38	1.35	1.92	0.058	0.041
	1.91	35.8	60	0.61	0.35	6.3	0.20	0.13	21.5	12.38	5.41	2.24	0.072	0.047
1,700	0.54	35.5	50	0.45	0.20	5.6	0.17	0.12	17.8	12.96	1.42	1.99	0.060	0.043
	1.91	37.5	60	0.61	0.35	6.1	0.19	0.13	22.5	12.96	5.66	2.27	0.073	0.048
1,800	0.54	37.0	50	0.45	0.20	5.6	0.17	0.12	18.5	13.53	1.48	2.06	0.063	0.045
	1.91	39.1	60	0.61	0.35	5.9	0.19	0.13	23.5	13.53	5.91	2.30	0.074	0.050

[a]For bulls that are at least 12 months of age and weigh more than 50 percent of their mature weight.

[b]Vitamin A requirements per pound of dry feed are 1,000 IUs for growing bulls and 1,770 IUs for breeding bulls.

Appendix Table 9 continues on page 341.

Appendix Table 9. Nutrient Requirements of Yearling and Breeding Bulls[a,b] (Continued)

			Diet Nutrient Density						Daily Nutrients per Animal					
Body Wt. (lbs.)	ADG (lbs.)	DMI (lbs./day)	TDN (% DM)	NEm (Mcal/lb.)	NEg (Mcal/lb.)	CP (% DM)	Ca (% DM)	P (% DM)	TDN (lb.)	NEm (Mcal)	NEg (Mcal)	CP (lb.)	Ca (lb.)	P (lb.)
2,300-lb. Mature Weight														
1,900	0.54	38.6	50	0.45	0.20	5.5	0.17	0.12	19.3	14.09	1.54	2.13	0.065	0.047
	1.91	40.8	60	0.61	0.35	5.7	0.18	0.13	24.5	14.09	6.16	2.33	0.075	0.051
2,000	0.54	40.1	50	0.45	0.20	5.5	0.17	0.12	20.1	14.64	1.60	2.19	0.067	0.050
	1.91	42.3	60	0.61	0.35	5.6	0.18	0.13	25.4	14.64	6.40	2.36	0.075	0.053
2,100	0.54	41.6	50	0.45	0.20	5.4	0.17	0.13	20.8	15.19	1.66	2.26	0.070	0.052
	1.91	43.9	60	0.61	0.35	5.5	0.17	0.12	26.3	15.19	6.64	2.40	0.076	0.054
2,200	0.54	43.1	50	0.45	0.20	5.4	0.17	0.13	21.6	15.73	1.72	2.32	0.072	0.054
	1.91	45.5	60	0.61	0.35	5.3	0.17	0.12	27.3	15.73	6.87	2.42	0.077	0.056
2,300	0.00	44.5	46	0.39	0.00	5.2	0.16	0.12	20.5	16.26	0.00	2.30	0.071	0.054
	0.54	47.0	50	0.45	0.20	5.1	0.16	0.12	23.5	16.26	1.78	2.39	0.075	0.056

[a]For bulls that are at least 12 months of age and weigh more than 50 percent of their mature weight.

[b]Vitamin A requirements per pound of dry feed are 1,000 IUs for growing bulls and 1,770 IUs for breeding bulls.

Appendix Table 10. Mineral Requirements of Beef Cattle

Mineral	Units	Growing and Finishing	Gestating Cows	Lactating Cows	Max. Tolerable Concentration
Chromium	ppm	--.--	--.--	--.--	1,000.00
Cobalt	ppm	0.10	0.10	0.10	10.00
Copper	ppm	10.00	10.00	10.00	40.00
Iodine	ppm	0.50	0.50	0.50	50.00
Iron	ppm	50.00	50.00	50.00	500.00
Magnesium	percent	0.10	0.12	0.20	0.60
Manganese	ppm	20.00	40.00	40.00	2,000.00
Molybdenum	ppm	--.--	--.--	--.--	5.00
Nickel	ppm	--.--	--.--	--.--	100.00
Potassium	percent	0.60	0.60	0.70	2.00
Selenium	ppm	0.10	0.10	0.10	5.00
Sulfur	percent	0.15	0.15	0.15	0.30 in high concentrate diets and 0.50 in high roughage diets
Zinc	ppm	30.00	30.00	30.00	500.00

Based on *Nutrient Requirements of Beef Cattle, 1996* and *Mineral Tolerance of Animals,* National Research Council, 2005.

Appendix Table 11. Typical Composition of Feeds For Cattle

(All values except dry matter are shown on a dry matter basis.)

Feedstuff	DM	Energy				Protein		Fiber				EE	ASH	CA	P	K	CL	S	Zn
		TDN	NEm	NEg	NEl	CP	UIP	CF	ADF	NDF	eNDF								
		%	Mcal/cwt			%	%	%	%	%	%	%	%	%	%	%	%	%	ppm
Alfalfa Cubes	91	57	57	25	57	18	30	29	36	46	40	2.0	11	1.30	0.23	1.9	0.37	0.33	20
Alfalfa Dehydrated 17% CP	92	61	62	31	61	19	60	26	34	45	6	3.0	11	1.42	0.25	2.5	0.45	0.28	21
Alfalfa Fresh	24	61	62	31	61	19	18	27	34	46	41	3.0	9	1.35	0.27	2.6	0.40	0.29	18
Alfalfa Hay Early Bloom	90	59	59	28	59	19	20	28	35	45	92	2.5	8	1.41	0.26	2.5	0.38	0.28	22
Alfalfa Hay Midbloom	89	58	58	26	58	17	23	30	36	47	92	2.3	9	1.40	0.24	2.0	0.38	0.27	24
Alfalfa Hay Full Bloom	88	54	54	20	54	16	25	34	40	52	92	2.0	8	1.20	0.23	1.7	0.37	0.25	23
Alfalfa Hay Mature	88	50	50	12	49	13	30	38	45	59	92	1.3	8	1.18	0.19	1.5	0.35	0.21	23
Alfalfa Leaf Meal	89	60	60	30	60	26	15	16	24	34	35	3.0	10	2.88	0.34	2.2		0.32	39
Alfalfa Seed Screenings	91	84	92	61	87	34		13	15			10.7	6	0.30	0.67				
Alfalfa Silage	30	55	55	21	55	18	19	28	37	49	82	3.0	9	1.40	0.29	2.6	0.41	0.29	26
Alfalfa Silage Wilted	39	58	58	26	58	18	22	28	37	49	82	3.0	9	1.40	0.29	2.6	0.41	0.29	26
Alfalfa Stems	89	47	47	7	46	11	44	44	51	68	100	1.3	6	0.90	0.18	2.5			
Almond Hulls	89	56	56	23	56	3	60	16	29	36	100	3.1	7	0.24	0.10	2.0	0.03	0.07	20
Ammonium Chloride	99	0	0	0	0	163	0	0	0	0	0	0.0		0.00	0.00	0.0	66.00	0.00	0
Ammonium Sulfate	99	0	0	0	0	132	0	0	0	0	0	0.0						24.15	
Apple Pomace Wet	20	68	70	41	69	5	10	18	27	36	27	5.2	3	0.13	0.12	0.5		0.04	11
Apple Pomace Dried	89	67	69	40	68	5	15	18	28	38	29	5.2	3	0.13	0.12	0.5		0.04	11
Apples	17	70	73	44	71	3	10	7	9	25	10	2.2	2	0.06	0.60	0.8			
Artichoke Tops (Jerusalem)	27	61	62	31	61	6		18	30	41	40	1.1	10	1.62	0.11	1.4			
Avocado Seed Meal	91	52	52	16	51	20		19	24			1.2	16						
Bahiagrass Hay	90	53	53	18	53	6	37	32	41	72	98	1.8	7	0.47	0.20	1.4		0.21	
Bakery Product Dried	90	90	100	68	94	11	30	3	9	30	0	11.5	4	0.16	0.27	0.4	2.25	0.15	33
Bananas	24	84	92	61	87	4		4	5			0.8	3	0.03	0.11	1.5			8
Barley Bran	91	59	59	28	59	12	28	21	27	36	6	4.3	7						
Barley Hay	90	57	57	25	57	9		28	37	65	98	2.1	8	0.30	0.28	1.6		0.19	25
Barley Grain	89	84	92	61	87	12	28	5	7	20	34	2.1	3	0.06	0.38	0.6	0.18	0.16	23
Barley Grain 2-row	87	84	92	61	87	12		6	8	24	34	2.3	2	0.05	0.31	0.6	0.18	0.17	
Barley Grain 6-row	87	84	92	61	87	11		6	8	24	34	2.2	3	0.05	0.36	0.6	0.18	0.15	
Barley Grain Lt. Wt. (42-44 lb/bu)	88	78	83	54	80	13	30	9	12	30	34	2.3	4						
Barley Grain Screenings	89	71	74	46	73	12		9	11			2.6	4	0.35	0.33	0.9		0.15	
Barley Grain, Steam Flaked	85	90	100	70	100	12	39	5	7	20	30	2.1	3	0.06	0.35	0.6	0.18	0.16	23

Appendix Table 11 continues on page 344

Appendix Table 11. Typical Composition of Feeds For Cattle (Continued)
(All values except dry matter are shown on a dry matter basis.)

Feedstuff	DM	Energy				Protein		Fiber											
		TDN %	NEm	NEg Mcal/cwt	NEl	CP %	UIP %	CF %	ADF %	NDF %	eNDF %	EE %	ASH %	CA %	P %	K %	CL %	S %	Zn ppm
Barley Grain Steam Rolled	86	84	92	61	87	12	38	5	7	20	27	2.1	3	0.06	0.41	0.6	0.18	0.17	30
Barley Feed Pearl Byproduct	90	74	78	49	76	15	25	12	15			3.9	5	0.05	0.45	0.7		0.06	
Barley Silage	35	59	58	26	58	12	22	34	37	58	61	3.0	9	0.46	0.30	2.4		0.22	28
Barley Silage Mature	35	58	58	26	58	12	25	30	34	50	61	3.5	9	0.30	0.20	1.5		0.15	25
Barley Straw	90	44	44	1	43	4	70	42	55	78	100	1.9	7	0.32	0.08	2.2	0.67	0.16	7
Beans Navy Cull	90	84	92	61	87	24	25	5	8	20	0	1.4	5	0.15	0.60	1.4	0.06	0.26	45
Beet Pulp Dried	91	76	81	52	78	9	44	21	26	46	33	0.7	5	0.65	0.08	0.9	0.40	0.22	21
Beet Pulp Wet	17	77	82	53	79	9	35	20	25	45	30	0.7	5	0.65	0.08	0.9	0.40	0.22	21
Beet Pulp Dried with Molasses	92	77	82	53	79	11	34	17	23	40	33	0.6	6	0.60	0.10	1.8		0.42	11
Beet Pulp Wet with Molasses	24	77	82	53	79	11	25	16	21	39	33	0.6	6	0.60	0.10	1.8		0.42	11
Beet Root (Sugar)	23	80	86	56	83	4		5	7	16		0.4	3						
Beet Tops (Sugar)	19	58	58	26	58	14		11	14	25	41	1.3	24	1.10	0.22	5.2	0.20	0.45	20
Beet Top Silage	25	52	52	16	51	12		12				2.0	32	1.38	0.22	5.7		0.57	20
Bermudagrass Coastal Dehydrated	90	62	63	33	63	16	40	26	29	40	10	3.8	7	0.40	0.25	1.8	0.72	0.23	18
Bermudagrass Coastal Hay	89	56	56	23	56	10	20	30	36	73	98	2.1	6	0.47	0.21	1.5	0.70	0.22	16
Bermudagrass Hay	89	53	53	18	53	10	18	29	37	72	98	1.9	8	0.46	0.20	1.5	0.70	0.25	31
Bermudagrass Silage	26	50	50	12	49	10	15	28	35	71	48	1.9	8	0.46	0.20	1.5	0.72	0.25	31
Birdsfoot Trefoil Fresh	22	66	68	38	67	21	20	21	31	47	41	4.4	9	1.78	0.25	2.6		0.25	31
Birdsfoot Trefoil Hay	89	57	57	25	57	16	22	31	38	50	92	2.2	8	1.73	0.24	1.8		0.25	28
Biuret	99	0	0	0	0	248	0	0	0	0	0	0.0	0	0.00	0.00	0.0	0.00	0.00	0
Blood Meal, Swine/Poultry	91	66	68	38	67	92	82	1	2	10	0	1.4	3	0.32	0.28	0.2	0.30	0.70	22
Bluegrass KY Fresh Early Bloom	36	69	71	43	70	15	20	27	32	60	41	3.9	7	0.37	0.30	1.9	0.42	0.19	25
Bluegrass Straw	93	45	45	3	44	6		40	50	78	90	1.1	6	0.20	0.10				
Bluestem Fresh Mature	61	50	50	12	49	6		34				2.5	5	0.40	0.12	0.8		0.05	28
Bone Meal Steamed, Swine/Poultry	95	16	27	0	11	13		1	0	0	0	11.6	77	27.00	12.74	0.2		2.50	290
Bread Byproduct	68	90	100	68	94	14	24	1	2	3	0	3.0	3	0.10	0.18	0.2	0.76	0.15	40
Brewers Grains Dried	92	84	92	61	87	25	54	14	24	49	18	7.5	4	0.30	0.58	0.1	0.15	0.32	78
Brewers Grains Wet	23	85	93	62	88	26	52	13	21	45	18	7.5	4	0.30	0.58	0.1	0.15	0.32	78
Brewers Yeast Dried	94	79	85	55	81	48		3				1.0	7	0.10	1.56	1.8		0.41	41
Bromegrass Fresh Immature	30	64	65	36	65	15	22	28	33	54	40	4.1	10	0.45	0.34	2.3		0.21	20
Bromegrass Hay	89	55	55	21	55	10	33	35	41	66	98	2.3	9	0.40	0.23	1.9	0.40	0.19	19

Appendix Table 11 continues on page 345

Appendix Table 11. Typical Composition of Feeds For Cattle (Continued)

(All values except dry matter are shown on a dry matter basis.)

Feedstuff	DM	Energy				Protein		Fiber				EE	ASH	CA	P	K	CL	S	Zn
		TDN %	NEm Mcal/cwt	NEg Mcal/cwt	NEI Mcal/cwt	CP %	UIP %	CF %	ADF %	NDF %	eNDF %	%	%	%	%	%	%	%	ppm
Bromegrass Haylage	35	57	57	25	57	11	26	36	44	69	61	2.5	8	0.38	0.30	2.0		0.20	19
Buckwheat Grain	88	75	79	50	77	12		13	17			2.8	2	0.11	0.36	0.5	0.05	0.16	10
Buttermilk Dried	92	88	98	65	91	34	0	5	0	0	0	5.0	10	1.44	1.00	0.9		0.09	44
Cactus,Prickle Pear	23	61	62	31	62	5		16	20	28		2.1	18	4.00	0.10	1.5		0.20	
Calcium Carbonate	99	0	0	0	0	0		0	0	0	0	0.0	99	38.50	0.04	0.1		0.00	0
Canarygrass Hay	91	53	53	18	53	9	26	32	34	67	98	2.7	8	0.38	0.25	2.7		0.14	18
Canola Meal Solvent Ext.	90	72	75	47	74	41	30	11	19	29	23	2.0	8	0.74	1.14	1.1	0.07	0.78	68
Carrot Pulp	14	62	63	33	63	6		19	23	40	0	7.8	9						
Carrot Root Fresh	12	83	90	60	86	10		9	11	20	0	1.4	10	0.55	0.32	2.5	0.50	0.17	
Carrot Tops	16	73	77	48	75	13		18	23	45	41	3.8	15	1.94	0.19	1.9			
Cattle Manure Dried	92	38	40	0	36	15		35	42	55	0	2.5	14	1.15	1.20	0.6		1.78	240
Cheatgrass Fresh Immature	21	68	70	41	69	16		23				2.7	10	0.60	0.28				
Citrus Pulp Dried	90	78	83	54	80	7	38	13	20	21	33	2.9	7	1.81	0.12	0.8	0.04	0.08	14
Clover Ladino Fresh	19	69	71	43	70	25	20	14	33	35	41	4.8	11	1.27	0.38	2.4		0.20	20
Clover Ladino Hay	90	61	62	31	61	21	25	22	32	36	92	2.0	9	1.35	0.32	2.4	0.30	0.20	17
Clover Red Fresh	24	64	65	36	65	18	21	24	33	44	41	4.0	9	1.70	0.30	2.0	0.60	0.17	23
Clover Red Hay	88	55	55	21	55	15	28	30	39	51	92	2.5	8	1.50	0.25	1.7	0.32	0.17	17
Clover Sweet Hay	91	53	53	18	53	16	30	30	38	50	92	2.4	9	1.27	0.25	1.8	0.37	0.46	
Coconut Meal Mech. Ext.	92	76	81	52	78	21	56	13	21	56	23	6.8	7	0.40	0.30	1.0	0.33	0.04	
Coffee Grounds	88	20	36	0	16	13		41	68	77	10	15.0	2	0.10	0.08				
Corn and Cob Meal	87	82	89	59	85	9	52	9	11	26	56	3.7	2	0.06	0.27	0.5	0.05	0.13	16
Corn Bran	91	76	81	52	78	11		10	17	51	0	6.3	3	0.04	0.15	0.1	0.13	0.08	18
Corn Cannery Waste	29	68	70	41	69	8	15	28	36	59	0	3.0	5	0.10	0.29	1.0		0.13	25
Corn Cobs	90	48	48	9	47	3	70	36	39	88	56	0.6	2	0.12	0.04	0.8		0.27	5
Corn Fodder	80	65	66	37	66	9	45	25	29	48	100	2.4	7	0.50	0.25	0.9	0.20	0.14	
Corn Germ Full-fat	97	135	198	160	198	12	55	6	11	36	20	44.9	2	0.02	0.28	0.1	0.02	0.17	60
Corn Gluten Feed	90	80	86	56	83	22	25	9	12	38	36	3.2	7	0.11	0.84	1.3	0.25	0.47	84
Corn Gluten Meal 41% CP	91	85	93	62	88	46	63	5	9	32	23	3.2	3	0.13	0.55	0.2	0.07	0.62	35
Corn Gluten Meal 60% CP	91	89	99	67	93	67	65	3	6	11	23	2.5	2	0.06	0.54	0.2	0.10	0.90	40
Corn Grain High Moisture	74	93	104	71	97	10	42	2	3	9	0	4.0	2	0.02	0.30	0.4	0.06	0.14	20

Appendix Table 11 continues on page 345

Appendix Table 11. Typical Composition of Feeds For Cattle (Continued)
(All values except dry matter are shown on a dry matter basis.)

Feedstuff	DM	Energy				Protein		Fiber				EE	ASH	CA	P	K	CL	S	Zn
		TDN %	NEm	NEg Mcal/cwt	NEl	CP %	UIP %	CF %	ADF %	NDF %	eNDF %	%	%	%	%	%	%	%	ppm
Corn Grain, High Oil	88	91	102	69	95	8	54	2	3	8	60	6.9	2	0.01	0.30	0.3	0.05	0.13	18
Corn Grain Hi-Lysine	92	87	96	64	90	12	58	4	4	11	60	4.4	2	0.03	0.24	0.4	0.05	0.11	18
Corn Grain Rolled	88	88	98	65	91	9	54	2	3	9	34	4.3	2	0.02	0.30	0.4	0.05	0.14	18
Corn Grain, Steam Flaked	85	93	104	71	97	9	59	2	3	9	40	4.1	2	0.02	0.27	0.4	0.05	0.14	18
Corn Grain Whole	88	88	98	65	91	9	58	2	3	9	60	4.3	2	0.02	0.30	0.4	0.05	0.14	18
Corn Screenings	86	91	102	69	95	10	52	3	4	9	20	4.3	2	0.04	0.27	0.4	0.05	0.12	16
Corn Silage Milk Stage	26	65	66	37	66	8	18	26	32	54	60	2.8	6	0.40	0.27	1.6		0.11	20
Corn Silage Mature Well Eared	34	72	75	47	74	8	28	21	27	46	70	3.1	5	0.28	0.23	1.1	0.20	0.13	22
Corn Silage Sweet Corn	24	65	66	37	66	11		20	32	57	60	5.0	5	0.24	0.26	1.2	0.17	0.16	39
Corn Stover Mature (Stalks)	80	54	54	20	54	5	30	35	43	70	100	1.3	7	0.45	0.15	1.2	0.30	0.14	22
Corn Whole Plant Pelleted	91	63	64	34	64	9	45	21	24	40	6	2.4	6	0.50	0.24	0.9		0.14	
Cottonseed Hulls	90	45	45	3	44	5	45	48	70	87	100	1.8	3	0.15	0.08	1.0	0.02	0.05	10
Cottonseed Meal, Mech. 41% CP	92	79	85	55	81	46	50	13	19	31	23	5.0	7	0.21	1.18	1.6	0.05	0.42	64
Cottonseed Meal, Solvent 41% CP	90	77	82	53	79	47	42	13	18	25	23	1.5	7	0.22	1.23	1.6	0.05	0.44	66
Cottonseed, Whole	91	95	107	73	99	23	38	27	37	47	100	19.4	5	0.16	0.64	1.0	0.06	0.24	34
Cottonseed, Whole, Delinted	90	95	107	73	99	24	39	19	28	40	100	22.9	5	0.12	0.54	1.2		0.24	36
Cottonseed, Whole, Extruded	92	87	98	67	91	26	50	32	44	53	33	9.5	5	0.17	0.68	1.3		0.24	38
Cotton Gin Trash (Burrs)	91	42	43	0	40	9		35	50	70	100	2.0	14	1.40	0.18	1.9		0.14	25
Crab Waste Meal Solvent Ext.	91	29	37	0	30	32	65	11	13			3.0	43	15.00	1.88	0.5	1.63	0.27	107
Crambe Meal, Mech.	92	88	98	65	91	28	50	24	33	42	25	17.0	7	1.22	0.78	1.0	0.65	1.18	41
Crambe Meal, Solvent Ext.	91	81	88	58	84	31	45	25	35	47	23	1.4	8	1.27	0.86	1.1	0.70	1.26	44
Cranberry Pulp Meal	88	49	49	11	48	7		26	47	54	33	15.7	2						
Crawfish Waste Meal	94	25	36	0	29	35	74	12	15				42	13.10	0.85				
Curacao Phosphate	99	0	0	0	0	0		0	0	0	0	0.0	95	34.00	15.00				
Defluorinated Phosphate	99	0	0	0	0	0		0	0	0	0	0.0	95	32.60	18.07	1.0			100
Diammonium Phosphate	98	0	0	0	0	115	0	0	0	0	0	0.0	35	0.52	20.41	0.0		2.16	
Dicalcium Phosphate	96	0	0	0	0	0		0	0	0	0	0.0	94	22.00	18.65	0.1		1.00	70
Distillers Dried Solubles	93	87	96	64	91	31	47	4	7	22	4	13.0	8	0.35	1.20	1.8	0.28	1.10	91
Distillers Corn Stillage	7	92	103	70	96	22	55	8	10	21	0	8.1	5	0.14	0.72	0.2		0.60	60
Distillers Grain, Barley	90	75	79	50	77	30	56	16	20	44	4	8.5	4	0.15	0.67	1.0	0.18	0.43	50
Distillers Grain, Corn, Dry	91	95	106	72	99	30	58	8	16	44	4	9.5	4	0.09	0.75	0.9	0.14	0.70	65

Appendix Table 11 continues on page 347

Appendix Table 11. Typical Composition of Feeds For Cattle (Continued)
(All values except dry matter are shown on a dry matter basis.)

Feedstuff	DM	Energy				Protein		Fiber				EE	ASH	CA	P	K	CL	S	Zn
		TDN %	NEm Mcal/cwt	NEg Mcal/cwt	NEl Mcal/cwt	CP %	UIP %	CF %	ADF %	NDF %	eNDF %	%	%	%	%	%	%	%	ppm
Distillers Grain, Corn, Wet	36	97	109	74	102	30	47	8	16	44	4	9.5	4	0.09	0.75	0.9	0.14	0.70	65
Distillers Grain, Corn with Solubles	89	98	111	76	103	30	54	8	16	38	4	11.9	6	0.20	0.75	0.9	0.18	0.80	85
Distillers Grain, Sorghum, Dry	91	84	92	61	87	33	62	13	20	44	4	10.0	4	0.20	0.68	0.3		0.50	50
Distillers Grain, Sorghum, Wet	35	86	95	63	89	33	55	13	19	43	4	10.0	4	0.20	0.68	0.3		0.50	50
Distillers Grain, Sorghum with Solubles	92	85	93	62	88	33	53	12	18	42	4	10.0	4	0.23	0.70	0.5		0.70	55
Distillers Grains, Wet	25	91	102	69	95	28	52	8	18	40	4	9.6	5	0.10	0.70	1.0	0.20	0.60	95
Elephant (Napier) grass hay, chopped	92	55	55	21	54	9		24	46	63	85	2.0	10	0.35	0.30	1.3		0.10	
Fat, Animal, Poultry, Vegetable	99	195	285	230	285	0		0	0	0	0	99.0	0	0.00	0.00	0.0			
Feather Meal Hydrolyzed	93	67	69	40	68	87	68	1	14	42	23	7.0	3	0.48	0.45	0.1	0.20	1.82	90
Fescue KY 31 Fresh	29	64	65	36	65	15	20	25	32	64	40	5.5	9	0.48	0.37	2.5		0.18	22
Fescue KY 31 Hay Early Bloom	88	60	60	30	60	18	22	25	31	64	98	6.6	8	0.48	0.36	2.6		0.27	24
Fescue KY 31 Hay Mature	88	52	52	16	51	11	30	30	42	73	98	5.0	6	0.45	0.26	1.7		0.14	22
Fescue (Red) Straw	94	43	44	0	41	4		41				1.1	6	0.00	0.06				
Fish Meal	90	74	78	49	76	66	60	1	2	12	10	9.0	20	5.55	3.15	0.7	0.76	0.80	130
Flax Seed Hulls	91	38	40	0	36	9		32	39	50	98	1.5	10						
Garbage Municipal Cooked	23	80	86	56	83	16		9	50	59	30	20.0	10	1.20	0.43	0.6	0.67		
Glycerol (Glycerin)	88	90	100	68	94	0	0	0	0	0	0	0.0	6				4.00		
Grain Dust	92	73	77	48	75	10		11				2.2	10	0.30	0.18				42
Grain Screenings	90	65	66	37	66	14		14				5.5	9	0.25	0.34				30
Grape Pomace Stemless	91	40	42	0	38	12	45	32	46	54	34	7.6	9	0.55	0.07	0.6	0.01		24
Grass Hay	88	58	58	26	58	10	30	33	41	63	98	3.0	6	0.60	0.21	2.0		0.20	28
Grass Silage	30	61	62	31	61	11	24	32	39	60	61	3.4	8	0.70	0.24	2.1		0.22	29
Guar Meal	90	72	75	47	74	39	34	16				3.9	5						
Hominy Feed	90	89	99	67	93	11	48	5	8	21	9	6.5	3	0.04	0.55	0.6	0.06	0.10	32
Hop Leaves	37	49	49	11	48	15		15				3.6	35	2.80	0.64				
Hop Vine Silage	30	53	53	18	53	15		21	24			3.1	20	3.30	0.37	1.8		0.22	44
Hops Spent	89	35	39	0	33	23		26	30			4.6	7	1.60	0.60				
Kelp Dried	91	32	38	0	29	7		7	10			0.5	39	2.72	0.31				
Kenaf Hay	92	48	48	9	47	10		31	44	56	98	2.9	12						
Kochia Fresh	29	55	55	21	55	16		23				1.2	18	1.10	0.30				
Kochia Hay	90	53	53	18	53	14		27				1.7	14	1.00	0.20				

Appendix Table 11 continues on page 348

Appendix Table 11. Typical Composition of Feeds For Cattle (Continued)

(All values except dry matter are shown on a dry matter basis.)

Feedstuff	DM	Energy				Protein		Fiber											
		TDN %	NEm	NEg Mcal/cwt	NEl	CP %	UIP %	CF %	ADF %	NDF %	eNDF %	EE %	ASH %	CA %	P %	K %	CL %	S %	Zn ppm
Kudzu Hay	90	54	54	20	54	16		33				2.6	7	3.00	0.23				
Lespedeza Fresh Early Bloom	25	60	60	30	60	16	50	32				2.0	10	1.20	0.24	1.1		0.21	
Lespedeza Hay	92	54	54	20	54	14	60	30				3.0	7	1.10	0.22	1.0		0.19	29
Limestone Dolomitic Ground	99	0	0	0	0	0	0	0	0	0	0	0.0	98	22.30	0.04	0.4			
Limestone Ground	98	0	0	0	0	0	0	0	0	0	0	0.0	98	34.00	0.02			0.03	
Linseed Meal Mech Ext.	91	82	89	59	85	37	40	10	17	24	23	6.0	6	0.42	0.90	1.4	0.04	0.46	59
Linseed Meal Solvent Ext.	91	77	82	53	79	38	36	10	18	25	23	1.7	6	0.43	0.91	1.5	0.04	0.47	60
Meadow Hay	90	50	50	12	49	7	23	33	44	70	98	2.5	9	0.61	0.18	1.6		0.17	24
Meat Meal, Swine/Poultry	93	71	74	46	73	56	64	2	7	48	0	10.5	24	9.00	4.42	0.5	1.27	0.48	190
Meat and Bone Meal, Swine/Poultry	93	72	75	47	74	56	24	1	5	34	0	10.0	29	13.50	6.50				
Milk, Dry, Skim	94	87	96	64	90	36	0	0	0	0	0	0.9	8	1.36	1.09	1.7	0.96	0.34	41
Mint Slug Silage	27	55	55	21	55	14		24				1.8	16	1.10	0.57				
Molasses Beet	77	75	79	50	77	8	0	0	0	0	0	0.2	12	0.14	0.03	6.0	1.64	0.60	18
Molasses Cane	77	74	78	49	76	6	0	0	0	0	0	0.5	14	0.95	0.09	4.2	2.30	0.68	15
Molasses Cane Dried	94	74	78	49	76	9	0	2	3	7	0	0.3	14	1.10	0.15	3.6	3.00		30
Molasses Citrus	65	75	79	50	77	9	0	0	0	0	0	0.3	8	1.84	0.15	0.2	0.11	0.23	137
Molasses, Cond. Fermentation Solubles	43	69	71	43	70	16	0	0	0	0	0	1.0	26	2.12	0.14	7.5	2.73	0.93	30
Molasses Wood, Hemicellulose	61	70	73	44	71	1	0	1	2	4	0	0.6	7	1.10	0.10	0.1		0.05	
Monoammonium Phosphate	98	0	0	0	0	70	0	0	0	0	0	0.0	24	0.30	24.70	0.0		1.42	81
Mono-Dicalcium Phosphate	97	0	0	0	0	0		0	0	0	0	0.0	94	16.70	21.10	0.1		1.20	70
Oat Grain	89	76	81	52	78	13	18	11	15	28	34	5.0	4	0.05	0.41	0.5	0.11	0.20	40
Oat Grain, Steam Flaked	84	88	98	65	91	13	26	11	15	30	32	4.9	4	0.05	0.37	0.5	0.11	0.20	40
Oat Groats	91	91	102	69	95	18	15	3				6.6	2	0.08	0.47	0.4	0.10	0.20	
Oat Hay	90	54	54	20	54	10	25	31	39	63	98	2.3	8	0.40	0.27	1.6	0.42	0.21	28
Oat Hulls	93	38	40	0	36	4	25	33	41	75	90	1.6	7	0.16	0.15	0.6	0.08	0.14	31
Oat Middlings	90	91	102	69	95	16	20	4	6			6.0	3	0.07	0.48	0.5		0.23	
Oat Mill By-product	89	33	38	0	30	7		27	37			2.4	6	0.13	0.22	0.6		0.24	
Oat Silage	35	60	60	30	60	12	21	31	39	59	61	3.2	10	0.34	0.30	2.4	0.50	0.25	27
Oat Straw	91	48	48	9	47	4	40	41	48	73	98	2.3	8	0.24	0.07	2.5	0.78	0.22	6
Orange Pulp Dried	89	79	85	55	81	9		9	16	20	33	1.8	4	0.71	0.11	0.6		0.05	
Orchardgrass Fresh Early Bloom	24	65	66	37	66	14	23	30	32	54	41	4.0	9	0.33	0.39	2.7	0.08	0.20	21

Appendix Table 11 continues on page 348

Appendix Table 11. Typical Composition of Feeds For Cattle (Continued)

(All values except dry matter are shown on a dry matter basis.)

Feedstuff	DM	Energy				Protein		Fiber											
		TDN %	NEm Mcal/cwt	NEg Mcal/cwt	NEl Mcal/cwt	CP %	UIP %	CF %	ADF %	NDF %	eNDF %	EE %	ASH %	CA %	P %	K %	CL %	S %	Zn ppm
Orchardgrass Hay	88	59	59	28	59	10	27	34	40	67	98	3.3	8	0.32	0.30	2.6	0.41	0.20	26
Pea Vine Hay	89	59	59	28	59	11		32	50	62	92	2.0	7	1.25	0.24	1.3		0.20	20
Pea Vine Silage	25	58	58	26	58	16		29	44	55	61	3.3	8	1.25	0.28	1.6		0.29	32
Pea Vine Straw	89	51	51	14	50	7		41	49	72	98	1.4	7	0.75	0.13	1.1		0.15	
Peanut Hulls	91	22	36	0	18	7		63	65	74	98	1.5	5	0.20	0.07	0.9			
Peanut Meal Solvent Ext.	91	77	82	53	79	51	27	9	16	27	23	2.5	6	0.26	0.62	1.1	0.03	0.30	38
Peanut Skins	92	0	0	0	0	17		13	20	28	0	22.0	3	0.19	0.20				
Pearl Millet Grain	87	82	89	59	85	13		2	6	18	34	4.5	3	0.03	0.36	0.5			
Peas Cull	88	85	93	62	88	23	22	7	9	12	0	1.4	4	0.14	0.46	1.1	0.06	0.26	30
Pineapple Bran	89	71	74	46	73	5		20	33	66	20	1.5	3	0.26	0.12				
Pineapple Greenchop	17	47	47	7	46	8		24	35	64	41	2.4	7	0.28	0.08				
Pineapple Presscake	21	71	74	46	73	5		24	35	69	20	0.8	3	0.25	0.09				
Potato Vine Silage	15	59	59	28	59	15		26				3.7	19	2.10	0.29	4.0		0.37	
Potato Waste Dried	89	85	93	62	88	8	0	7	9	15	0	0.5	5	0.16	0.25	1.2	0.39	0.11	12
Potato Waste Wet	14	82	89	59	85	7	0	9	11	18	0	1.5	3	0.16	0.25	1.2	0.36	0.11	12
Potato Waste Filter Cake	14	77	82	53	79	5	0	2				7.7	3	0.10	0.19	0.2			
Potato Waste Wet with Lime	17	80	86	56	83	5	0	10	12	16	0	0.3	9	4.20	0.18				
Potatoes Cull	21	80	86	56	83	10	0	2	3	4	0	0.4	5	0.03	0.24	2.2	0.30	0.09	
Poultry By-product Meal	93	79	85	55	81	62	49	2				14.5	17	4.00	2.25	0.5	0.58	0.56	129
Poultry Manure Dried	89	38	40	0	36	28	22	13	15	35	0	2.1	33	10.20	2.80	2.3	1.05	0.20	520
Prairie Hay	91	50	50	12	49	7	37	34	47	67	98	2.0	8	0.40	0.15	1.1	0.06	0.06	34
Pumpkins, Cull	11	80	86	56	83	15		14	21	30	0	8.9	9	0.24	0.43	3.3			
Rice Bran	91	71	74	46	73	14	30	13	18	24	0	16.0	11	0.07	1.70	1.8	0.09	0.19	40
Rice Grain	89	79	85	55	81	8	30	10	12	16	34	1.9	5	0.07	0.32	0.4	0.09	0.05	17
Rice Hulls	92	13	35	0	8	3	45	44	70	81	90	0.9	20	0.12	0.07	0.5	0.08	0.08	24
Rice Mill By-product	91	39	41	0	37	7		32	50	60	0	5.7	19	0.25	0.48	2.2		0.30	31
Rice Polishings	90	90	100	68	94	14		4	5			14.0	9	0.05	1.34	1.2	0.12	0.19	28
Rice Straw	91	40	42	0	38	4		38	47	72	100	1.4	13	0.23	0.08	1.2		0.11	
Rice Straw Ammoniated	87	45	45	3	44	9		39	53	68	100	1.3	12	0.25	0.08	1.1		0.11	
Rye Grain	89	80	86	56	83	14	20	3	9	19	34	2.5	3	0.07	0.55	0.5	0.03	0.17	33
Rye Grass Hay	90	58	58	26	58	10	30	33	38	65	98	3.3	8	0.45	0.30	2.2		0.18	27

Appendix Table 11 continues on page 350

Appendix Table 11. Typical Composition of Feeds For Cattle (Continued)

(All values except dry matter are shown on a dry matter basis.)

		Energy				Protein		Fiber											
Feedstuff	DM	TDN %	NEm	NEg Mcal/cwt	NEl	CP %	UIP %	CF %	ADF %	NDF %	eNDF %	EE %	ASH %	CA %	P %	K %	CL %	S %	Zn ppm
Rye Grass Silage	32	59	59	28	59	14	25	22	37	59	61	3.3	8	0.43	0.38	2.9	0.73	0.23	29
Rye Straw	89	44	44	1	43	4		44	55	71	100	1.5	6	0.24	0.09	1.0	0.24	0.11	
Safflower Hulls	91	14	35	0	34	4		58	73	90	100	3.7	2						
Safflower Meal Dehulled Solvent Ext.	91	75	79	50	77	47		11	20	27	30	0.6	7	0.38	1.60	1.2	0.18	0.22	36
Safflower Meal Solvent Ext.	91	56	56	23	56	24		33	41	57	36	1.3	6	0.35	0.79	0.9	0.21	0.23	65
Sagebrush Fresh	50	50	50	12	49	13		25	30	38		9.2	10	1.00	0.25			0.22	
Sanfoin Hay	88	61	62	31	62	14	60	24				3.1	9						
Shrimp Waste Meal	90	48	48	9	47	50	60	11				5.5	25	8.50	1.75		1.15		
Sodium Tripolyphosphate	96	0	0	0	0	0		0	0	0	0	0.0	96	0.00	25.98	0.0		0.00	
Sorghum Grain (Milo) Flaked	82	90	100	68	94	11	62	3	6	15	38	3.1	2	0.04	0.28	0.4	0.10	0.14	18
Sorghum Grain (Milo) Ground	89	82	89	59	85	11	55	3	6	15	5	3.1	2	0.04	0.32	0.4	0.10	0.14	18
Sorghum Silage	32	59	59	28	59	9	25	27	38	59	70	2.7	6	0.48	0.21	1.7	0.45	0.11	30
Sorghum Stover	87	54	54	20	54	5		33	41	65	100	1.8	10	0.50	0.12	1.2			
Soybean Hay	89	52	52	16	51	16		33	40	55	92	3.5	8	1.28	0.29	1.0	0.15	0.24	24
Soybean Hulls	90	77	82	52	79	13	28	39	48	62	28	2.3	5	0.60	0.19	1.3	0.02	0.12	38
Soybean Meal Solvent 44% CP	89	84	92	61	87	49	35	7	10	15	23	1.5	7	0.36	0.70	2.2	0.07	0.41	62
Soybean Meal Solvent 49% CP	89	87	96	64	90	54	36	4	6	8	23	1.3	7	0.28	0.71	2.2	0.08	0.45	61
Soybean Straw	88	42	43	0	40	5		44	54	70	100	1.4	6	1.59	0.06	0.6		0.26	
Soybean Mill Feed	90	50	50	12	49	15		36	46			1.9	6	0.46	0.19	1.7		0.07	
Soybeans Whole	88	92	103	70	96	41	28	8	11	15	100	18.8	5	0.27	0.64	1.9	0.03	0.34	56
Soybeans Whole, Extruded	88	93	104	71	97	40	35	9	11	15	100	18.8	5	0.27	0.64	2.0	0.03	0.34	56
Soybeans Whole, Roasted	88	93	104	71	97	40	48	9	11	15	100	18.8	5	0.27	0.64	2.0	0.03	0.34	56
Spelt Grain	88	75	79	50	77	13	27	10	17	21	34	2.1	4	0.04	0.40	0.4		0.15	47
Sudangrass Fresh Immature	18	70	73	44	71	17		23	29	55	41	3.9	9	0.46	0.36	2.0		0.11	24
Sudangrass Hay	88	57	57	25	57	9	30	36	43	67	98	1.8	10	0.50	0.22	2.2	0.80	0.12	26
Sudangrass Silage	31	58	58	26	58	10	28	30	42	64	61	3.1	10	0.58	0.27	2.4	0.52	0.14	29
Sunflower Seed Meal Solvent Ext.	92	65	66	37	66	40	27	18	22	36	23	2.8	8	0.44	0.97	1.1	0.15	0.33	55
Sunflower Seed Meal with Hulls	91	57	57	25	57	31	35	27	32	44	37	2.4	7	0.40	1.03	1.0		0.30	85
Sunflower Seed Hulls	90	40	42	0	38	4	65	52	63	73	90	2.2	3	0.00	0.11	0.2		0.19	200
Sugar Cane Bagasse	91	39	41	0	37	1		49	60	86	100	0.6	4	0.90	0.29	0.5		0.10	
Tapioca Meal,Cassava By-product	89	82	89	59	85	1		5	8	34		0.8	3	0.03	0.05				

Appendix Table 11 continues on page 351

Appendix Table 11. Typical Composition of Feeds For Cattle (Continued)

(All values except dry matter are shown on a dry matter basis.)

Feedstuff	DM	Energy				Protein		Fiber				EE	ASH	CA	P	K	CL	S	Zn
		TDN %	NEm Mcal/cwt	NEg Mcal/cwt	NEl Mcal/cwt	CP %	UIP %	CF %	ADF %	NDF %	eNDF %	%	%	%	%	%	%	%	ppm
Timothy Fresh Pre-bloom	26	64	65	36	65	11	20	31	36	59	41	3.8	7	0.40	0.28	1.9	0.57	0.15	28
Timothy Hay Early Bloom	88	59	59	28	59	11	22	32	39	63	98	2.7	6	0.58	0.26	1.9	0.51	0.21	30
Timothy Hay Full Bloom	88	57	57	25	57	8	30	34	40	65	98	2.6	5	0.43	0.20	1.8	0.62	0.13	25
Timothy Silage	34	59	59	28	59	10	25	34	45	70	61	3.4	7	0.50	0.27	1.7		0.15	
Tomato Pomace Dried	92	64	65	36	65	23		26	50	55	34	10.6	6	0.43	0.59	3.6			
Tomatoes	6	69	71	43	70	16		9	11			4.0	6	0.14	0.35	4.2			
Triticale Grain	89	85	93	62	88	14	25	4	5	22	34	2.4	2	0.07	0.39	0.5		0.17	37
Triticale Hay	90	56	56	23	56			34	41	69	98			0.30	0.26	2.3			25
Triticale Silage	34	58	58	26	58	10		30	39	56	61	3.6		0.58	0.34	2.7		0.28	36
Turnip Roots	9	86	95	63	89	16	0	11	34	44	40	1.6	9	0.65	0.31	3.1	0.65	0.43	40
Turnip Tops (Purple)	18	68	70	41	69	18		10	13			2.6	14	3.10	0.40	3.0	1.80	0.27	
Urea 46%N	99	0	0	0	0	12	0	0	0	0	0	0.0	0	0.00	0.00	0.0	0.00	0.00	0
Vetch Hay	89	58	58	26	58	288	14	30	33	48	92	1.8	8	1.25	0.34	2.4		0.13	
Wheat Bran	89	70	73	44	71	12	28	11	14	46	4	4.4	7	0.13	1.32	1.4	0.05	0.24	96
Wheat Fresh, Pasture	21	71	74	46	73	18	16	18	30	50	41	4.0	13	0.35	0.36	3.1	0.67	0.22	
Wheat Grain	89	88	98	65	91	9	23	3	4	12	0	2.3	2	0.05	0.43	0.4	0.09	0.15	40
Wheat Grain Hard	89	88	98	65	91	14	28	3	6	14	0	2.0	2	0.05	0.43	0.5		0.16	45
Wheat Grain Soft	89	88	98	65	91	14	23	3	4	12	0	2.0	2	0.06	0.40	0.4		0.15	30
Wheat Grain Sprouted	86	88	98	65	91	14	18	3	4	13	0	2.0	2	0.04	0.36	0.4		0.17	45
Wheat Grain, Steam Flaked	85	91	102	69	95	12	29	3	4	12	0	2.3	2	0.05	0.39	0.4		0.15	40
Wheat Hay	90	57	57	25	57	20	25	29	38	66	98	2.0	8	0.21	0.22	1.4	0.50	0.19	23
Wheat Middlings	89	80	86	56	83	18	22	8	11	36	2	4.7	5	0.14	1.00	1.3	0.05	0.20	98
Wheat Mill Run	90	75	81	52	78	17	28	9	12	37	0	4.5	6	0.11	1.10	1.2	0.07	0.22	90
Wheat Shorts	89	78	83	54	80	19	25	8	10	30	0	5.3	5	0.10	0.93	1.1	0.08	0.20	118
Wheat Silage	33	59	59	28	59	9	21	28	37	62	61	3.2	8	0.40	0.28	2.1	0.50	0.21	27
Wheat Straw	91	43	44	0	41	12	60	43	57	81	98	1.8	8	0.17	0.06	1.3	0.32	0.17	6
Wheat Straw Ammoniated	85	50	50	12	49	3	25	40	55	76	98	1.5	9	0.15	0.05	1.3	0.30	0.16	6
Wheatgrass Crested Fresh Early Bloom	37	60	60	30	60	20	25	26	28	50	41	1.6	7	0.46	0.32	2.4			
Wheatgrass Crested Fresh Full Bloom	50	55	55	21	55	11	33	33	36	65	41	1.6	7	0.39	0.28	2.1			
Wheatgrass Crested Hay	92	54	54	20	54	10	33	33	36	65	98	2.4	7	0.33	0.20	2.0			32
Whey Dried	94	82	89	59	85	10	15	0	0	0	0	0.9	10	0.98	0.88	1.3	1.20	0.92	10
Yeast, Brewer's	92	79	85	55	81	47	30	3	4		0	0.9	7	0.13	1.49	1.8			

Index

D

E

F